Making IT Work

History of Computing

William Aspray and Thomas J. Misa, editors

Janet Abbate, *Gender in the History of Computing: Reimagining Expertise, Opportunity, and Achievement through Women's Lives*

John Agar, *The Government Machine: A Revolutionary History of the Computer*

William Aspray and Paul E. Ceruzzi, *The Internet and American Business*

William Aspray, *John von Neumann and the Origins of Modern Computing*

Charles J. Bashe, Lyle R. Johnson, John H. Palmer, and Emerson W. Pugh, *IBM's Early Computers*

Martin Campbell-Kelly, *From Airline Reservations to Sonic the Hedgehog: A History of the Software Industry*

Paul E. Ceruzzi, *A History of Modern Computing*

I. Bernard Cohen, *Howard Aiken: Portrait of a Computer Pioneer*

I. Bernard Cohen and Gregory W. Welch, editors, *Makin' Numbers: Howard Aiken and the Computer*

Thomas Haigh, Mark Priestley, and Crispin Rope, *ENIAC in Action: Making and Remaking the Modern Computer*

John Hendry, *Innovating for Failure: Government Policy and the Early British Computer Industry*

Marie Hicks, *Programmed Inequality: How Britain Discarded Women Technologists and Lost Its Edge in Computing*

Michael Lindgren, *Glory and Failure: The Difference Engines of Johann Müller, Charles Babbage, and Georg and Edvard Scheutz*

David E. Lundstrom, *A Few Good Men from Univac*

René Moreau, *The Computer Comes of Age: The People, the Hardware, and the Software*

Arthur L. Norberg, *Computers and Commerce: A Study of Technology and Management at Eckert-Mauchly Computer Company, Engineering Research Associates, and Remington Rand, 1946–1957*

Emerson W. Pugh, *Building IBM: Shaping an Industry and Its Technology*

Emerson W. Pugh, *Memories That Shaped an Industry*

Emerson W. Pugh, Lyle R. Johnson, and John H. Palmer, *IBM's 360 and Early 370 Systems*

Kent C. Redmond and Thomas M. Smith, *From Whirlwind to MITRE: The R&D Story of the SAGE Air Defense Computer*

Alex Roland with Philip Shiman, *Strategic Computing: DARPA and the Quest for Machine Intelligence, 1983–1993*

Raúl Rojas and Ulf Hashagen, editors, *The First Computers—History and Architectures*

Dinesh C. Sharma, *The Outsourcer: A Comprehensive History of India's IT Revolution*

Dorothy Stein, *Ada: A Life and a Legacy*

John Vardalas, *The Computer Revolution in Canada: Building National Technological Competence, 1945–1980*

Maurice V. Wilkes, *Memoirs of a Computer Pioneer*

Jeffrey R. Yost, *Making IT Work: A History of the Computer Services Industry*

Making IT Work

A History of the Computer Services Industry

Jeffrey R. Yost

The MIT Press
Cambridge, Massachusetts
London, England

© 2017 Massachusetts Institute of Technology

All rights reserved. No part of this book may be reproduced in any form by any electronic or mechanical means (including photocopying, recording, or information storage and retrieval) without permission in writing from the publisher.

Set in ITC Stone Sans Std and ITC Stone Serif Std by Toppan Best-set Premedia Limited. Printed and bound in the United States of America.

Library of Congress Cataloging-in-Publication Data

Names: Yost, Jeffrey R., author.
Title: Making IT work : a history of the computer services industry / Jeffrey R. Yost.
Description: Cambridge, MA : MIT Press, [2017] | Series: History of computing | Includes bibliographical references and index.
Identifiers: LCCN 2017002542 | ISBN 9780262036726 (hardcover : alk. paper)
Subjects: LCSH: Computer service industry--History.
Classification: LCC HD9696.2.A2 Y67 2017 | DDC 338.4/700409--dc23 LC record available at https://lccn.loc.gov/2017002542

10 9 8 7 6 5 4 3 2 1

To Linda, John, and Chris

Contents

Acknowledgments   ix

Introduction   1

I   The Industry's Origins

1   Constructing Expertise: Arthur Andersen, Diebold, and Canning-Sisson, 1953–1964   19
2   Processing Data: IBM-SBC and Automatic Data Processing, 1953–1964   45
3   Programming Machines: Computer Usage, C-E-I-R, and Computer Sciences, 1955–1967   63
4   Integrating Systems: System Development and Informatics, 1955–1969   89

II   The Industry's Identity

5   Cooperating Competitors: ADAPSO, 1961–1982   117
6   Managing Facilities: Electronic Data Systems, 1962–1984   135
7   Sharing Time: Tymshare, 1965–1984   153
8   Expanding Capabilities: IBM and Control Data, 1960–1988   177
9   Brokering Contractors: Gentry, Inc., Phyllis Murphy and Associates, COMSYS, and the NACCB, 1972–1998   211

III   Geographical and Organizational Change

10   Transforming Giants, Offshoring Work, and Creating Clouds, 1982–2016   231

Conclusion 273

Notes 287
Bibliographic Notes 347
Index 353

Acknowledgments

Above all I am grateful to my colleagues at the Charles Babbage Institute. Thomas Misa read the entire manuscript and offered many insightful suggestions. Additionally, he helped me free up substantial time (from my other CBI responsibilities) in 2016 to enable me to finish researching and writing the manuscript. Kathryn Charlet expertly proofed the whole manuscript, correcting errors and offering very useful stylistic suggestions. Past and present CBI archivists—Arvid Nelsen, Stephanie Crowe, and Amanda Wick—were also helpful to my research and in facilitating the use of CBI photographs. CBI Senior Research Fellow James Cortada has been an extremely valuable sounding board for ideas on this book project, as well as many other book and article efforts, and he read and commented on chapter 10 of the manuscript. I am also very grateful to my colleagues on the faculty in the History of Science, Technology, and Medicine at the University of Minnesota, a terrific and collegial group of scholars who have always been encouraging. Professor Emeritus from the program and CBI's founding director Arthur Norberg provided very useful guidance, as did the late Michael Mahoney, in my early years as a computer history specialist, to whom I will always be deeply indebted.

Coeditor of the History of Computing series for the MIT Press (along with Misa), William Aspray has been very helpful to me on this book project and throughout my career. He has frequently and expertly offered comments and guidance on my scholarly work. Similarly, Martin Campbell-Kelly has generously provided important feedback and encouragement for nearly two decades. Nathan Ensmenger has also been an invaluable colleague. I had a number of discussions with all three of these scholars regarding our revision to the jointly authored *Computer: A History of the Information Machine, 3rd Edition*. These discussions led to concentrated reflection on the past two decades of IT history, which was useful to me in thinking about the context of the recent history of computer services.

My book primarily is based on archival research, which I supplemented with some oral histories, trade literature, and other sources. Burton Grad and Luanne Johnson (industry veterans who were founders of Software History Center and longtime leaders of CHM Software Industry SIG) were helpful in facilitating donations of archival materials to CBI on the computer services industry (and also on software products). Among other collections, they assisted with ITAA's donation of the ADAPSO organizational records to CBI. The records of this association were invaluable to my writing of chapter 5. Further, Burt and Luanne organized and ran multiple history workshops on the IT services industry. They invited me to participate (as a moderator and oral historian) in all of these events, which was extremely beneficial to my research. Burt and Luanne donated many of the oral histories from these workshops to CBI (and the others to CHM). And Burt generously read the entire manuscript of this book prior to my submission of it to the MIT Press. I am very grateful to both Burt and Luanne. I am also very grateful to Kohichiro Hotta, Director of Fujitsu's Heritage Hall, for his guidance regarding resources on Fujitsu's services history.

The history of computing academic community is fast growing but remains very friendly and close-knit—quite simply, a terrific environment for scholars. One key factor to this camaraderie is the thriving Society for the History of Technology's (SHOT) Special Interest Group in Computers, Information, and Society (SIGCIS). Thomas Haigh's skillful leadership of SIGCIS for many years led to this organization's growth in membership and its increased scholarly engagement (especially the Sunday SHOT SIGCIS workshop). Andrew Russell took over from Tom in recent years and with his dedicated leadership the organization has continued to reach new heights. I am so grateful to both Tom and Andy. There are many members of SIGCIS who have commented on my written work and conference papers, and/or who I have had fruitful conversations with on the history of computing. Among these scholars, I would especially like to thank Janet Abbate, Atsushi Akera, Gerard Alberts, Ross Bassett, David Brock, Paul Ceruzzi, Gerardo Con Diaz, David Grier, Lars Heide, Marie Hicks, Peggy Kidwell, David Kirsch, Chigusa Kita, Jennifer Light, Ken Lipartito, Eden Medina, Mara Mills, Joe November, Joy Rankin, Corinna Schlombs, Rebecca Slayton, Hong-Hong Tinn, Ksenia Tatarchenko, Steven Usselman, and David Walden. In SHOT, but outside of SIGCIS, I wish to offer special thanks to Carroll Pursell and David Hochfelder. I am also deeply grateful to family and friends who have always been gracious and helpful at every step.

Introduction

The computer services industry—which includes consulting services, programming services, systems integration, management of data centers, and time sharing (a forerunner of cloud computing)—is one of the most important industries, and historically one of the least understood. The industry, which in 2014 had worldwide annual revenue of $954.8 billion and which employs millions of people, played a critical role in ushering in the information technology revolution of the second half of the twentieth century.[1] It now is helping to define and redefine IT (and the global economy). Despite its size, it has received far less attention—from scholars or from journalists or other writers—than the computer hardware industry or the software products industry.

The true importance of the computer services industry lies not in its impressive revenue, but in what it has accomplished in shaping the information technology landscape. Since shortly after the advent of digital computing, this industry has been fundamental to the organizational adoption of computers, and to making these machines and systems useful. It has long provided customized solutions for the IT wants and needs of customer organizations.

This book is at the intersection of the history of technology and business history. It seeks to understand information technology services within their larger technological systems consisting of (but not limited to) tools of the trade (computers, operating systems, networking, programming languages, and tacit knowledge), labor, sites of production, and various types of intermediate and end-user organizational customers. What this industry produces varies greatly in form and function, from the IBM customer engineer who "carries a bag" and keeps mainframes up and running to the systems analyst designing a new business application and overseeing a team of programmers writing code—the ethereal "ghost in the machine," that like the

Figure I.1
The Burroughs Corporation salesman shown in this photo carried a bag and helped keep installations of Burroughs equipment running in the mid to late 1950s. Courtesy of Charles Babbage Institute, University of Minnesota.

product of poets, is "created by the exertion of the imagination" and "only slightly removed from pure thought stuff."[2]

There is also great variance of scale and scope of computer services, ranging from an independent contractor doing network administration services for the staff of a small church or a consultant advising on simulation software tools for an energy exploration client to hundreds of programmers, analysts, and managers engaged in developing and deploying a real-time airline reservation system, or the efforts of an even larger systems integration team to create and refine a billion-dollar system of networked computers and software to manage logistics for the US Air Force in real time. With regard to business history, the book seeks to contribute to what historian Kenneth Lipartito recently has characterized as the "new business and economic history," in which the "material" and the "mental" are simultaneously examined to understand the past.[3] Using such an approach, I examine trade patterns, globalization, organizational structures, value chains, managerial philosophies, financial and business strategies, professional organizations, political economy, mergers and acquisitions, regulation, anti-trust lawsuits, and lobbying. At the same time, I also consider discourses, imagery, icons, language, memes, and metaphors—gender representation, gendered language and spaces, corporate cultures, notions about expertise, marketing of "solutions," and visualizing clouds.

Design and development of electronic digital computers in the United States first occurred during World War II, and commercial systems were on the market by the early 1950s. In the computer industry's first decade, most scientific and defense-oriented programming was done internally, by personnel within the organizations that had acquired digital computers. Scientists and engineers in government laboratories and defense-related organizations largely taught themselves and their colleagues to program, either in machine code or with programming tools (compilers) that translated more accessible notation into machine code. As systems grew increasingly complex, especially with the advent of early computer networking, the emerging computer services industry provided critical systems integration work.

Meanwhile, outside of the big science/defense sector, the computer services industry was fundamental to the origin (in the mid 1950s) and the phenomenal growth of computer-based business data processing. With large-scale systems integration, and particularly business data processing, the computer services industry also played a fundamental and continuous role in circulating IT knowledge, reducing the disadvantages to organizations with less internal IT infrastructure and uplifting the overall performance of information technology.

Client organizations and the trade and popular presses often have been critical of computer services companies and the broader industry, especially for real and perceived failures on specific, large-scale, high-profile government projects. Early problems with the roll-out and use of Healthcare.gov (the Web-based platform often referred to as "Obamacare")—an effort that cost hundreds of millions of dollars and involved sixty IT services firms—represents only one of the more recent debacles (real or perceived) of large-scale systems integration services.⁴ For contracting specialists and project managers at client firms and government departments and agencies, the presence of failures, delays, and cost overruns by their computer services contractors have at times seemed to be more the rule than the exception. These problems, however, have been at least as prevalent on comparable internal IT projects and speak more to the technical and managerial challenges of designing, developing, integrating, and maintaining software and systems than the computer services industry's underperformance or lack of expertise. Although large efforts in programming and systems integration always introduce major technical and organizational challenges, even small programming projects can be difficult. Some of the problems stem from the complexity of programming, others from difficulties with the identification and clear communication of the needed business attributes and functionality for systems to technically oriented programmers, and still others from clients' changing specifications in the middle of a project.⁵ Since the earliest days of computer programming, specialists and the trade press have often referred to programming as a "black art."⁶ Despite decades of efforts at software engineering, programming remains largely a craft enterprise that requires significant creativity and is deeply resistant to efforts to impose factory practices in order to scale up. Quite simply, creating software is hard and managing large-scale programming and systems integration is much harder.

Throughout the history of the computer services industry, the boundaries of the industry have been varied, rapidly changing, and difficult to define. The industry's complexity, rapid evolution, and sometimes stealthy nature (from classified military projects and corporate client preferences for inconspicuous services contractors to the industry's narrow marketing efforts relative to the computer hardware and software products trades) have contributed to its being largely ignored by historians and other scholars. The absence of high-quality sources of material on the history of the industry has also been problematic.

Martin Campbell-Kelly has published the most significant piece of scholarship on the early years of the computer services industry, a chapter in his book *From Airline Reservations to Sonic the Hedgehog: A History of the Software Industry*.⁷ In that highly insightful book, Campbell-Kelly focuses on the history of the software products industry, not on computer services. In detailing the history of software products, Campbell-Kelly rarely discusses the services industry beyond its first decade, or outside of two chapters. His chapter on the early days of the computer services industry essentially serves as a pre-history to software products; as he notes, some of the early services firms attempted to reap scale benefits by selling identical code (products) to multiple customers.[8] There are also several articles on the history of the industry, and a few internally produced company histories, but the literature is sparse compared to the computer services industry's long-standing technical, economic, social, and cultural importance.[9]

Though the scale benefits of selling products proved attractive, the services industry certainly did not diminish with the advent of software products. The software products industry emerged in the early 1960s, less than a decade after the start of the computer services trade, and the two industries have grown rapidly side by side ever since. The need for customized solutions, not just standardized products, remained paramount for many customers, and the computer services industry has always been the larger of the two industries—despite far greater public recognition of software products, especially in the era of the personal computer.

For decades, individuals have been customers of software products, whereas the computer services industry has long targeted primarily organizational customers and clients.[10] The emergence of one clear software products leader in the personal computer era (in revenue and as a provider of the most common applications and operating systems), Microsoft, has contributed to the high profile of products relative to services. Journalists have created a cottage industry writing books and articles about Microsoft and its co-founder, Bill Gates (now the wealthiest person in the world, with a net worth of approximately $75 billion).[11] Though Microsoft is targeting services as never before under its new CEO Satya Narayana Nadella's push to cloud computing, it is still widely perceived to be (and to a large extent it still is) principally a software products company. The only widely recognized public figure from the computer services industry, H. Ross Perot, is known far more for his battle with General Motors' then Chairman and CEO Roger Smith, for orchestrating a rescue of two of his employees from Iran in 1979, and for his 1992 campaign for the presidency of the US than as the visionary founder and longtime leader of Electronic Data Systems (EDS).[12]

The boundaries of the computer services industry, as an agent of and a respondent to our rapidly changing IT world and its surrounding institutions, practices, and people, have been continuously in flux. It is an enigmatic industry, characterized by diversity and dynamism, that sometimes defies clear definition. In addition to its seemingly amorphous nature, the lack of strong primary sources has long posed major challenges to the historical study of this industry. These have been among the reasons why the scholarly community (and other writers) have overlooked this fundamentally important industry for many years.

This book was made possible by newly available archival resources (particularly at the Charles Babbage Institute), by tapping long existing records in new ways (especially from the IBM Corporate Archives), and by oral history interviews and extensive mining of trade literature.[13]

A few major computer services corporations have been in operation for more than 40 years; among them are Automatic Data Processing, Computer Sciences Corporation, and International Business Machines. Big Eight accounting firms (including Arthur Andersen and Price Waterhouse) and management consulting firms (such as McKinsey and Company, and Deloitte, Haskins, and Sells, now known as Deloitte Touche) also have participated, to varying degrees, in this industry. In recent decades, large corporations long known for their leadership in computer hardware—IBM, Hewlett-Packard (HP), Fujitsu, and Xerox—have focused heavily on computer services as a primary profit center. IBM is currently the worldwide revenue leader of the computer services industry; and since its acquisition of EDS in 2008, HP has been second. Despite the presence of giants, many small enterprises—independent contractor brokerages, specialized consultancies, and self-employed individual contractors—have contributed mightily to this industry.

In short, the computer services industry (which recently has also been identified as "IT services") has been fundamental to making computers, software, networking, and overall IT systems *work* (that is, operate reliably, securely, and efficiently). It also has been important to the circulation of knowledge and best practices in IT, to the identification of new applications for computers, and to creating IT *work* (in the sense of extending the growth of the international IT labor force).

For many years the industry was largely national, and some segments and firms operated primarily or even exclusively in particular metropolitan areas or regions. The industry originated in the United States (easily the largest computer market throughout the second half of the twentieth century and beyond), and the US is still by far the global leader in computer services. The largest US computer services firms—IBM, HP, Accenture, and Computer Sciences Corporation—now have substantial facilities and presence overseas, and compete with international corporations such as Japan-based Fujitsu, France-based Cap Gemini, and India-based Tata Consultancy, Infosys, and Wipro. The Indian computer services industry has grown more rapidly in the past 15 years than its American or European counterparts, and Tata Consultancy has broken into the global top ten in

IT services industry revenue. With all of the largest computer services companies operating on an international basis, the location of origin, or the corporate headquarters, has become less significant over time. Cognizant (based in Teaneck, New Jersey) is often listed with the Indian giants, since the vast majority of its staff of over 200,000 is based in India. Yet alongside the industry's globalization and the rise of large multinational companies, many thousands of small computer services firms continue to operate exclusively within a 50-mile radius.[14]

In the early to mid 1950s, when pioneering firms launched computer services, they focused on distinct types of services. Martin Campbell-Kelly concentrates heavily on Computer Usage Corporation (CUC) and on C-E-I-R, both of which were programming services enterprises, in characterizing the launch and early history of the computer services industry. (CUC formed in 1955 and C-E-I-R entered the computer services market in 1956.) Although these firms were important, and although the programming services segment has long been one of the industry's most important segments, computer services really originated in two other segments: consulting services and data processing (service bureaus). Both of those segments began in 1953 (service bureaus have a much older pre-computer history), two years before the programming services segment. Prior to contracting with providers of programming services, client companies and organizations had to decide to acquire mainframe digital computers and to decide what mainframe system to buy or lease. Some took their cues strictly from the mainframe firms—especially if they had been longtime customers of the computer provider for pre-computer punch card tabulation machinery (as many had been with IBM or Remington Rand). Many companies and organizations, however, wanted independent expertise on making the important decision whether to invest in entering the computer age, the timing for this move, what computer system to acquire, whether to buy or to lease, what applications to focus on, and what programming to do in house and what programming to outsource; they also wanted detailed cost–benefit analyses of different possibilities. Consulting firms also provided thorough studies of competitors in individual industries to help firms understand threats and opportunities for IT in their particular trade. In 1953 and 1954, several businesses—among them the accounting firm Arthur Andersen and Company and startup companies John Diebold and Associates and Canning, Sisson, and Associates—launched the computer services industry by providing IT consulting services. In this book, by providing case histories of those three firms, I show how they provided customers with the know-how they needed in order to use digital computers effectively.

With the advent of computer networking, systems became increasing complex. Firms emerged to help organizations (primarily departments and agencies of the federal government) integrate systems by bringing hardware, software, and networking together to create effective and efficient systems. System Development, founded in 1956, was the first systems integration firm.

For companies and for organizations, outsourcing computer-based data processing became an alternative to acquiring digital computers, greatly reducing upfront costs. Data processing using pre-computer punch card tabulation dated back to the late nineteenth century, and often services were bundled with purchases or rentals of tabulation machines. In 1932, IBM formally launched fee-based punch card data processing services by selling time on tabulation machines and expertise from IBM machine operators through its newly formed service bureaus in major cities around the world. IBM's first computer-based service bureau was launched in 1953.[15]

Whereas IBM provided a broad array of data processing services, specialty data processing providers emerged after World War II. Of these specialized data processing providers, none became more significant than Automatic Payrolls, Inc., founded in 1949. In the late 1950s, the firm—still specializing in the involved, regular, and mundane task of preparing payrolls for clients—switched to incorporate first punch card tabulation equipment and then (at the start of the 1960s) digital computing technology and had changed its name to Automatic Data Processing (ADP). Consulting services, programming services, systems integration, and data processing outsourcing (or sale of machine/system time) characterized the formative years of the computer services industry—the focus of part I of this book.

Throughout the book, I use case studies of individual firms to introduce particular segments of the computer services industry, highlighting the visionary leaders who launched different areas of the industry and in the process forever transformed computing and software. In addition to the case studies, I discuss other firms briefly to provide further context. The companies (cases) were chosen both for their prominence and as representative examples of currents in the broader industry.

Early on, virtually all computer services firms specialized in a particular segment: consulting, programming services, systems integration, or business data processing/service bureau operations. These first four segments originated in the 1950s. Two additional segments, facilities management and time sharing, emerged by the mid 1960s. To a certain degree, time sharing (as a services business) was merely an innovative way of extending the functions of service bureaus to take advantage of computer networking and the sharing of resources across different geographies (computer processing no longer needed to be in the same location as the computer user). It also fostered more direct interaction with systems by clients than in batch processing environments.

The requisite knowledge, skills, practices, marketing, and contracting strategies differed significantly among these various segments of the computer services industry. Eventually, most of the large-scale computer services companies invested in developing the resources and organizational capabilities to provide a range of different types of computer services, gradually adding additional industry segments to their repertoire of businesses.[16] For instance, within five years of its formation in 1959, the programming services specialist Computer Sciences Corporation developed a highly successful systems integration business focusing heavily on government clients.

The book concentrates mostly on developments in the United States and on firms based there. In large part, this reflects the industry. Three of the today's four global leaders among IT services companies (IBM, HP Enterprise, and Accenture) are based in the US; Japan-based Fujitsu rounds out the top four. All four of these firms have personnel and offices spanning much of the world. In the more distant past, all of the industry's top ten companies were based in the US. Though the book unquestionably has a US focus (and is in part a product of resources available in archives located in the US), I also explore the overseas expansion of US-based industry giants (IBM, Andersen/Accenture, Control Data, Tymshare, and others). Of equal significance, I analyze the history of the IT services powerhouses Fujitsu (based in Japan), Cap Gemini (based in France), and India's Big Five (Tata Consultancy, Wipro, Infosys, HCL Technologies, and Cognizant) to provide further international perspective on this global industry.

Chapter 1 concentrates on Arthur Andersen and Company, John Diebold and Associates, and Canning, Sisson, and Associates, and on the industry segment of computer consulting services that those firms pioneered. Arthur Andersen and Company was a large accounting firm that, like several of its competitors, also engaged in management consulting. It probably was the first company in the world to contract to provide computer consulting services. In 1953 it was hired by General Electric to provide advisory services on computer-based payroll and inventory management systems for GE's pioneering Appliance Park facility in Louisville. John Diebold and

Associates and Canning, Sisson, and Associates, both launched in 1954, may well have been the two earliest startup computer consulting firms.[17] All three of the case studies in chapter 1 concentrate on how computer services companies constructed and conveyed expertise at a time when computer-based business data processing was entirely new. These case studies also introduce the fundamental role of computer consulting companies, and later of all computer services firms, in circulating knowledge and best practices among clients.

Although many larger corporations had acquired computers by the late 1950s, others had not. Some found it more efficient to rent computer time or to outsource data processing. Computer-based service bureaus that sold computer time and provided certain other computer data processing services emerged in 1953 with IBM's Technical Service Bureau, which later became part of IBM's Service Bureau Division. In 1956, as part of an IBM settlement of a US Justice Department anti-trust lawsuit, IBM transformed the Technical Service Bureau into the Service Bureau Corporation (SBC), a wholly owned subsidiary. All of the service bureaus of the 1960s maintained, or brokered the use of, mainframe systems and sold computer time; some of them also offered associated programming and data processing services (processing the bureaus would do for the client). Chapter 2 details the activities and the growth of computer-based service bureaus (including their pre-history in punch card tabulation bureaus) and of data processing specialists such as Automatic Data Processing (ADP).

Chapter 3 analyzes three of the earliest programming services companies: Computer Usage Corporation (CUC), C-E-I-R, Inc., and Computer Sciences Corporation (CSC). In 1955, Computer Usage and C-E-I-R pioneered programming services, which allowed customers to make use of their new digital mainframe computers to advance their enterprises. One important element of success in programming services was recruiting and retaining talented programmers and systems analysts. Programmers could vary widely in productivity, and often there were shortages of quality programmers. Many in the industry and trade presses spoke of a "software crisis," in which software and its creators were the bottleneck to getting the most from computing systems. The case study of C-E-I-R in chapter 3 benefits from the Charles Babbage Institute's archive of the firm's records. CSC began nearly five years later than CUC and C-E-I-R, but thanks to important early contracts it grew faster than these companies. Chapter 3 also introduces the topic of gender in the industry.

IBM was the primary computer hardware contractor for the Semi-Automatic Ground Environment (SAGE) radar and computerized air defense system in the mid 1950s, and was solicited by the Air Force to take on SAGE programming and systems integration work. Ultimately, IBM's leaders declined the opportunity to take the lead in this new and highly challenging programming and systems integration task, and the Air Force turned to nonprofit research corporation, RAND, for a solution. In the mid 1950s, the RAND Corporation established System Development Division— soon spun off and renamed System Development Corporation (SDC)—to do the complicated programming, debugging, and systems integration work on SAGE. By the late 1950s, SDC had more programmers and systems analysts on staff (over 500) than any other organization in the world, and the company grew larger than its parent, RAND. There was high turnover at SDC, and over the succeeding ten years former SDC employees helped to populate the computer industry, the computer services industry, and the emerging software products industry.

Informatics, a for-profit corporation founded in 1962, provides a contrast to the (initially) nonprofit SDC. For a number of years the federal government was the primary source of contracts for systems integration, but soon more and more large corporations (among them airlines seeking real-time reservation systems) became important clients. Most of the major mainframe computer firms launched divisions to serve the federal government on systems integration projects by the 1960s—though this services work, from a revenue standpoint, was sometimes invisible as it was bundled in with the cost of hardware. Chapter 4 focuses on the histories of SDC and Informatics.

The computer services industry originated with consulting services, soon followed by the data processing, programming services, and systems integration segments. Part II of this book explores the emergence and evolution of the computer services industry's identity in the 1960s, the 1970s, and the 1980s.

Unfortunately, relatively few scholars have followed business historian Louis Galambos in analyzing trade organizations.[18] In considering the computer, software, and services industries, scholars have conducted in-depth studies of professional organizations (including the Data Processing Management Association) and software user groups (including SHARE, Inc.), but have paid far less attention to trade organizations.[19] An exception is Larry Browning and Judy Shetler's valuable study of a prominent semiconductor trade association, Sematech, but there is nothing comparable in computing, software, or services.[20] Not only do trade associations help competitors cooperate to advance an industry (and segments within it); they help define an industry's identity. This was certainly true with the Association of Data

Processing Services Organizations (ADAPSO),[21] launched in 1961 by service bureau companies to represent and advance the interests of their industry.[22] ADAPSO quickly broadened to serve other types of computer services businesses, and in time to serve software products enterprises as well. ADAPSO's history and the emerging identity of the computer services industry in the early 1960s are the topics of chapter 5, which benefits greatly from use of ADAPSO's organizational records in the archives of the Charles Babbage Institute.

Chapter 6 and 7 focus on two new segments of the computer services industry that originated in the 1960s: facilities management and time sharing. Both of those segments sought to advance efficiencies in computing and data processing, but they had different origins and paths. Facilities management, as with all computer services, required technical skill, but primarily was an organizational invention (outsourcing that combined programming services, data processing services, and maintenance services) to improve efficiency. In contrast, computer time sharing consisted in part of technical advances in networking and operating systems to aid efficiency, enhance the user experience, and make computing more economically and geographically accessible.

H. Ross Perot, the founder of Electronic Data Systems, astutely recognized that many firms that had been early adopters of digital computers had peaks and troughs in workload (processing time and labor) in their data processing departments. Some of these organizations had gotten in over their heads in rapidly transitioning from pre-computer data processing to computer-based data processing. Perot and EDS established a business model of contracting to completely take over an organization's IT facilities on a fixed-price basis, and utilized downtime for workload and computer processing resources (excess capacity) at one client to serve other clients.

In the early 1960s, scientists at the Massachusetts Institute of Technology invented time sharing, later reinvented at other institutions and then modified and commercialized by General Electric, Tymshare, and other for-profit corporations.[23] Time sharing rapidly rotated (at split-second intervals) the processing resources of mainframe computers among many users (at basic terminals or smaller computers) to give users the experience of having their own mainframe.[24] In its early years it operated through either local networking or telecommunications-based remote networking. In the late 1960s several time-sharing enterprises built computer networks connecting facilities within many cities throughout the nation. For a number of years, Tymshare was second only to General Electric (GE) in the computer time-sharing industry—both of which built sizable computer networks. Some computer services providers that focused on other business areas quickly added time-sharing and facilities-management businesses. Time sharing was a major force in bringing computing resources to new customers and to new regions of the nation and the world, and was a partial realization of efforts to make computing a broad-based, accessible "utility."

To varying degrees, the mainframe firms of the 1960s provided computer services to their customers. In some cases it was bundled as part of the package to hardware customers; in others it was established as a distinct business division. While IBM was by far the largest computer producer, controlling 60 percent or more of the industry from the late 1950s through the 1970s, it also engaged in major services efforts, such as the programming and systems integration of the Semi-Automatic Business Research Environment (SABRE)—a project completed in partnership with its customer American Airlines. IBM also provided its hardware customers with extensive educational services. IBM Customer Engineers (CEs) kept machines operable with hardware maintenance and debugging assistance. Beginning in 1960, IBM System Engineers (SEs) provided clients with systems analysis consultation. Further, IBM's Federal Systems Division provided extensive computer services to the federal government throughout the 1960s, the 1970s, and the 1980s.

IBM, as the largest mainframe producer, had far more services personnel than any competitor in the computer industry. In terms of the proportion of services revenue to overall revenue, however, no mainframe manufacturer was more deeply involved in selling computer services before 1990 than Control Data Corporation (CDC). Although CDC historical literature focuses heavily on the company as the supercomputer industry leader, by 1964, when its first supercomputer (the CDC 6600) hit the market, CDC was firmly committed to growing its already considerable computer services business. It did this both through internal expansion and through acquisitions—it acquired the programming services pioneer C-E-I-R in 1967 and (from IBM) the Service Bureau Corporation in 1973. Chapter 8 documents IBM's pre-1970s work in the services field and CDC's transition to become primarily a services company. The analyses of the services work and services businesses of these two firms were made possible by resources in the IBM Corporate Archives (at IBM) and the Control Data Corporation Records (housed at the Charles Babbage Institute).

By the late 1960s the computer services industry consisted of a number of large corporations—including IBM, CSC, EDS, and CDC—but the industry has always had many small enterprises (including many businesses serving only one local metropolitan area). Some of these small firms competed with the large corporations for contracts, but more often they focused on providing computer services to small to mid-size companies, local governments, and nonprofit organizations. Without these smaller providers of computer services, many enterprises would not have been able to effectively and efficiently adopt computing technology for their data processing needs. Some clients relied on self-employed independent contractors for computer programming. Many independent contractors, however, lacked either the ability or the desire to network and to successfully market themselves to potential clients. In addition, marketing could be time consuming and expensive for individual contractors, reducing billable hours. Clients often needed multiple contractors for programming projects, and there was much administrative effort and worrisome risk with hiring numerous independent contractors. Meanwhile, traditional computer services companies, in keeping salaried employees on staff ("on the bench") when they were not out servicing clients, created inefficiencies. In 1972, seeing all this, Grace Gentry established a new segment of the computer services industry: independent contractor brokerages. Her company—Gentry, Inc.—pioneered the industry segment of firms brokering relationships between independent contractors and customer organizations/clients. Carefully screening and assessing the skills of experienced programmers and systems analyst contractors, Gentry placed qualified contractors in client firms, taking a modest percentage of what the client paid for each person-hour of work.

Chapter 9 analyzes Gentry, Inc., and two other early independent contractor brokerages: Phyllis Murphy and Associates and COMSYS. These companies worked with between 50 and 200 contractors at a time, placing them in both short-term and long-term projects with clients. The chapter also explores the trade association for this industry segment, the National Association of Computer Consultant Businesses. The NACCB—launched solely to combat changes successfully lobbied into federal tax code by ADAPSO, whose members consisted of larger services companies that wanted to reduce or eliminate competition from independent contractors and brokerages—quickly evolved into a broad-based trade association for independent contractor brokerages. The chapter also analyzes gender in the computer and computer services industries. Grace Gentry, Phyllis Murphy, and other female entrepreneurs had faced substantial barriers to advancement while working for large organizations in computing and data processing earlier in their careers; becoming entrepreneurs, their own boss, was a way to potentially move past that. The chapter was made possible by oral histories and by NACCB records provided to me by Grace Gentry.

Since the late 1980s, computer services companies and the broader industry have gone through fundamental organizational and geographic changes. Chapter 10 focuses on those changes, surveying three themes—transforming giants (IBM, HP, Fujitsu, Capgemini, and Andersen/Accenture), offshoring enterprises and labor (India's Big Five IT services providers and US-based IT services companies changing geographical deployment of their workforces), and cloud services—over the past 35 years. Cloud computing, a mid-1990s re-invention and reformulation of time sharing, is a dynamic, paradigm-changing, fast-growing phenomenon that has attracted startup companies seeking to define the field for particular applications such as Salesforce.com (in customer relationship management software), existing IT services giants (IBM, HP, Fujitsu, Capgemini, CSC, and India's Big Five), and large firms from other areas of IT (among them the e-commerce powerhouse Amazon and software products leader Microsoft). With such a cast of providers, cloud computing's momentum is strong and the IT services industry is evolving rapidly.

The case-focused approach of this book provides a look into the tremendous diversity of the computer services industry over the past six decades and the evolving strategies of many firms. The book's structure also facilitates analysis of a number of core themes. First and foremost is the role of the computer services industry in shaping information technology and making IT work over the past 60 years (creating applications for computers that operate effectively). Second, the industry has become a strong contributor to the US economy and making IT work (in terms of national and international jobs creation—the global computer services industry labor force is composed of millions of workers). The falling price for processing power and memory (Moore's Law), coupled with the growth of computer networking knowhow and infrastructure, has enabled considerable offshoring of some types of computer services work (particularly to India—both to US multinationals overseas as well as to overseas-based firms), while other services work is strongly tied to place (either by necessity or preference/opportunity). Third, the computer services industry has long had a fundamental role in the circulation of IT knowledge and skill. This has helped to level the IT playing field, partially offsetting disadvantages faced by organizations with less internal IT infrastructure.

Fourth, the computer services industry has played a major role in boosting efficiency and flexibility. It allows firms to outsource IT work to true specialists for particular types of applications, and to balance the inevitable uneven needs for IT labor and computer processing—cloud computing appears to further advance such economies and efficiencies. Only time will tell whether cloud computing becomes the new dominant IT services model. It is currently approaching 15 percent of overall IT services industry revenue, and is growing much more rapidly than the overall IT services industry.

I The Industry's Origins

1 Constructing Expertise: Arthur Andersen, Diebold, and Canning-Sisson, 1953–1964

On February 14, 1946, nearly six months after the end of World War II, the US Department of War officially dedicated the Electronic Numerical Integrator and Computer (ENIAC) at the University of Pennsylvania. The machine, a 30-by-60-foot behemoth containing about 18,000 vacuum tubes, had been operating for three months. Various types of analog and electromechanical digital computers had preceded it, from Vannevar Bush's Differential Analyzer and George Stibitz's relay-based Complex Number Computer in the 1930s to Howard Aiken's Mark I and Konrad Zuse's Z3 in the early 1940s. But the ENIAC was the first meaningful electronic digital computer.[1]

The US Army provided more than $400,000 for the design and development of the ENIAC, a project led by the physicist John Mauchly and the electrical engineer J. Presper Eckert at the University of Pennsylvania's Moore School of Electrical Engineering. The Department of War authorized the funds to create a machine of unprecedented speed and accuracy for calculating ballistic firing tables. From the launch of the project through the end of the war, the ENIAC effort was little known, having been classified as "confidential."[2] The *New York Times* reported that at the dedication officials of the Department of War characterized the ENIAC as a "tool with which to begin to rebuild scientific affairs on new foundations."[3] In December 1945, unbeknownst to spectators, the ENIAC had been used to calculate equations on thermonuclear reactions in support of research for developing a hydrogen bomb.[4]

Scientists, engineers, government officials, and the press all understood the ENIAC and other electronic digital computers of the second half of the 1940s to be advanced scientific calculating machines. Common metaphors for the ENIAC and other digital computers of the time were "giant brain" and "electronic brain."[5]

In 1946, Eckert and Mauchly left the Moore School and founded the Electronic Control Company (soon renamed Eckert-Mauchly Computer Corporation) to design and manufacture digital computers. That same year, in St. Paul, former US Navy cryptographers and investors launched Engineering Research Associates to engage in naval electronics contracting and to design and build digital computers. In their first years, these pioneering computer companies focused solely on building computational machines and other electrical equipment for the scientific, engineering, and defense markets.

At a session of the 1949 annual meeting of the American Institute of Electrical Engineers (AIEE) that lasted nearly four hours, engineering professionals discussed the future possibilities for digital computers. Among the list of potential applications, the engineers postulated that digital computers could be useful for flying airplanes, tuning televisions, forecasting

Figure 1.1
Frances Bilas (right) and Betty Jennings (left) adjusting program settings on the ENIAC master programmer in 1946. Courtesy of Charles Babbage Institute, University of Minnesota.

the weather, and automating manufacturing. This distinguished group also debated what was required for a computer to "have a will of its own" and what constituted a "thinking" device.[6] At the meeting, Jay Forrester, an engineer from MIT's Servomechanisms Laboratory who was orchestrating the expansion of an analog flight simulator project into a successful effort to build the first real-time digital computer, declared "present efforts in digital computing are only exploratory, the engineering effort which will ultimately be justified may exceed that devoted to the wartime development of radar."[7] While Forrester and others understood the limitations of comprehending a new technology in its infancy, this group of distinguished AIEE engineers nevertheless correctly identified important future applications, recognized the importance of computers to questions being raised in the young field of cybernetics, and anticipated issues that would be central to the future field of artificial intelligence. Absent from these engineers' reported discussions was the possibility of applying digital computers to data processing for businesses.[8]

In the second half of the 1940s and in the early 1950s, the state of the art in business data processing was the use of electromechanical punch card tabulating machines from the office machines giants International Business Machines (IBM) and Remington Rand. IBM began under the name Computer-Tabulating-Recording Company (C-T-R) in 1911. It was a leading producer of punch card tabulation machines from its origin well into the 1960s (when selling and leasing computers became its primary business).

The transition from punch card tabulation to computers took time and was uneven among various data processing machine user industries. Before 1960, Remington Rand was IBM's primary competitor in the market for punch card tabulation machines. As historian Thomas Haigh has pointed out, in the early 1950s it was far from clear that the electronic digital computer would be the key electronic product for business data processing.[9]

Like IBM's leaders, top executives at IBM's competitors (Remington Rand, Burroughs, and National Cash Register) fully understood by the late 1940s that the future of office machines lay in electronics.[10] Those firms, however, lacked the extensive financial resources that IBM was able to devote to greatly expand its electronics research and development. Recognizing the market for scientific computing and hopeful of applications of digital computers in business, Remington Rand acquired the Eckert-Mauchly Computer Corporation (in 1950) and Engineering Research Associates (in 1952) in stock-based deals.[11]

While Remington Rand's Universal Automatic Computer, (UNIVAC)—first made available in 1952—was designed and marketed as a machine for both science and business, in most configurations the price exceeded $1 million, an amount few corporations could justify spending for new data processing technology. The much smaller IBM 650 sold for between $200,000 and $400,000, and leased for as little as $3,500 a month.[12] IBM had more than 2,000 IBM 650 installations (sales and leases) during the 650's eight-year production run, compared with fewer than 50 installations for Remington Rand's UNIVAC. IBM also continued to generate substantial revenue from selling and leasing its pre-computer punch card tabulation systems throughout the second half of the 1950s and into the 1960s.[13]

In the 1950s, a decision to acquire a digital computer (or multiple computers) appeared daunting for most corporations and other organizations. Mainframe computers were extremely expensive and were technological and institutionally disruptive. Moreover, the economic payoff digital computers might provide for business data processing was uncertain. Pre-computer punch card tabulation equipment, which most major American corporations had acquired between the late 1920s and the early 1940s, also had been costly, but significantly less costly than digital computers. Many corporate decision makers likely had some knowledge of the challenges of getting punch card tabulation departments up and running. By the advent of the first viable commercial computers on the market (in 1952), established routines were in place within corporate data processing departments—from the technology itself and its spatial location to the personnel, their knowledge and skills, and organizational culture. The thought of completely uprooting these organizational structures and routines in favor of new technology tended to be overwhelming—and represented just the start of the many challenging questions to answer: What would be the first applications? What formal training procedures would be launched? What new positions would be created? What would be done in house? How would corporate or organizational reporting responsibilities change?

In opposition to the many reasons for corporate leaders' apprehension with taking the leap to an expensive and unproven new technology for business data processing was the fear of being left behind. By the mid 1950s, computer manufacturers' sales staffs, journalists, consultants, and others had fueled this fear by serving as evangelists for the inevitable electronics-based (and sometimes specifically computer-based) technological revolution in business data processing.[14]

Three particularly important and very different works of historical scholarship have been published to date on early corporate adoption of computers in the United States: Thomas Haigh's article on the first five years of the "electronic revolution" in business data processing, JoAnne Yates's book on punch card tabulation and computer data processing in the life insurance industry, and James W. Cortada's trilogy surveying computing in more than forty different industries and economic sectors. Haigh highlights the revolutionary rhetoric and provides a meaningful look into the occupations and power struggles in corporations.[15] Yates conveys the importance of continuity and gradual migration to new information technology in the life insurance industry and the influence of this trade in helping to shape punch card tabulation and computing.[16] Cortada's three-book survey offers substantial breadth in documenting major currents in the usage of computing over decades in many industries, and government and educational sectors. The focus of all three of these scholars' works lies outside of the role consulting firms played in the organizational adoption of computers.[17]

Haigh quotes several 1950s consultants espousing revolutionary rhetoric on computing.[18] This accurately conveys a general pro-computer perspective among some consultants in the early computer industry. There were a number of self-identified individual consultants by the mid to late 1950s—some with minimal knowledge but anxious to contribute to and capitalize from a created revolutionary moment in electronics and computing. Other computer consultants, however, began the process of developing *true* expertise, engaged in exhaustive research and analysis, and became consultancies that client organizations returned to time and again. The remainder of this chapter explores three of those businesses: Arthur Andersen and Company (whose accounting division no longer exists, but whose consulting division is now Accenture, the third-largest IT services firm in the world), Diebold and Associates (which grew throughout the 37-year tenure of its visionary leader John Diebold), and Canning, Sisson, and Associates (which served clients for nearly ten years before Richard Canning left to "scale up" the expertise he sold by entering the technical/managerial advisory publishing business). These three businesses, which were among the first multi-person computer consulting enterprises in the United States, represent the proper beginning of the computer services industry. In all three cases, the themes of constructing expertise (both in terms of building real expertise and in terms of projecting expertise as they learned in a new and rapidly evolving field) come to the fore. The three short case studies presented below also exemplify the computer consulting industry's role in circulating knowledge and best practices in computing and programming—a defining theme to this day, and one that is evident throughout the computer services industry. The case of Arthur Andersen and Company provides a first glimpse into the somewhat porous and increasingly diminishing boundaries between accounting consulting and managerial consulting, on the one hand, and IT consulting, on the other.

Arthur Andersen and Company

In 1913, Arthur Andersen and Clarence DeLany left the accounting firm Price Waterhouse to purchase the Audit Company of California to form Chicago-based Andersen, DeLany and Company (renamed Arthur Andersen and Company in 1918).[19] Since the late nineteenth century, engineers and cost accountants had worked side by side at accounting firms conducting administrative examinations for their clients. Like its industry peers, Arthur Andersen and Company began to perform such services regularly shortly after the launch of the new firm—services that increasingly took the form of modern management consulting.

As historian Christopher McKenna convincingly argues in his book *The World's Newest Profession*, cost accounting and the nature of administrative examinations by accounting firms transformed with the rise of "associationalism" in the 1920s as future president Herbert Hoover took control of the Commerce Department. With his engineering mindset, Hoover sought to solve the problem of excessive competition without resorting to heavy government intervention. During Hoover's tenure as Secretary of Commerce (1921–1928) and as the 31st president of the United States (1929–1933), administrative examinations by cost accountants and engineers provided a conduit for industry-wide sharing of cost information and for conducting managerial audits. Such organizational and strategic audits became the core of modern management consulting. As McKenna relates, in 1933 (early in President Franklin D. Roosevelt's administration) the Glass-Steagall Act forced accounting firms (and banks) out of this activity and spawned a thriving independent management consulting industry.[20]

McKenna argues that the federal government took (managerial) consulting away from accounting firms with the Glass-Steagall Act, but enabled a specialized form of it (a gift of sorts), computer consulting, by preventing IBM from entering the field with a 1956 consent decree. While McKenna's book is a much needed high-quality scholarly history of managerial consulting (largely focused on strategy consulting), his quite brief discussion of computer consulting is distorting. McKenna draws from historian Robert Sobel in relating that IBM sales and service staff would aid with computer installations to make sure everything went right, but fails to recognize the importance of these bundled services IBM provided. He also exaggerates or overemphasizes a counterfactual (if the IBM consent decree with the Justice Department did not exist, no significant independent computer consulting business would have emerged to compete with IBM), and he is off in his chronology. Andersen and Company bundled data processing advisory services before it became a focused Administrative Services Division, and IBM bundled software and services for decades before it became a major profit center for the firm. An important element of the consulting business was impartial outside advice. Many firms would not have been comfortable securing all their advice and guidance from one source—IBM, the leading hardware firm (or from any of IBM's mainframe-producing competitors). The notion that IBM could have controlled all or even most computer services (and especially consulting advisory services) seems dubious. IBM is the largest computer services company in the world today and is fully committed to this area as its primary business, yet its share of the global market is less than 6 percent.[21]

In the early postwar years, a decade before the IBM consent decree, Arthur Andersen and Company expanded its Administrative Accounting Division, which specialized in various administrative services, including punch card tabulation. In the 1930s and the early 1940s many of its professionals had learned about business data processing technology from working with punch card tabulation systems at client firms. More significantly, one technology-focused manager and two future Andersen employees who had gained deep expertise in state-of-the-art punch card systems in the military during the war led to Andersen and Company's revitalizing administrative advisory work for clients on punch card tabulation systems and, later, digital computers.

Leonard Spacek, who upon Arthur Andersen's death in 1947 succeeded him as managing partner, began his tenure with Andersen and Company in 1928.[22] During his 45 years at the firm, Spacek served countless clients, but estimated he did fewer than ten accounting audits.[23] His specialty was administrative services, a division of Andersen he expanded after gaining leadership of the company. Shortly after World War II, Spacek hired John Higgins and Joseph Glickauf. Higgins had worked at Andersen and Company previously before joining IBM and later serving in the Navy during the war (specializing in management of mechanization projects). Glickauf joined the Navy in 1942 and worked on developing more efficient payroll and inventory systems at the Naval Bureau of Supplies and Accounts in Cincinnati. Spacek hired these two naval punch card tabulation system specialists not for accounting (Glickauf, in particular, had little knowledge of accounting), but to advise clients on the area of expertise they had gained during the war: data processing systems.[24]

In 1950 Spacek met with Willis Gale, the chief executive of Andersen's largest client, Commonwealth Edison. Commonwealth Edison was not merely an accounting client; it also retained Andersen and Company's advisory services, the Administrative Accounting Division, to improve management systems to track its extensive property records. At the meeting, Gale, who had a deep technical interest and curiosity, asked Spacek about possible digital computer applications for data processing. Spacek assured Gale he would look into that possibility right away and assigned Joe Glickauf to the task. Glickauf met with J. Presper Eckert and John Mauchly and received a demonstration of the UNIVAC later that year. In 1951 Glickauf and other Andersen employees built a small digital computer, dubbed the Arthur Andersen Demonstration Computer (AADEC), to advance their digital computing knowledge and skills. Spacek, Glickauf, and Higgins selected five of the most technically astute Andersen employees to study digital computing—one of whom, Joe Carrico, spent six months with Eckert and Mauchly (both of whom had become friends with Spacek). By 1952 the Administrative Accounting Division was renamed Administrative Services Division to more accurately represent its activities (primarily data processing system advisory services).[25]

On March 27, 1953, General Electric executives, having heard about the commitment Arthur Andersen and Company had made to researching digital computer applications, hired the company for a $64,000 feasibility study for a UNIVAC I installation to run payroll processing and other systems for GE's new Appliance Park facility in Louisville. Later that year, in anticipation of the UNIVAC that would arrive at Appliance Park in a matter of months, Andersen and Company's Administrative Services Division conducted its first "Programming and Installation School" (held at the downtown Chicago campus of Northwestern University) for Carrico and other pioneers to teach 26 new hires or assignees to the Administrative Services Division.[26]

John Higgins and Joseph Glickauf, the leaders of Andersen and Company's Administration Services Division, published an article titled "Electronics Down to Earth" in the March-April 1954 issue of the *Harvard Business Review*.[27] It highlighted the General Electric Appliance Division's planned UNIVAC installation, but more broadly it raised questions for businesses contemplating acquiring computers for business data processing. As the title suggests, Higgins and Glickauf tried to convey balance in the face of the grandiose, revolutionary rhetoric that promised the computer as the cure for all business ills. In the article they mentioned a second client for which Andersen and Company had conducted a computer feasibility study—a firm that was significantly smaller than GE. That unidentified firm had 5,000 employees and a complex payroll because of its excessively high number of job and benefit classifications. Andersen and Company's suggestion to the management of the firm was to reduce unneeded complexity by having fewer classifications so as to be able to run payroll economically using their existing pre-computer punch card tabulation equipment. While Higgins and Glickauf's article is optimistic on the future of digital computers for business data processing, it focuses on the importance of deep analysis in finding appropriate solutions for individual organizations and their circumstances. Higgins and Glickauf also argued that the growing preoccupation with high-speed printing was unwarranted, as most organizations could not benefit significantly enough to offset the higher costs of the equipment. Virtually all computer services consulting companies sought to follow the lead of Higgins and Glickauf and to publish business advisory articles and books early and often. Though some resorted to far less realistic assessments by blindly promoting the coming revolution for business (probably reasoning that it would result in more consulting revenue), others, Higgins and Glickauf among them, provided thorough, deeply reasoned, and customized research-based studies.

The manager of the Business Procedures Section of General Electric's Louisville plant, Roddy Osborn, was in charge of the company's UNIVAC installation project. Not missing an opportunity to promote GE, he published a company-aggrandizing article titled "GE and UNIVAC: Harnessing the High-Speed Computer" in the *Harvard Business Review*. "[W]hile scientists and engineers have been wide-awake in making progress with these remarkable tools [digital computers]," Osborn stated early in the article, "business, like Rip Van Winkle, has been asleep. GE's installation of a UNIVAC may be business's first blink."[28]

As became commonplace in the years and decades that followed, the outsourced advisory, training, and system-building services were downplayed by the client, saving brief mention of Andersen and Company for the next-to-last page of the nine-page article.[29] Though what Osborn did write went further than many future clients in publicly recognizing computer services contractors, he commented that "because of the pioneering aspect of the program, we found it wise to employ an independent management consultant firm ... Arthur Andersen ... experienced in computer logic and developments."[30]

More than many other early computer consulting firms that focused on advisory services, Andersen and Company quickly contracted to provide computer programming services as well. It contributed critical programming work for the GE payroll application in 1954, and it probably was the first firm in the United States to engage in contract work for computer programming services. Running into problems, the Andersen and Company team had to engage in reprogramming the entire payroll system, which Glickauf promised that Andersen and Company would do without additional charges. Roddy Osborn and other executives at Appliance Park understood the challenges of pioneering in the new field of applications programming, wanted to be fair to Andersen and Company, and agreed to pay for the reprogramming on a cost-plus basis.[31]

The high-profile GE project helped establish Andersen and Company's computer consulting expertise and led to a rapid succession of additional clients in the mid 1950s, including computer installations for the International Shoe Company in St. Louis and for Andersen's longtime client Commonwealth Edison. The latter contract demonstrated how Andersen and Company successfully practiced account control to transition clients from pre-computing administrative services to computer services. Andersen's Administrative Services advised Commonwealth Edison to use IBM's first entry into the electronic digital computing market, the IBM 701, and completed this installation at the start of 1956. Advising and installing successive IBM models became commonplace for Andersen in the years that followed—with Commonwealth Edison, Field Enterprises, and others.[32] Beginning with the installation at GE's Appliance Park, John Higgins established a standard practice of recording, indexing, and sharing computer consulting experiences throughout Andersen's Administrative Services Division. This facilitated and advanced a culture of circulating knowledge, which greatly aided many clients and proved important to the division's success.

By 1960, Andersen and Company's Administrative Services Division had grown to 150 employees. Around that time, the division also began providing large-scale systems integration services to commercial clients, and by the mid 1960s it had extended such services to the federal government. Throughout the 1960s, the 1970s, and the 1980s the enterprise grew rapidly. Andersen Consulting Services (as the computer services division was called by the late 1980s) came to exceed the company's accounting business in profitability. In 2000 it was split from the company, and it soon adopted

Constructing Expertise 29

the name Accenture. By 2001, Accenture, with 65,000 employees, was the largest exclusively consulting firm in the world.[33]

Diebold and Associates

Arthur Andersen and Company was unusual in the early computer consulting services industry. Far more than its peers in accounting services and management consulting, it devoted substantial resources to quickly building and expanding other computer services offerings (particularly programming services). Most pioneering computer consulting operations began as startup firms with one or several associates launching and slowly growing enterprises that focused on advisory services. (These businesses either did not venture into programming services, or did so far more gradually than Andersen and Company.) One such firm—(John) Diebold and Associates, later Diebold Group, Inc.—stood out for helping to launch the consulting services segment of the industry, and for its charismatic and visionary leader.

John Diebold was born in 1926 in Weehawken, New Jersey. After graduating from Weehawken High School, the 18-year-old Diebold entered

Figure 1.2
John Diebold, circa late 1960s. Courtesy of Charles Babbage Institute, University of Minnesota.

Swarthmore College, but soon took a hiatus to enter the US Merchant Marine Academy, where he developed a fascination with machines and particularly with radar-controlled automatic tracking and firing systems. In 1946 he returned to Swarthmore, where he got a BS degree in engineering. Upon graduating, he entered the Harvard Business School's MBA program. At HBS he worked closely with George Doriot, a pioneering venture capitalist who had co-founded American Research and Development Corporation (ARDC) in 1946.[34]

Both Diebold's experience at the Merchant Marine Academy and his studies at the Harvard Business School contributed to his growing fascination with automatic machinery. A close faculty colleague of Doriot's at HBS, Curtis Tarr, proved influential. In 1950, Tarr served as faculty advisor to Diebold's project to research and write a report titled "Making the Automatic Factory a Reality."[35] Diebold's fascination with digital computers grew rapidly, and he studied the work and writings of John von Neumann and Norbert Weiner, both of whom he contacted and met with in the early 1950s.[36] Diebold would go on to substantially revise and expand his HBS report into his soon-to-be-famous 1952 book *Automation: The Advent of the Automatic Factory*.[37]

In 1951, Diebold completed his MBA program. He then sought employment as a management consultant and was hired by a small Chicago consulting firm led by E. O. Griffenhagen. Diebold incorporated his factory automation studies into certain consulting assignments and was successful working for Griffenhagen. Nevertheless, within several years Diebold wanted the freedom of his own consulting operation. In 1954 he returned to Weehawken to found a consulting enterprise.[38]

Highlighting the tremendous importance of publishing to establish name recognition and conveying expertise in consulting, Diebold published frequently throughout in his career. Diebold did not invent the term "automation" with his 1952 book, but that book and his subsequent publications and work did much to advance both the use of the new term and computer-based automating efforts of various kinds.[39] The subtitle of Diebold's book suggests that it is focused solely on automating production machinery and processes within factories, and indeed the first two thirds of the book concentrates on this topic. The final third, however, considers the future of automation of information in offices and other settings and how it might impact organizations and society. Though Diebold clearly advocated the use of computer applications for the betterment of business, government, and society, he was not blindly optimistic; he recognized the importance of deep organization-level and industry-specific analysis.

Intensive research was the guiding force of John Diebold and Associates, the computer consulting business he founded in Weehawken. (Later it relocated to New York City.)

Most of the small, privately owned computer consulting firms that went out of existence twenty years ago left little or no trace for researchers interested in their past activities. In contrast, John Diebold left behind a rich archive (housed at the Charles Babbage Institute) detailing his firm's work—a complete set of more than 500 consulting reports the company prepared for corporate and organizational clients throughout the firm's 37-year existence.[40]

Diebold and Associates contracted with their first clients in 1956. That attracting the first customers proved difficult is understandable insofar as the very first movers hired the well-established Arthur Andersen and Company in 1954 and 1955. Thereafter, the growing attention to Diebold's book helped with name recognition and attracting clients. Table 1.1 lists the firms that Diebold and Associates served from the mid 1950s to 1960. From the early 1960s on, the company expanded steadily and produced far more reports each year.[41]

Table 1.1 illustrates the diversity of Diebold and Associates' early clients. Most of the studies that the firm conducted were contract work for corporate or organizational clients. By 1960 the firm had already consulted on potential acquisitions and implementations of computing and software systems at American corporate giants, smaller enterprises in the US and overseas, governments, public utilities, hospitals, and other types of organizations. Applications ranged from business data processing and competitor research to factory automation and developing aerospace control systems. Diebold and Associates provided expert advice not only on technology and systems, but also on a myriad of organizational and managerial issues involved with IT transformations. A typical consulting project lasted from six weeks to six months and yielded a report of 80–250 pages presenting extensive research findings.[42] Among Diebold's clients between 1956 and 1959 were Alwac, Mead Johnson, and the American Hospital Association.

Alwac is not a widely recognized name in computing, despite the involvement of the highly successful Swedish entrepreneur Axel Lennart Wenner-Gren.[43] Wenner-Gren had made his early fortune in the vacuum cleaner and refrigerator industries. In 1952 he became involved in computing, partnering with an ex-Northrop Aircraft engineer to form Alwac in Redondo Beach, California. Like several of other mainframe digital computer companies that formed in the Los Angeles metropolitan area (a

Table 1.1
Companies and organizational clients of Diebold and Associates, 1956–1960.

Alwac Corporation
Alweg Monorail
American Hospital Association
AMF Pinspotters, Inc.
Arco Auto Carriers
Bear, Stearns & Company
Boeing Airplane Company
Bulova Watch Company
Carl M. Rhodes & Company
Coopers Inc.
Department of Highways, Ontario
Federal Electric Corporation
General Electric Company
Gladding McBean & Company
Illinois Agricultural Association
Kellogg Switchboard & Supply Company
Kleber-Colombes
Mead Johnson & Co.
Owens-Illinois Glass Company
Pepperell Manufacturing Company
Radio Corporation of America
Remington Rand
Shuron Optical Company
Socony-Mobil Oil Company
Stromberg-Carson
Swiss Bank Corporation
Teleautograph Corporation
The Peoples Gas Light and Coke Company
The State Insurance Fund
Western Virginia Pulp Paper Company
Westinghouse Electric Corporation

hotbed of aerospace and defense contracting), the firm struggled as a business. In contrast to several computer industry leaders—that transitioned from pre-computer office machine specialists to become dominant in the computer industry, or startups fortunate enough to be acquired by such firms—most new ventures in digital computing of the early 1950s had periods of growth, but their existence was always precarious.

As Alwac struggled to contain costs in the mid and late 1950s, it brought in Diebold and Associates to do a thorough analysis of the firm and its mechanisms, competitive environment, and product and market opportunities. Diebold and Associates found a firm with few controls in place that was not basing its strategy on sound evaluations of the market and of its competitors. Diebold consultants put together a series of recommendations for controlling costs and improving the organization of the firm. Alwac had two programming groups that did not coordinate or communicate effectively. The programmers had little understanding of the internal circuitry of the ALWAC I computer, the company's initial product. Diebold consultants developed a plan to put the programmers in a single large programming room and to educate them on the requisite computer circuitry. That a computer startup in the 1950s would have difficulty organizing programming efforts is understandable—programming would prove a major challenge for even the largest computer firms in the 1950s and the 1960s. For instance, IBM experienced delays and challenges with FORTRAN, OS/360, SABRE, and other major internal and external programming projects. Diebold and Associates brought the combined wisdom of studying and learning from a number of programming efforts by the time they consulted for Alwac in the late 1950s.[44]

Despite sound advice from Diebold and Associates, Alwac continued to suffer severe cash-flow problems and ultimately could not compete effectively with the office machine and electronics leaders (IBM, NCR, Remington/Sperry Rand, GE, RCA, and Burroughs) that, along with two well-funded and well-organized startups from 1957, Control Data Corporation and Digital Equipment Corporation, would dominate the computer industry in the 1960s. Nonetheless, it is telling that Alwac contracted with Diebold and Associates to work continuously for two years to examine and advise on every aspect of the firm: engineering, design, product development, programming, marketing, and accounting. Much of the expertise that enabled Diebold and Associates to aid Alwac, as well as more successful computer industry participants (Burroughs, IBM, RCA, GE), came out of its experience at teaching and learning from office and factory automation users and emerging users, who made up the bulk of its clientele.[45]

In 1895, Edward Mead Johnson (one of the three brothers who had founded Johnson & Johnson in 1886) launched a side business, American Ferment Company, to produce and sell digestive aids. Two years later, he left Johnson & Johnson to focus on American Ferment, which was renamed Mead Johnson & Company in 1905. The company became a significant producer of infant formula, diet aids, and other products. In 1957, it contracted with Diebold and Associates to do a preliminary study of possibilities for administrative data processing.[46]

Although Diebold and Associates reported preliminary assessments of how an electronic digital computer system could help schedule production and minimize total costs (material, purchasing, manufacturing, and distribution), achieve faster and more accurate customer invoicing, provide current sales analysis reports, and assist the Research Division with the development of new products, the true focus of the six-week study was on the existing Tabulating Department and its operations. That department's technology consisted of pre-computer IBM punch card tabulation equipment. Diebold consultants determined that the firm had major organizational and managerial problems. Although Mead Johnson & Company had been hiring more and more employees and had been increasing its output for decades, the Diebold consultants argued its Tabulating Department was failing to meet the company's needs. They found that the department's manager lacked the ability that was needed to exercise control over personnel and equipment. They noted that he failed to educate users of his department's services, and also that he promised to complete reports on impossibly short time schedules and then failed to deliver them. At the same time, this manager's supervisors missed opportunities for the department to serve the corporation on account of a short-sighted personnel policy, and they failed to require coordination between the Tabulating Department and other departments. The Tabulating Department's manager was not being invited to appropriate meetings in other areas of the firm and was being kept in the dark about initiatives that would require his department to provide new services in the future.[47]

Computer technology came at a high cost, and organizations had to be creative, cooperative, and proactive in seeking to fully take advantage of new systems. The skill sets of managers and operators in tabulating departments often were insufficient for a larger-scale computerized operation. Diebold consultants recognized and recorded the inadequate technical understanding of Mead Johnson's Tabulating Department manager, but put far greater emphasis on the organizational and managerial skill set a new manager would need to fully take advantage of computing within the company, and the way different sectors and divisions of the corporation would need to cooperate in new ways. Ultimately, Diebold consultants provided not only technical evaluations, but also thoughtful recommendations for recruiting, training, promotion, and knowledge sharing to facilitate a more competent Tabulating Department regardless of whether Mead Johnson invested in computer technology.[48] More generally, Diebold and Associates' evaluations assessed existing and potential new technology within Mead Johnson's managerial, organizational, and personnel contexts. Mead Johnson & Company's management found the study so helpful that the firm contracted with Diebold for three additional studies over the succeeding decade.[49]

The American Hospital Association (AHA) was founded in 1898 to promote good practices related to hospital health care and health administration. In 1958 it hired Diebold and Associates to analyze and make recommendations for Baylor University Hospital (in Dallas)—as a test bed—in order to establish standards and best practices in health administration and data processing that could be utilized at many other (AHA-member) hospitals. Diebold consultants analyzed current practices and evaluated options and opportunities for computing applications to the following functions: accounts receivable, accounts payable, general accounting, payroll, inventory control, purchasing, and medical records and statistics. While the research focused on Baylor University Hospital, Diebold and Associates gathered extensive data on several other hospitals for comparative purposes, including Memorial Hospital for Cancer in New York and Barnes Hospital in St. Louis.[50]

Diebold consultants provided a complete plan for how a data processing infrastructure based on digital computers could be implemented at Baylor University Hospital. The consultants performed an extensive cost analysis of the existing record-keeping and retrieval mechanism, and projected costs for a computerized record system. They concluded that a general-purpose computer with large random-access memory and relatively low arithmetic and logic speeds would suffice for processing a hospital's records. They stipulated that such a system wasn't likely to bring savings if installed right away (early 1959), but that cost trajectories on systems and memory were such that computers could probably bring savings to most large hospitals, including Baylor University Hospital, within three to five years. (Computing costs would continue to decline relative to processing power and memory, and existing clerical costs would continue to rise.) Diebold and Associates provided a highly detailed plan for utilizing punched card tabulation and digital computer systems to organize and use data.[51]

In the late 1950s, the field of medical computing was gaining attention (focused on patient records) because of the work of Robert Ledley and Lee Lusted, radiologists who published extensively on how computing could benefit diagnosis. Their ideas would not be realized until ten years later, when the cardiologist Homer Warner instituted some of them at Latter Day Saints (LDS) Hospital in Salt Lake City. LDS Hospital was acquired by Intermountain, which remains at the forefront of the still somewhat underdeveloped fields of computerized diagnostic tools.[52] On the other hand, computer applications for hospital administration grew significantly in the 1960s, in the 1970s, and afterward. Diebold and Associates' recommendations, and the associated test bed system at Baylor University Hospital, contributed to this phenomenon.

Over time, Diebold's firm supplemented client-based work with producing and selling industry-specific analysis reports, or position papers on IT trends. This was secondary, as client contracts were always the focus of the firm. Many of Diebold's clients became repeat customers over years and decades, hiring Diebold and Associates to evaluate and assess new IT systems, new applications, and organizational issues associated with information technological change. Some industries that appear underrepresented on the list of the first thirty clients but which we know were early adopters of computers, such as insurance and banking, had long-established relationships with IBM, Burroughs, and other computer manufacturers, and greater internal expertise with computing at an earlier date. Banking and insurance firms, however, became frequent Diebold clients in the 1960s, 1970s, and 1980s—including Bankers Trust, Carteret Savings Bank, Dollar Savings Bank, Bear Stearns, Blue Cross and Blue Shield, American Insurance Services Group, American International Group, Central National Insurance, and Travelers. From natural gas, oil, aerospace, investment, agribusiness, and pharmaceuticals to paper, paint, electronics, airline, earth-moving machines, and tobacco companies, Diebold's industrial client base was extremely broad. It even served the trade union Communication Workers of America to help this organization understand the impact of automation on labor issues in their industry. In fact, this organization commissioned nine studies from Diebold between 1964 and 1983 making it one of the firm's longest-term clients.[53] Beginning in the late 1950s, it also served client companies and organizations around the world, with a particularly rich base of Japanese firms and Japanese government organizations—including Mitsubishi, Japan Information Processing Development Center, Japan Trade Center, and the Japanese Ministry of Industry and Trade. Likewise, the firm consulted for many different local governments, and federal departments and agencies. The mix of industries and government, and geographic diversity of clients, helped to insulate Diebold's firm from economic downturns in particular regions and economic sectors.[54]

To judge from the number of client contracts, completed reports, the length and depth of studies, and the fact that most projects involved sending a small team, Diebold and Associates must have had at least a dozen consultants on staff by 1960, and many dozens by the late 1960s (by that time under the new name, Diebold Group, to reflect its larger size). The number grew steadily in succeeding decades. At its peak, in the 1980s, the firm had more than 300 employees.[55] Soon after launching the enterprise, John Diebold became more of a manager of consultants and programming specialists than a consultant himself.

John Diebold, as well as many of his consultants on staff, published frequently. After the publication of Diebold's highly popular book *Automation*, publications had a major effect on the growth of his business. Not only was publishing a free or inexpensive marketing tool, it was a particularly effective one because articles and books in respected journals and published by distinguished presses conveyed expertise—exactly what Diebold's firm sold. Diebold and Associates/Diebold Group conducted dozens of consulting projects for clients in the second half of the 1950s, and more than 100 in each of the three succeeding decades. In all it served more than 300 unique client organizations, introducing many to digital computing, new applications, better organizational processes and structures, and more effective and efficient use of information technology. In 1991 John Diebold sold Diebold Group, Inc. to Daimler-Benz.[56]

Canning, Sisson, and Associates

Though small compared to Andersen and Company's consulting operation, Diebold and Associates was a substantially larger enterprise than many emerging computer consultant firms in the early industry—some of which were sole proprietors or partnerships. Perhaps the earliest partnership in computer consulting services was Canning, Sisson, and Associates, which began in 1954, the same year as Diebold and Associates.[57] Canning, Sisson, and Associates—founded by Richard Canning and Roger Sisson—had a somewhat different business model than either Andersen and Company's Administrative Services Division or Diebold's firm. A core activity of Canning and Sisson's company was spreading expertise in computer systems and applications by holding fee-based educational seminars.

Richard Canning, who had worked on a new radar bombing set during World War II while serving overseas, joined IBM immediately after the war and was assigned to production engineering. Less interested in the production of IBM machines than in their industrial application, he soon took an IBM course on manufacturing control at the firm's facility in Endicott, New York. The rigidity of IBM and the experience of raising a family in upstate New York, where winters were harsh, led Canning to seek greener pastures with a smaller and more flexible firm in Southern California. Canning would be the first of many who gained experience at IBM, later to leave for computer consulting services or other types of careers in computer services.[58]

After writing some letters to Southern California firms, Canning was offered and accepted a position with Electronic Engineering Company in February 1950 (located in Santa Ana). About two years later, to augment his income and pursue an interest in computer applications to manufacturing that he had been unable to at IBM, Canning joined a research group as part of the Management Sciences Research Project (funded by the Office of Naval Research) at the University of California at Los Angeles, where he had the title Associate Research Engineer. The group was focused on applying computers and operations research to aircraft manufacturing, and on advising the ONR's Logistics Branch. Roger Sisson, who had a background in electrical engineering and had worked at National Cash Register previously, joined the UCLA project on the same basis. Soon thereafter, in fall 1954, Canning and Sisson decided to form a computer consulting firm together. Its name was Canning, Sisson, and Associates.[59]

Back in 1953, before the installation of the very first data processing computer to be used in business in the United States (a UNIVAC at GE's Appliance Park), Canning began teaching a night school course on computer applications for business in Los Angeles. He also taught a one-week condensed version of the course at the University of Chicago. There was no textbook on the subject, so Canning began to write one. Ultimately it became *Electronic Data Processing for Business and Industry*, a seminal work in the early digital computer business applications field, published in 1956.[60] The latter portion of Diebold's 1952 book *Automation* had suggested the possibility of widespread computer applications in business, but Canning produced the first high-quality textbook detailing state-of-the-art knowledge about applying computers to a range of such applications— from department store inventory control and airline reservations to managerial decision making. Context and understanding provided by Canning's early participation in the pioneering ONR study at UCLA enabled him to write this deeply researched text.[61] What stands out about the volume is that it does not present a wildly optimistic utopian vision; rather, it presents a sober in-depth evaluation of business processes, operations research, and existing equipment and methods of applications. Canning notes that the shift to electronics and digital computing did not have to be wholesale, and that in most instances a wholesale shift was unwise. He emphasizes that firms and organizations can gradually incorporate electronics equipment to complement and improve functions and run hybrid electromechanical and electronics data processing systems. He discusses ways to make optimal use of digital computers, but argues that firms and organizations could (and often should) make gradual transitions that advance operations while containing costs.[62] This is also true of Canning's second book, *Installing Electronic Data Processing Systems* (1957), in which he offers advice on opportunities and challenges by drawing on specific cases of installations, hiring and organizing of data processing personnel, general programming, programming computer runs, costs of operations, and other topics.[63]

From the start, Canning and Sisson decided to avoid any real or perceived conflict of interest by refusing to simultaneously consult for multiple companies in the same industry. Both men had started out in operations research and computer applications to aircraft production control and were optimally situated (in terms of knowledge and geography) to serve aerospace firms. Limiting themselves to one aerospace firm at a time, they quickly conducted research and developed broader expertise to provide quality advisory services on electronics and computer data processing applications for other industries.

Knowing the substantial tuition that University of Chicago charged for his popular week-long course, Canning, the more talented educator of the two, began teaching one-week fee-based seminars on computer applications to business data processing and other industrial applications in Los Angeles, in New York, and in Chicago. Sisson would sometimes co-teach the Los Angeles seminars with Canning, but in order to keep travel costs down he seldom traveled with Canning for the New York and Chicago seminars. The five-day events carried a $200 fee, which included lunches, and were held in luxury hotels. Canning and Sisson downplayed their roles as consultants at these events, so attendees did not feel they were paying a substantial fee just to get a sales pitch—and when asked about consulting, they often recommended major accounting and management consulting firms such as Andersen and Company, Coopers and Lybrand, and McKinsey and Company. Nonetheless, attendees often suggested getting together for dinner or drinks, which led to site visits to companies, and ultimately significant consulting contracts for Canning and Sisson, particularly on the East Coast. One such client was the Associated Merchandizing Corporation, a purchasing association for member US department stores. Recording, processing, and analyzing information quickly was (and is) critical to department stores' purchasing and cost containment. In the late 1950s, AMC decided to move forward with a project in which AMC members would support a single computer project to analyze information and better understand how computers could be applied, using one store (a Higbee's department store in Cleveland) as a test site. Canning and Sisson advised on hardware, on accounts payable, on payroll, and on other software applications for the test site.[64]

Over time, publications—Canning's first book, and both Canning's and Sisson's ongoing publication of articles and books—were important to the continuing success of the business. As with all consulting firms, getting off the ground with initial clients was the most difficult part. Canning's and Sisson's work on the ONR project, their publications, and holding high-level seminars all brought name recognition, leading to early clients—and successful projects produced invaluable word-of-mouth marketing when data processing managers came together at professional meetings such as the National Machine Accountants Association/Data Processing Management Association (NMAA/DPMA), or other events and venues.[65]

For Canning, Sisson, and Associates, in contrast with most computer consultants, publishing was not merely a critical tool to display expertise and establish a brand; it was also a business. To be successful, consultants had to keep abreast of current literature on computer applications in industry. This included the occasional article on the topic in well-known business publications such as the *Harvard Business Review*, but also many trade journals for individual industries as well as technical publications. Roger Sisson reasoned that data processing managers (potential clients) had to do the same. With this in mind, Sisson and Canning decided to launch *Data Processing Digest* as a monthly subscription-based publication to synthesize useful information for busy managers. Sisson had supervised Margaret Milligan, a technical writer at National Cash Register, who was hired away from NCR by Canning and Sisson to write the content for and oversee *Data Processing Digest*. She requested and obtained blanket permission to synthesize or summarize articles from most of the relevant publishers, and would spend endless hours in the UCLA library looking for appropriate articles to summarize. The first issue (April 1955) consisted of ten pages. While Canning and Sisson would provide some advice and occasionally write, Milligan did most of the writing, editing, formatting, and managing of the publication. Sending out subscription forms to 3,000 contact addresses Canning and Sisson had assembled, they quickly secured 300 annual subscribers—a base that would grow steadily. Milligan networked within the emerging computer community and was often the only woman at meetings of the Los Angeles chapter of the Association for Computing Machinery.[66] As with data processing managers, emerging computer consultants, and others, Canning and Sisson benefited from the publication as a time-efficient tool to stay current on the literature and to make wise decisions on what articles to read in full.

Canning and Sisson were well aware of Diebold and Associates in the mid 1950s, and Sisson visited Diebold once in his New York office, but the two computer consultancies generally did not compete for contracts. Canning and Sisson usually sought shorter-term contracting assignments and lacked the geographical scope of Diebold's firm. From the beginning, Diebold aimed to build a steadily growing consulting corporation, while Canning and Sisson were focused on running a partnership where they were the primary consultants. Over time, however, Canning and Sisson differed on their vision for the future of the company. Roger Sisson wanted to expand into programming services and become a manager of a broader-based computer consulting company, along the lines of Diebold and Associates or Arthur Andersen and Company's Administrative Services Division. Richard Canning, in contrast, was more content with advisory consulting, teaching seminars, and operating a very small firm. With these differences, the business partners decided to part ways in 1958. Canning continued to consult and hold seminars, and in time, sought to leverage his expertise by starting a new publication, while Sisson abandoned consulting and joined the Ford Motor Company as an aeronautical engineer.[67] In 1961, Canning and Sisson sold *Data Processing Digest* to Milligan, who expanded it to include commissioned book reviews and articles and published it for another 25 years.[68]

Data Processing Digest alerted Canning to the potential business of publishing on computer applications, and in September 1962 he had a career-changing idea. Whereas *Data Processing Digest* gave a quick synopsis of many of the relevant articles on computer-based business data processing, Canning envisioned a new publication to focus on original content and to offer greater depth. Most articles on business data processing were quite short, usually no longer than a few pages. Canning's plan was to launch a publication, to be named *EDP Analyzer*, in which he would write a single in-depth article of 10–16 pages concentrating on a significant theme in the field for each monthly issue. The subscriber base was primarily companies, not individuals, and *EDP Analyzer* was priced as an expert journal for companies/institutions at $36 (approximately $280 in today's dollars) a year. With the boldness of a true entrepreneur, Canning invested all the money he had—$4,000—in direct mail marketing for the first issue (and offered a money-back guarantee to initial subscribers). He wrote all the issues from 1963 through 1977, sharing duties with his daughter until 1987, when publication ceased.

Throughout 1963, Canning continued consulting, but subsequently he devoted full time to the journal. In his years publishing *EDP Analyzer*, Canning traveled extensively to spark ideas for future issues' themes. His consulting days had provided him with a rich network of friends and contacts. In that sense, his post-consulting publishing career mirrored many aspects of consulting—traveling extensively, meeting many data processing and computing professionals, engaging in research, and circulating knowledge. *EDP Analyzer* leveraged this dissemination of expertise, as the subscriber base grew substantially and eventually reached a peak of 10,000—a remarkable feat insofar as many organizations had only one subscription and shared or illegally photocopied the issues for all members of the data processing staff.[69] The same down-to-earth advice Canning gave as a consultant—emphasizing deep analysis and cost-effective evolutionary rather than revolutionary change—radiated throughout the pages of *EDP Analyzer*.

Conclusion

The computer consulting sector was the origin of the computer services industry, and Arthur Andersen and Company's Administrative Services Division was probably the first computer consulting enterprise. Most computer consulting companies were small—indeed, many were one- or two-person operations, like Canning, Sisson, and Associates. These sole proprietorships, partnerships, or very small corporations tended to leave little or no documentary record of their existence behind, despite being enormously influential in aggregate in providing expertise in the emerging field of computer-based business data processing.

John Diebold almost singlehandedly popularized the term "automation" for manufacturing and business data processing, and in 1954 he founded the most significant startup computer consulting corporation. Ultimately computer consultants sold their expertise and sold their time. From the standpoint of individual consultants, the time limitation (of perhaps 2,000 or 2,500 billable hours a year) is what led Andersen and Company to expand into programming services and become a large enterprise, led John Diebold to add consultants and programming professionals and become a manager of a midsized firm, and led Richard Canning to leverage his knowledge and expertise by entering the publishing business.

2 Processing Data: IBM-SBC and Automatic Data Processing, 1953–1964

Although the computer consulting services business—which began in 1953—was the origin of the computer services industry, other segments quickly emerged. As was noted in chapter 1, in the 1950s consulting services were often employed by organizations to get recommendations so as to make better decisions on acquiring a digital computer (or computers), or for advice with set-up and perhaps with initial applications programming. A substantial number of organizations, however, did not investigate acquiring a digital computer system in the early years of this new technology, or they did so and decided against it. In these instances, the decision might have signified an organization opting to continue to use its internal pre-computer business machine (generally punch card tabulation) operation (if it had one), or it might have represented a choice to continue to outsource data processing (again, throughout this book "outsource" refers to securing services outside a firm or an organization, not necessarily outside the country).[1]

Beginning in the early 1930s, organizations could outsource data processing to a new type of facility launched as a division of International Business Machines: the IBM Service Bureau. IBM's service bureaus of the 1930s—which quickly spread to major cities throughout the US and much of the developed world (and spawned smaller competitors)—had the latest in punch card tabulation equipment, and had knowledgeable employees who could get the most out of the machines so as to meet customers' data processing needs.

This chapter explores both the pre-computer service bureau business (in the years 1932–1953) and the advent and early history of digital-computer-based service bureaus in the mid 1950s and later, focusing in both eras on the industry's leader, IBM.[2]

The vast historical literature on IBM has almost entirely ignored IBM's Service Bureau Division (SBD) and its successor, the Service Bureau

Corporation (SBC). I say "almost" because many IBM histories include a perfunctory mention that IBM agreed to spin off its Service Bureau Division (as a wholly owned subsidiary, Service Bureau Corporation) in 1956 as part of a settlement, or consent decree, with the US Department of Justice. Most of these IBM histories have a second and final brief appearance for IBM's (or more precisely its subsidiary's) service bureau operations—a quick recounting of how IBM sold SBC to Control Data Corporation (CDC) in 1973 as part of a legal settlement.[3] In short, these treatments have severely neglected the service bureaus, in which IBM helped solidify organizational capabilities in data processing that proved valuable to all operations of the company.[4]

In addition to recounting the broad contours of IBM's SBD/SBC history, which was the heart of the data processing services segment of the computer services industry, this chapter also addresses the emergence of specialized data processing firms, through a case study of industry leader Automatic Data Processing, Inc., which early on was named Automatic Payrolls, Inc.

Before exploring SBD/SBC, it is important to provide brief context on the origin and early history of IBM, including the roles of two visionary leaders, Herman Hollerith and Thomas J. Watson.

The Emergence of C-T-R / IBM

Herman Hollerith, born in 1860, was an engineer and entrepreneur instrumental to transforming information technology. He invented punch card tabulation machines in the mid 1880s as a labor-saving data processing tool in response to the census data processing problem. Roughly 1,500 tally clerks manually processed the 1880 US Census, creating 21,000 pages of census reports, a task that took seven years.[5] Hollerith saw this first hand as a 19-year-old "special agent" employed on the 1880 US Census.[6] Far more data would be collected for the 1890 US Census (the largest data collection project in the world to date), and the leaders at the US Bureau of the Census recognized the need for a better processing mechanism and held a competition for the design and construction of assistive labor-saving machinery. Hollerith's "Electric Tabulating System" won the competition, and he secured the contract for the 1890 US Census. His machines read and tabulated information coded on punch cards.[7]

The idea of punched holes on cards instructing operations of a mechanical machine dated back to inventor Joseph Marie Jacquard's 1802 loom, a labor-saving device that facilitated high-precision complex weaving. In the 1830s, Charles Babbage—inventor, mathematician, and philosopher—designed a mechanical "computer" (the Analytical Engine) that included ideas for punch cards with machine instruction and calculation.[8]

Hollerith's machines, unlike Babbage's designs (none of which resulted in constructed machines during Babbage's lifetime), utilized the newly harnessed power of electricity. Holes punched in the appropriate positions on the cards allowed pins on the machine to make electrical connections and tally the data. Hollerith's system was the only one in the competition in which, once the cards had been punched (more than 62 million in the 1890 US Census, one for each citizen), the sorting and counting could be accomplished automatically. Hollerith leased rather than sold the Western Electric-manufactured (subcontracted) Hollerith census machines—which grew from an initial 56 tabulating machines to a hundred, and many hundreds of card punches.[9] Leasing was preferable to Hollerith to maintain service for the machines and to make certain they would work. As Hollerith's biographer Geoffrey Austrian explained, "Timely and expert service ... was as important as furnishing the machines."[10] The 1890 US Census was processed in two and a half years, generating more than 26,000 pages of census reports. The cost was approximately $11.5 million, with estimates that Hollerith machines resulted in savings of $5 million relative to the largely manual methods of the 1880 US Census.[11]

In 1896, Hollerith launched the Tabulating Machine Company to more fully reap the rewards of his invention. Hollerith secured the contract to supply the machines to process the 1900 US Census. But because the decision-makers at the US Bureau of the Census believed he charged an excessive amount, they found another solution, using very similar technology, for the 1910 US Census. This, however, was not the blow it might have been. By that time Hollerith had diversified his customer base. In addition to the 1900 US Census, Hollerith won contracts over the next 15 years with more than ten foreign census departments, as well as for a number of applications in the railroad, insurance, steel, automobile, textile and other industries.[12] In 1911 Hollerith sold his firm—businessman Charles Flint facilitated the combination of the Tabulating Machine Company with three other companies—and the foremost pioneer of tabulation machines, having accumulated considerable wealth, for the most part retired.[13]

The resulting enterprise of Flint's orchestrated merger was the Computing-Tabulating-Recording Company (C-T-R), a seller of punch card tabulating machines, clocks, industrial scales, and a range of other machines. While the other businesses gradually diminished in significance relative to tabulating machines in the 1920s and the 1930s, this merger brought important manufacturing know-how to the former Tabulating Machine Company, which previously had outsourced manufacturing. C-T-R was in effect the launch of International Business Machines, the name the firm took in 1924 (originally the name of C-T-R's Canadian operations, which had been launched in 1917) to reflect its rapidly growing international operations and sales.

C-T-R would change forever when Thomas J. Watson signed on as general manager in 1914 and became the firm's president the following year. Watson was an experienced business machine salesman and sales manager, having spent nearly twenty years at National Cash Register. With a talented and controversial longtime leader, John Patterson, NCR had developed one of the leading sales organizations in the world. Watson quickly instituted and refined at C-T-R many of the sales management processes and techniques developed by NCR, including sales territories, quota and point systems, sales conventions, rewarding successful salesmen, educating users, instituting user feedback to help shape product and service innovation, and a conservative dress code.[14] He also brought with him and promoted to his employees a motto—"Think" that he had come up with late in his tenure with NCR.[15]

Services from the Start at C-T-R

Watson took quick action in his first five years of leadership of C-T-R. He integrated manufacturing between the constituent companies and divisions, authorized the hiring of talented engineers away from established technology companies, established an aggressive research and development organization, and set a foundation for advanced training to achieve an unparalleled sales and services infrastructure. To Watson, sales and services were extremely important and completely intertwined. As he saw it, they were critical to making the firm a permanent enterprise.[16] Like Hollerith's company, C-T-R leased its tabulating machines to properly maintain machines (protect reputation), create a steady revenue stream, help sell associated products (primarily punch cards), and enhance opportunities for frequent customer system upgrade cycles. As a 1917 article in the *T.M. Business Record* (a publication of the C-T-R/Tabulating Machine Company) put it, "Service is the backbone of our business. You have here what might be termed a perpetual motion machine [sales to service to sales to service] ... . If there is any one thing that will continue to keep this company closely linked to its customers, it is the constant fulfillment of service."[17]

Far more than just the initial installation of tabulating machines and keeping these systems in optimal working order at customer sites, C-T-R's field services personnel were also regularly counseling customers on methods and best practices for operations—from optimizing efficiency with the routine to pioneering new applications. Field services personnel often were involved in helping to configure and reconfigure the many wires on panels of C-T-R/IBM punch card tabulation machines to "program" the machines for particular data processing work.[18]

In a talk before a C-T-R 1920 Executive School "graduating class" a senior C-T-R leader, Clark Hayes, emphasized that the work of the "Tabulating Machine Organization" consisted of "calling on customers at least once a month" and "making sure their working plans are right."[19] Long before computers—and long before information technology consultants at Diebold and Associates or Canning, Sisson and Associates—C-T-R was providing expertise in optimizing data processing for customers using tabulating machines.[20]

C-T-R/IBM field services specialists, along with regular salesmen, sometimes pioneered knowledge in new data processing techniques. Equally if not more significant, they frequently circulated knowledge and best practices through a fast-growing number of organizational and industrial users on an ever-increasing scale and scope of applications.[21] Larger companies were called on far more frequently than the minimum once a month in C-T-R's early years, and eventually major installations had C-T-R/IBM field engineers as almost permanent fixtures of their tabulating machine operations. These services were included with leases of tabulating systems.

In 1925, the newly named International Business Machines recognized and enhanced the organization of and the future foundation for advanced tabulation machine services to customers with the formation of the Business Services Department, headed by Frederick W. Nichol. This department was focused on expertise and excellence with enhancing customer data processing applications. Nichol joined C-T-R in 1914 as a salesman. In 1917 he enlisted in the US Army and rose to the rank of major. In 1924 he returned to IBM and became the Assistant General Sales Manager of IBM Canada before moving back to IBM's US operations the following year, becoming the founding leader of the Business Services Department and an assistant to Watson.[22] He would go on to become a vice president and one of Watson's most trusted advisors. He also was a frequent distinguished speaker at IBM conventions and sales schools in the late 1920s and the 1930s. His many talks emphasizing the interrelationship of sales and services highlighted

IBM as *both* a business machine products and a services-focused company throughout the interwar years and beyond.[23]

IBM's Service Bureau Division

Back in the early 1920s, Thomas Watson commissioned a major study investigating the possibility of forming a new division to provide organizations with a data processing outsourcing option—what would later be referred to as a "service bureau." This proposed division would serve organizations too small to efficiently utilize a leased punch card tabulation machine system, as well as to potentially help existing customers with peak demand for data processing (partial outsourcing to balance out loads). The existing field services workforce would be drawn from and would be used interchangeably (as needed) with IBM data processing services operations (service bureaus).

On April 25, 1922, a formal data processing services business plan—with extensive analysis and financial projections—was sent to Watson. The plan called for launching an initial C-T-R processing bureau in New York City containing multiple tabulation and related data processing machines. It estimated the startup cost for this proposed new division would be $16,000 in the first year, with expected revenue at $12,000. According to the plan, with both startup and operating costs, this would result in an initial-year loss of about $8,000. With great optimism, the estimate for the second year was for revenue of $62,000, with profit of $31,000, leading to a cash accumulation position of $23,000. IBM did launch a division closely matching important elements of the plan; however, it did not do so for an entire decade. It would be years before it would achieve that level of profits, even in nominal terms, and margins were always far short of these initial expectations.[24]

Clearly defined reasons for this considerable delay appear lost to history, but in all likelihood economic conditions and uncertainty were a meaningful factor. World War I demand (both government contracts and economic stimulus from heavy government expenditures) created a boom time for the firm. Between 1914 and 1919 the company's revenues, profits, and workforce all roughly doubled.[25] While C-T-R's worldwide workforce grew slightly in the late teens and early 1920s, the US workforce of the company declined.[26] In 1922, the year of the proposed "Services Division," there was a recession in the US. Only with the financial bubble in the second half of the 1920s did substantial growth in revenue, profits, and workforce levels return. Then in October 1929 the New York Stock Exchange crashed. This initial shock was met by continuing economic decline in the early 1930s.

Watson famously maintained and even substantially increased IBM's domestic and international workforce in the 1930s Great Depression years, as other office machine manufacturers, including rival tabulation machine maker Remington Rand, retrenched.[27] IBM was fourth of four in profitability in the US office machine oligopoly in 1928, behind Remington Rand, NCR, and accounting machine specialist Burroughs Corporation. A mere 11 years later (having grown substantially), IBM was by far the most profitable office machine manufacturer.[28] IBM anticipated and took advantage of government data processing needs with the advent of New Deal initiatives in 1933, and expanded New Deal programs later in the decade. Of immense importance, in the year after the Social Security Act of 1935, IBM won the major contract for data processing with the Social Security Administration.[29]

While the number of IBM employees declined significantly inside and outside the US in 1934, all other years in the 1930s saw year-to-year growth in employee count. And both the number of domestic and world IBM employees roughly doubled from 1930 to 1940. This included growth of 14 percent of the domestic workforce in 1932, the year the firm launched Tabulating Machine Service Bureau, which was also referred to as the IBM Service Bureau Division (initially just a domestic operation).[30]

Thomas Watson's timing in launching IBM's Service Bureau Division—punch card tabulation data processing outsourcing operations in IBM branch offices in major US cities, and later throughout much of the world—may appear odd. National unemployment had grown to nearly 25 percent in 1932, and along with 1933, it was the nadir of the Great Depression. Watson felt that tough times called for increasing effort and investment in sales and services, and he substantially added to the sales and services workforce that year and throughout much of the 1930s. IBM increased government sales, but many business customers had to cut back on their internal tabulating machine investments in the lean economic times. Their data processing needs, however, did not disappear. The timing of the launch of IBM Service Bureau Division was to give such customers greater options as well as to go after *all* data processing business, no matter the size. Watson wanted every potential customer pursued aggressively.[31]

IBM Service Bureau Division made it possible for some smaller companies and organizations—for the first time—to take advantage of the speed and efficiency of punch card tabulation, and the expertise of IBM's staff. It also optimized IBM's sales and services workforce utilization, as field sales and services staff could be used interchangeably with Service Bureau Division personnel. Further, the increased size, number of machines, and activity at branches served as an advertisement for IBM in larger cities and towns—potential customers could see the busy data processing work in operation.[32] In short, rather than expansion in good times—which were the guiding thoughts of such a division when the preliminary study/business plan was initiated just before the 1922 recession—Service Bureau Division was launched in 1932 in the broader context of expansion during bad times. It was created to take advantage of near-term rapidly increasing government business, to seek any business opportunities no matter the size, to avoid losing business with organizations forced to downsize data processing infrastructure, and to position the firm for the anticipated broader economic recovery in the private sector (that as it turned out, did not fully emerge until the stimulus of the wartime economy of the early 1940s, and substantial postwar demand for business machines).

The roll-out of IBM Service Bureaus was rapid in 1932. After the first such bureau's launch in New York City early in the year, by late October there were very active IBM Service Bureau operations in major cities coast to coast, including San Francisco, Washington, Philadelphia, Cleveland, Pittsburgh, Denver, Minneapolis, and Providence. Industries and other organizations using these services included publishers, retailers, manufacturers, advertisers, government, and utilities. Applications ranged from accounting and inventory control to human resources recordkeeping, and industrial and customer surveys. As anticipated by IBM, in some instances, the business was with organizations new to punch card tabulation, in others it was existing customers that either wanted to apply tabulation machines for new applications (perhaps adding inventory management data processing on an outsourced—service bureau—basis to the accounting data processing done internally), or those organizations that had downsized in response to economic conditions.[33]

The number of IBM Service Bureaus continued to grow steadily throughout the 1930s and the 1940s, as did the volume of data processing work completed and revenue generated by these facilities. Both private enterprise and various levels of government increasingly employed IBM data processing services—whether on site or at the service bureaus. During these two decades, IBM Service Bureau operations were extended to facilities in major European cities, as well as the many parts of the world where IBM had sales offices. In 1949 IBM formed its World Trade Corporation, a wholly owned subsidiary for all of its non-US offices and operations, which included all of its non-US service bureaus.

IBM Service Bureau Corporation

As was noted earlier, IBM was the leading business machine company in the world in 1939, the year World War II broke out after Germany's invasion of Poland at the start of September. With Japan's bombing of Pearl Harbor and the United States' entrance into the war in December 1941, IBM soon turned its facilities and infrastructure to focus on the war effort. IBM's annual revenue more than doubled during the war (from roughly $60 million in 1940 to $138 million in 1945), but its net earnings remained between $8 million and $11 million, where it had been throughout the second half of the 1930s. Wartime expenditures and the near-term diminished industrial capacity in war theaters resulted in the US standing alone as a dominant economic power when the war concluded in September 1945. With IBM fully returning to regular production and operations in late 1945 and 1946, its strained margins ended and net earnings rose rapidly—to roughly $19 million in 1946 and $24 million in 1947.[34]

IBM's top competitor, Remington Rand, which had acquired the Eckert-Mauchly Computer Corporation in 1950, came out with the Universal Automatic Computer (UNIVAC) in 1952. The first significant commercial digital computer to hit the market, that large-scale mainframe—"universal" signifying its uses for both science and business—received much attention and achieved several dozen installations in the succeeding five years.[35]

Meanwhile IBM took a different path to its first fully electronic digital computer, the Defense Calculator, or IBM 701. The firm had been investing heavily in electronics research and development during the war as well as partnering with elite universities—Columbia University on scientific computation machines and Harvard University with the electro-mechanical Mark I computer. While proactive on research, Watson was cautious on entering the digital computer market. But advisors (among them James Birkenstock, Cuthbert Hurd, and his son Thomas Watson, Jr.) helped sway him to authorize the IBM 701, as did eighteen letters of intent for purchase or lease (secured by Birkenstock and Hurd) from government and defense contractors. The first 701 was delivered in 1952.[36] As the Cold War escalated in the early 1950s, Watson's patriotism played into his authorization of the Defense Calculator.

As Watson Jr. was assuming more of the company's leadership responsibilities before officially taking over from his father in 1956, IBM announced a smaller-scale, less expensive, data-processing-oriented digital computer, the IBM 650 in 1953. This smaller computer was first delivered in December 1954 (to John Hancock Mutual Life Insurance Company, Boston,

Massachusetts).[37] In all, more than 2,000 IBM 650s were installed (far exceeding any computer model from any firm to date, and earning the moniker "Model T of computing") before the last was built in 1962, which was the same year that computer revenue for IBM first exceeded that of rental and sales income from pre-computer punch card tabulation machine equipment.[38] This fact highlights the success of IBM's pre-computer tabulation equipment, as well as IBM's measured practices (based on customer operations and their wants and needs), as it maintained account control and migrated its tabulation machine customer base to digital computers in the second half of the 1950s and the early 1960s.

Likewise, during these years it shifted its Service Bureaus to install digital computers in place of, or for a time at least, in addition to, tabulating machines. In 1953, shortly after its completion, the firm installed an IBM 701 at its New York City World Headquarters—the first digital computer at an IBM services facility. This advanced processing installation, which was specifically designated as a "Scientific Computer Center," was also the first IBM Service Bureau to acquire the follow-on large-scale scientific-application focused IBM 704 computer. IBM began to install business data-processing-oriented IBM 650 computers at a number of larger Service Bureaus in the mid 1950s.[39] In this time frame, Thomas Watson Jr. emphasized to his most senior sales executives the great opportunity with IBM Service Bureau Division and the fact that he felt it was underutilized and under-marketed. He indicated the IBM 650 "electronic workhorse" was or soon would be at 16 of the firm's 82 Service Bureau facilities, stressing the importance of sales and service "technical methods and know-how" and the training that many existing IBM Service Bureau personnel had recently completed, and proclaimed "These machines solve with ease such problems as pipeline design production technology, developing characteristics of advanced aircraft, and earth movement calculations for road builders, in addition to such routine jobs as production control and cost accounting."[40]

In 1956, the Department of Justice–IBM consent decree (a settlement to a suit filed against the firm by the Department of Justice in 1952), in part, stipulated that IBM Service Bureau Division had to be spun off as a wholly owned subsidiary and could not share workforces (an employee had to either work at IBM or Service Bureau). While this change—carried out officially on January 1, 1957—eliminated the flexibility in shifting technical and sales staff to meet immediate needs, the expertise and know-how gained at customer sites by field services personnel and IBM service bureau technical staff probably continued to diffuse back and forth between

IBM and its wholly owned services contracting subsidiary, Service Bureau Corporation.[41]

At the start of the 1960s IBM began delivering IBM 1401 systems to 25 of its largest US service bureau facilities. This was IBM's first transistorized computer. During its production run of about ten years, more than 10,000 were produced and installed (primarily at customer locations). IBM also began to link smaller computers to larger SBC computers through tele-processing. As the decade progressed, an increasing percentage of IBM Service Bureau facilities had 1401 computers, and by the mid 1960s, increasingly they installed IBM System/360 series computers. As the firm continually invested in upgrading its service bureaus to its newest digital computers (making them far more efficient), IBM gradually closed some smaller, poorer-performing bureaus—by 1964 the number of IBM's US Service Bureaus declined to 71.[42]

Increasingly IBM's SBC not only sold data processing services, but also computer time, advisory services, and programming services. There were other service bureaus, some run by IBM's small number of computer industry competitors, as well as independent operations (which often had just one or several locations).[43] By the mid to late 1960s, a specialized form of bureau—focused on remotely selling just computer time, or computer time and certain services and software products—emerged: time-sharing companies (the focus of chapter 7).

In 1973, IBM's wholly owned Service Bureau Corporation was sold to competitor Control Data Corporation (CDC) on very favorable terms ($16 million with some guaranteed business for five years) as a partial settlement to a drawn-out lawsuit CDC filed against IBM in 1968 for various anti-competitive practices. At that time SBC's revenues were approximately $63 million a year, a relatively small piece of IBM's roughly $9.5 billion in overall revenue.[44] SBC's full importance, however, cannot be understood merely by the income it generated. A substantial portion of its strategy was to target small data processing customers and help them to grow into having tabulation machines—and later, IBM digital computers—of their own. Further, IBM Service Bureaus provided added incremental capacity to existing organizational customers of IBM systems to meet their data processing wants and needs efficiently and effectively.

Automatic Payrolls, Inc./Automatic Data Processing

IBM Service Bureau Division stood alone in scale and scope among data processing services using tabulation machines in the interwar period and

Figure 2.1
A data entry specialist for the Royal Globe Insurance Company (a British firm) keying in data on a terminal linked to the Service Bureau Corporation's time-sharing network, circa early 1970s. Courtesy of Charles Babbage Institute.

in the early postwar years, as did SBC in the mid 1950s through the early 1970s (partially at first, and soon overwhelmingly, using digital computers). Automatic Data Processing (ADP), which became a major player in the field of data processing services by the mid 1970s, could not have had a more different origin or a more different early history.

While IBM's Service Bureau was a division (and later wholly owned subsidiary) of a multinational juggernaut, Automatic Payrolls Inc. (ADP's name at its founding and through its early years) was initially a one-person startup firm that grew very slowly in its early years. From the 1920s on, IBM had strong organizational capabilities in data processing field service. API started 17 years after IBM's Service Bureau Division, and had no such operations or capabilities. API built none of its own machines. IBM built all machines/systems it used in data processing services. IBM sought out virtually every conceivable data processing services application, as API specialized in just one for a considerable time, and this would always be its largest area—payroll processing. In the era of punch card tabulation machines (before 1954), API did not even focus on using punch card tabulation machines for data processing. In fact, Automatic Payrolls was a misnomer (or anticipatory) in its first years—with the basic machines it used, the company was essentially manually processing payrolls. From the mid 1950s on, IBM's SBD/SBC was an early and rapid adopter with regard to installing digital computers, while ADP was a relatively late adopter of computers (not until 1961). IBM sales and field service personnel helped market, train, and sometimes work at the firm's Service Bureau Division (before SBC), and thus it had a sales and services force that ranged in the thousands in the 1950s (and in hundreds from the start of the division in 1932), API/ADP had a sales force of one or two for the first half of the 1950s. Nonetheless, sales were a common theme—there was an overwhelming commitment at both SBD/SBC and API/ADP to sales, and this, coupled with execution, made all the difference to these businesses' success over time.

Henry Taub, a Hungarian-American with a recent bachelor's degree in accounting from New York University, founded Automatic Payrolls, Inc. in a small space above Grinker's Ice Cream Parlor on Church Street in Paterson, New Jersey in 1949. Taub, born in 1927, had skipped ahead in the Paterson public school system and was only 19 years old when he got his degree from NYU in 1947. He worked for a couple years for an accounting firm, which had a client that, due to the illness of a bookkeeper in charge, had missed distributing employee payroll one payday. After witnessing the chaos this caused, Taub saw an opportunity for a payroll services outsourcing firm. In 1949 he launched Automatic Payrolls, Inc., to process payrolls for nearby small to mid-size businesses. For this enterprise he initially partnered with two other accountants from the firm where he worked, who provided the needed $6,000 in startup capital. In the first year, his two financiers became impatient with the lack of growth, and Taub established a payment plan to buy out his partners and become sole owner of the firm.[45]

The first customer was New Era Dye and Finishing, which was in the nearby town of Fair Lawn, New Jersey. In the first years, Taub and the company were without a vehicle; Taub picked up the paper payroll data and rode buses to deliver payroll checks or voucher envelopes.[46]

Within its first several years, the company, which moved to somewhat larger quarters in the basement of the nearby Carroll Plaza Hotel (on Market Street), slowly began to expand, gaining a larger and broader clientele. Henry Taub's brother Joseph, just out of high school, became API's first hire at the start of the 1950s. Henry and Joseph Taub and Frank Lautenberg (who signed on shortly after Joseph, initially as a part-timer in 1952) made up the triumvirate that would lead the firm for more than three decades.[47]

Frank Lautenberg, born in 1924 in Paterson, grew up in the same racially and ethnically diverse working class neighborhood as the Taubs. He too was a product of Paterson's public schools. Both families' fathers were laborers in the city's silk mills. Frank Lautenberg attributed his father's early death from cancer (when his father was in his early forties) to mill conditions, and this experience shaped his values and concern for employee rights, which he would champion as a five-term New Jersey senator, as well as earlier in his distinguished career helping to lead ADP as president and chairman of the board.[48]

After serving in the US Army Signal Corps during the war, Frank Lautenberg used the GI Bill to fund his completion of an undergraduate economics degree at Columbia University (in 1949). He became a salesman for Prudential Insurance, which for a time had adjacent office space to API. Fascinated with Henry Taub's idea, he offered to work part-time for only a sales commission. After Lautenberg proved himself (boosting business by 30 percent in short order to reach thirteen customers), he and Henry Taub came to an arrangement in early 1954, and Lautenberg became API's full-time sales and marketing manager. By 1957 he was making major contributions, and on a handshake agreement with Henry Taub, the two decided he would be one of the three partners in the business (and Lautenberg was considered by the Taub brothers from that point forth one of the founders).[49]

Lautenberg remembered Henry Taub as the visionary who would also be the front person in setting new customers up (making arrangement for data delivery), Joseph as a good organizer who managed the API/ADP office and staff, and himself as the full-time sales and marketing specialist. He had learned at Prudential that if he could "muster up enough courage to knock on the door, that he could get people's attention."[50] That is exactly what he did as he built up API's customer base in the 1950s.

Part of the challenge as Lautenberg recalled, "payroll records were considered sacred ... so I soon discovered that my job as the principal sales and marketing person was to convince the prospect of the wisdom of outsourcing."[51] Conveying and earning trust was absolutely critical as decision makers at (potential) customer organizations had to be convinced that information would never leak out to "their competitors, but also their own employees. ... Their [pay] scales weren't always exactly the same."[52] Efficiency was an easy sell if he could get by this hurdle, as a 100 person payroll might be done by API for one fourth of a bookkeeper's salary—and in a fraction of the time.[53]

None of the three of them were particularly technical, as Lautenberg pointed out, and before the late 1950s they did not use (relatively expensive) punch card tabulation machines. To add and multiply columns of numbers they used a second-hand Friden calculator, then, slightly later, a second-hand Comptometer accounting machine. For several years between the Comptometer and acquiring its first computer in 1961, the firm, beginning in 1957, used an IBM punch card tabulation system. At that time, the three spun off a company with a new name, Automatic Tabulating Services, doing some limited data processing services work other than payrolls— including calculating bowling scores for local leagues.[54] And with its first computer, an IBM 1401, API became Automatic Data Processing, or ADP, Inc., and Automatic Tabulating Services was merged back in to become part of ADP.[55]

As API/ADP was upgrading its technology it also expanded its facilities. Back in 1956, API moved to a larger facility—a former supermarket on Route 46 in neighboring Clifton, New Jersey.[56] With the arrival of the IBM punch card tabulation machine, the calculation task for the first time became substantially automated, that is after the major new task of punching data on cards for each payroll had been completed. Support staff had grown slowly to that point, but accelerated more rapidly with the growing customer base and size of customers and the need for key punchers. Following the employee-friendly ethos of the Taubs and Lautenberg, the company offered different schedule options. The firm provided flexible schedules so women could work part time while their husbands were at work and their children were at school. Then there was a second shift of full-time key punchers who came in at 4 or 5 PM and worked until midnight. Another job was pickup and delivery. At the start it had been Henry Taub riding the buses, but as sales rose the firm acquired a Volkswagen and, later, multiple vehicles and pickup/delivery drivers. Many customers in these years wanted a cash payroll (which was popular with employees). In these cases ADP would just calculate and print the payroll amount and other identifying information on voucher envelopes that the customer organization's payroll clerk would then fill with the appropriate amount of cash.[57]

The IBM tabulating machine brought some change and added efficiency, but nothing compared to adding digital computers, first the IBM 1401 in 1961, and then, IBM System/360 series computers in the middle of the decade. Lautenberg remembers in the early 1960s that the IBM sales force sought to convince them how much easier a digital computer would be than "tab" equipment.

The greatest advantage, as Lautenberg later reflected, was with programming, which helped facilitate replication of processes, and thus the firm's geographic reach (which had already grown a bit by then) through acquisitions. The company hired a combination of domain experts (accountants and brokerage specialists) and technical experts (tabulating machine operators, computer operators, and programmers). With none of the three principals being technical, Lautenberg recalled that they as a group had "the tenacity and strength to stumble through the mistakes." And that they deserved more credit "for starting the services business than ... for being in the computer business." Nonetheless, programming expertise to apply IT to domains was critical. Lautenberg emphasized that was: "the beginning of our ability to establish common production facilities in different places." He explained, "It wasn't realistic before when so much of it was manual ... . We could do the programming that was necessary, and then we made an acquisition in Miami, Florida, or Boston, Massachusetts, or Cleveland, Ohio, or wherever, we could send the service process out and have it working." As Lautenberg emphasized, this "standardized the market." "But," he continued, "it wasn't until 1965 when the IBM System/360 was produced that the world opened up."[58]

To grow the business, Henry Taub and Frank Lautenberg shared the task of scouting out acquisition possibilities all over the country throughout the early 1960s after they had acquired a leased IBM 1401, and this accelerated in the mid 1960s with the new IBM System/360 series. To do this they needed capital, and stock of clearly recognized and respected value. Hence, they took the company public in 1961. In that year ADP recorded more than $400,000 in revenue and had 300 clients and 125 employees.[59] The 1961 ADP initial public offering, handled by Oppenheimer & Company, yielded $300,000 in cash.[60] Using this capital, but primarily the firm's (now) well-respected company stock, ADP acquired a number of small payroll processing and other data processing shops with annual revenues of about $100,000 or $200,000. Then they would quickly expand these operations. In the early 1960s ADP acquired Arthur Kranseler's small Boston firm (which had been doing statistical processing work for Harvard University and MIT) and David Perlman's service bureau in Miami. These were followed by many more acquisitions that decade and in succeeding ones.[61]

Payroll was ADP's core, and remains so to this day, but in the 1960s it added other significant areas of data processing. Of these, none was more important to ADP than brokerage services. In the late 1950s, Jack Nash of Oppenheimer & Company asked Henry Taub to examine the brokerage's back-office operations. Taub concluded that his firm did not have the resources to provide any outsourcing. This changed with ADP's acquisition of the IBM 1401. In 1962, ADP launched its Brokerage Services Division with one client: Oppenheimer & Company. It helped the brokerage company process its many daily financial transactions and maintain records, in the early years processing more than 300 transactions per night. (During the 1990s, overall brokerage transactions by ADP would exceed a million a day.) Despite the odds, ADP succeeded in this business space that was occupied by data processing powerhouses like IBM's Service Bureau Corporation and Sperry Rand.[62] In 1972, ADP began to offer large-scale inventory and accounting data processing services to auto dealers with its Dealer Services Division. Two years later it expanded to Europe by purchasing a Dutch payroll services bureau, just the beginning of its worldwide expansion. By 1975 its revenues topped $150 million, the firm had 35,000 clients, and a labor force of 5,000.[63]

Conclusion

Even though C-T-R first investigated data processing services as a possible new business in 1922, IBM's Service Bureau Division was not launched until 1932. It was as an outsourcing option for customers, and one that for the first time partially monetized IBM's long-established work and capabilities in data processing services. Even with this division's steady growth, it was a small portion of the data processing services it provided to customers (most such services were "freely" bundled in with hardware contracts and provided at customer sites) in the pre-computer tabulation machine era as well as in the early computer era.

ADP (initially as API) started slowly, but by the mid 1970s the company had substantially more annual revenue than Service Bureau Corporation, which had recently been sold by IBM to Control Data Corporation.[64]

That ADP was larger than SBC by the mid 1970s is a testament to the importance of ADP's specialty, payroll processing, which it now does for the vast majority of Fortune 500 companies. ADP's leaders also demonstrated strategic insight into other business process outsourcing specialties, including brokerage transactions processing. Business process outsourcing in recent decades, with the growth of computer networking throughout the world and inexpensive computer memory and microprocessors, has in part shifted to countries with lower labor costs—in particular, India.[65]

3 Programming Machines: Computer Usage, C-E-I-R, and Computer Sciences, 1955–1967

In many respects, the most substantial and defining segment of the computer services industry is programming services, which, along with some computer customers hiring or using existing staff to program mainframes (and later, minicomputers), is fundamentally concerned with building the instruction-based applications (and, at times, compilers, specialized programming languages, and systems programs) that make computers useful.[1] Two startup firms pioneered this segment: Computer Usage Company (CUC) and Council/Corporation for Economic and Industry Research (C-E-I-R, Inc.). These two companies—quite different in their origins—began in the mid 1950s and grew rapidly in the 1960s. They, however, did not have as long a sustained influence as Computer Sciences Corporation (CSC), a major player in today's computer services industry that was launched in 1959, grew steadily in the first half of the 1960s by specializing in systems programming for computer manufacturers, and eventually offered other types of programming and systems integration services to the federal and local governments, corporations, and other organizational customers.[2]

All three, at various points and to differing degrees, advised on new computing installations, developed services to directly sell or broker computer time, and provided standard "back office" data processing (indicative of the partial overlap of segments of the computer services industry). CUC's and C-E-I-R's core was in applications programming, and this also became central to CSC during its first decade and beyond. This chapter explores each of these three firms and their efforts—through hiring, training, developing, and deploying individuals in the emerging occupations of programmers and systems analysts—to make computers useful to organizational customers.[3]

Computer Usage Company

John W. Sheldon and Elmer Kubie founded Computer Usage Company— the first independent firm in the computer services industry to focus primarily on programming services— in March 1955. Both founders had worked for IBM for several years during the first half of the 1950s. Sheldon left IBM in 1953 for consulting work on applied mathematics; Kubie left in 1954 for a position in operations research at General Electric.[4] Working at IBM at this critical juncture, just as some of its executives (including Cuthbert Hurd and Thomas J. Watson Jr.) surveyed the emerging market for computers and convinced the elder Watson to enter this industry with the IBM Defense Calculator/IBM 701, coupled with the IT giant's long-term commitment to data processing services (Service Bureau Division) and its rapidly growing presence with scientific calculation/computation, helped provide the experience and the motivation for Sheldon and Kubie's entrepreneurial venture.[5]

Born in New York City in 1923, Sheldon attended Phillips Exeter Academy and completed most of his undergraduate work at Yale University before serving as a radar and sonar specialist in the US Navy on a destroyer in the Pacific during the last two years of World War II. In 1945 and 1947 respectively, he received bachelor's and master's degrees in physics from Yale University. He worked for a couple years at Columbia University's Thomas J. Watson Scientific Computing Laboratory before joining IBM in 1949 to head its New York City Technical Computing Bureau (TCB). Sheldon led the effort to install the IBM 701 at the bureau in late 1952—an installation that was providing client services (primarily for government and industrial scientific and technical applications) by early 1953.[6]

Months before Sheldon undertook to install the TCB's IBM 701, Elmer Kubie joined Sheldon's group at the bureau. Kubie's background—in many respects analogous to Sheldon's—included a two-year enrollment as an electronics technician in the Navy and a BS in physics (financed under the GI Bill) from Columbia University. He came to the TCB as his first assignment upon joining IBM in 1952, overlapping with and working for Sheldon for several months.[7]

Early in 1955, Sheldon approached Kubie about forming a computer services company, one that would focus on "programming" rather than "consulting." After a few days of thinking it over Kubie was on board—signing on, at Sheldon's request, as president, while Sheldon took the title of Director of Technical Applications Development (both were original members of the board). Sheldon came up with $40,000 ($20,000 himself and an equal amount from a family member) in startup funds. In April 1955, Sheldon and Kubie hired a secretary and bookkeeper (Louise Greene) and four programmers (Barbara Lesser, Anne Shea, Rheena Trunk, and Dorothy Walsh).[8]

In the late 1940s the computer staff that would later be referred to as "programmers" were often called "operators"—and, with the advent of compilers in the early 1950s, "coders." By the mid 1950s the term "programmer" coexisted with "coder." The optimal educational background and skills for such work was uncertain, and frequently debated among hiring authorities within organizations. In 1955, IBM developed its famed Programmer Aptitude Test (PAT). Many computer firms, organizational users with substantial programming requirements, and programming services companies were employing versions of IBM's PAT, or a similar test, by the early 1960s. The IBM PAT—in its original form as well as in later versions—concentrated significantly on mathematics and logic skills. Computer Usage administered an aptitude test in the early 1960s that required applicants to solve logic problems by studying an IBM 1401 console light panel.[9] IBM, CUC, and other organizations often found testing helpful to assess programming aptitude, but educational background and test results were imperfect predictors of potential skill in the art of programming. Rarely was analysis done to try to correlate (or disprove) an association between high scores (or middle or low ones, if such hires were made) and job performance as a programmer. System Development Corporation probably was the first to formally do personality assessments in hiring programmers (they also used aptitude tests).[10] As historian Nathan Ensmenger emphasizes, the skyrocketing demand for programmers in the 1950s and the 1960s—which *Businessweek* in 1966 called a computing "crisis" resulting from a "software gap"—dominated the thinking of hiring authorities.[11] Self-reported "job satisfaction" being the best measure to try to correlate with personality assessments (in view of the general absence of systematic, objective performance evaluations), firms rarely were able to hire only those scoring the highest on tests or showing the "optimal" personality to have job satisfaction in the programming field.[12] Further, programming was an extremely wide domain—mathematical ability may have meant more for certain types of programming, such as scientific programming or systems programming, than for business programming.

During World War II, the ENIAC project had drawn on the expertise of the "ENIAC girls," six highly gifted women with backgrounds in mathematics: Frances Bilas, Betty Jennings, Ruth Lichterman, Kay McNulty, Betty Snyder, and Marilyn Wescoff. Historian Jennifer Light insightfully analyzed these women's "ambiguous entry into computing" during the wartime environment of disrupted gender roles. The antecedents of female human computers (pre-1945) on ballistics firing table calculation efforts and the original characterization of the new occupation as "operators," where programming was achieved by elaborate configuring of wires on plug boards (physically similar to the female-gendered occupation of telephone operators), played into the opportunity for these six principal programmers for this influential computer.[13] At the time, hardware engineering was of distinctly higher status relative to the equally demanding technical task of programming.

No documentation exists on the specific contributions of CUC's first four programmers, but a surviving "Roster of Personnel" from July 1961 indicates the hierarchy and gender mix of CUC technical staff and officers a half-dozen years after the firm's founding.[14] By 1961, Louise Greene had become the corporate secretary and controller, joining President Kubie, Vice President G. Liston Tatum, and Consul William Sloane as one of CUC's four corporate officers.[15]

CUC "analysts" (at three levels) oversaw projects and engaged in system design to ensure that the company met the needs of customer/client organizations. They oversaw the coding work (at two levels) of "programmers." Many early programming services firms would have similar hierarchies within the two broad categories of "systems analysts" and "programmers." In 1961 the hierarchy of managerial/technical staff in the New York City office (in descending order) consisted of: principal analysts, senior analysts, analysts, senior programmers, and programmers. All five principal analysts were male. Three of the seven senior analysts were women, including one of the original four programmers, Barbara Lesser. Of the seventeen analysts, seven were women, also including one of the original four, Reena Trunk (the other two original female programmers were not on the roster and presumably had left CUC). Four of the eight programmers and six of the fifteen senior programmers were women. As 42 percent of senior analysts and 33 percent of senior programmers (in mid 1961, out of CUC's main New York City office), women were unquestionably making major technical and managerial contributions, but there probably was a "glass ceiling" preventing their rise to the top managerial/technical job title: principal analyst.[16] Regarding corporate leadership, it appears there was less of a continuing opportunity for women as the company grew. By 1966, Louise Greene was no longer with CUC, and all five officers and all three board members were male.[17]

Throughout March and April of 1955, CUC had been run out of the living room of Sheldon's apartment on New York's Upper East Side, the most pronounced retrofit being a large wall-hung chalkboard. At the start of May the firm moved to occupy the fifth floor of a midtown Manhattan office building at 18 East 41st Street. That year, and for many additional ones, CUC did not possess a computer of its own; the company most commonly programmed on customer computers on site at customer locations or bought computer time from a service bureau. Throughout 1955—drawing on the two founders' backgrounds at IBM with the IBM 701—the customers' desired applications tended to be in scientific rather than business computing. The first project, for California Research Corporation, was to program an IBM 701 application for simulating the radial flow of fluids to an oil well, and computer time was purchased from Sheldon's and Kubie's former IBM worksite, the IBM New York City TCB. Kubie supervised the programmers on this project.[18] In 1956 Sheldon led a CUC effort, working alongside a customer (E. L. Wachpress of the Knolls Atomic Power Laboratory), to program an application to solve the "multi-group, steady-state, neutron-diffusion equation in three different two-dimensional geometries." Sheldon's underlying logic for the program was later applied within the atomic energy programs in the United States, in the United Kingdom, and in France.[19]

In late 1952, Kubie, along with other senior engineers at IBM, had advocated to the company's top executives to have the firm develop and market an intermediate-priced business-oriented computer. Once the project was approved and began at year's end, Kubie, then a member of the IBM Applied Science Department-Mathematical Group, assisted engineers in designing and developing the IBM 650.[20] Working for Kubie on this project was George Trimble, a mathematician and engineer who would leave IBM to join CUC in February 1956 as the startup's eighth employee. That year CUC had its first business programming application projects, which included creating an IBM 650 actuarial program that analyzed the relationship between blood pressure and body build and mortality and longevity for the Prudential Life Insurance Company and a sales and inventory management program for Hartfield Stores, a West Coast women's clothing retailer headquartered in New York City.[21]

The founding of CUC by two former IBM technical/managerial employees, the startup services firm's usage of the IBM TCB, and the hiring away of Trimble and other employees highlight the importance of the symbiotic relationship that existed between IBM and CUC. IBM executives undoubtedly hated to see technical talent like Trimble leave—he not only worked on the IBM 650 design, but also wrote early software packages for the machine and was a frequent speaker at IBM 650 classes at Endicott in 1954 and 1955. At the same time, the computer manufacturer benefited from CUC making IBM 701s, IBM 650s, IBM 1401s, and (in the mid 1960s and later) IBM System/360s useful to a broad range of IBM's corporate, government, and other organizational hardware systems customers.[22]

Moreover, IBM's (and later CUC's) Cuthbert Hurd estimated that at times (probably in mid 1960s with the System/360 project) 40 to 60 percent of Computer Usage's business came from (indirectly, but in partnership with) IBM.[23] CUC was asked by IBM to begin programming work on the System/360 in November 1963 in fulfillment of a request from the Federal Aviation Agency (FAA).[24] No System/360 hardware was available yet, and initial work was needed developing a JOVIAL compiler and an "elementary operating system" ideally to be ready as soon as the hardware—the System/360 Model 50—was available. (The FAA had ordered six of these systems.) The CUC software effort began by working on a simulation of a System/360 Model 50 developed on an IBM 7030 "Stretch" supercomputer located at the US Naval Weapons Laboratory in Dahlgren, Virginia.[25] By late 1966, the CUC team—drawing on roughly 60 person-years of experience—had written and checked approximately 200,000 lines of code for the system. Both CUC and IBM benefited from this successful project done for the FAA for air traffic control. In 1966 CUC staff involved in this programming effort published the book *Programming the IBM System/360*, which was officially authored by "The Staff of Computer Usage Company." Inside the book, twenty authors are named—ten men and ten women who worked on System/360 Model 50 JOVIAL compiler and operating system software. This CUC project helped IBM fulfill a major hardware system contract, allowed IBM and CUC software developers to learn from each other, and generated a significant portion (probably at least 40 percent) of CUC's revenue over the project's three-year life.[26]

IBM was late to focus on time sharing with System/360 and some leading time-sharing scientists at MIT's Project MAC (especially Multics project leader Fernando Corbató) chose to go with General Electric (using the GE 645 mainframe) as a partner/supplier on MIT's Multics time-sharing operating system project in the second half of the 1960s. On both technical and economic grounds Corbató and his Project MAC colleagues believed GE to be the better choice.[27] IBM helped support the MIT computer laboratory, which had successively installed an IBM 704, an IBM 709, and an IBM 7090—its powerful scientific computers—in the second half of the 1950s and at the start of the 1960s. In view of successful past partnering, IBM's leaders were disappointed by MIT Project MAC computer scientists' decision to go with GE for Multics. These IBM executives also perceived General Electric, with its substantial financial resources, as a potential major competitor. Ultimately, this competitive threat never meaningfully materialized. This largely was because GE top executives, far less committed to computing, deprived the division of resources in the 1960s and sold its computer division to Honeywell in 1970.[28]

In the mid 1960s, with GE a potentially fierce competitor, IBM scrambled to get time-sharing software out and available to customers and as part of this effort contracted with CUC to implement the command system for Time-Sharing System (TSS) 67. Another portion of the TSS 67 had been contracted to Computer Sciences Corporation (CSC).[29] As IBM was moving toward greater vertical integration on the semiconductor side in the early to mid 1960s, launching and rapidly growing its Components Division, by essentially partnering in various types of indirect outsourcing of software development (to CUC, CSC, and other firms) to enable optimal fulfilment of hardware contracts, the corporation's top executives were drawing the boundary of the firm (at times favoring a network or market model over vertical integration or hierarchy) somewhat differently on the software side.[30]

In the late 1950s and the early 1960s, CUC expanded geographically and added several new computer services business segments. In 1957 CUC established an Education Division, which ran educational workshops on operating systems (for IBM computers). In 1959 CUC opened a Washington office as contracts for the US Navy and other federal government business continued to grow after the start of a project in 1958 to develop information requirements for Naval Operational Control (OpCon) Centers.[31]

In 1960, CUC launched its Computer Time Sales Division (CTSD), which bought and sold computer time on customer machines to more efficiently utilize expensive computing resources. This business, initially just focused on the New York City area, sold time on three IBM 650s and soon thereafter sold time on several IBM 1401s. The enterprise not only aided customers and generated revenue for CUC, it also was an excellent tool to introduce new potential organizational customers to Computer Usage.[32]

Also of great significance in 1960 CUC had an initial public offering—raising a net of more than $185,000—to help finance the company's continuing growth.[33] That year CUC expanded to the West Coast opening an office in Los Angeles. By the end of 1961, CUC had 114 staff members—most of them at its New York City headquarters, about a fourth of them in Washington, and about a tenth of them at its new Los Angeles facility.[34]

Over the next several years, it also launched offices in Dallas, where it was developing an entire software package for Texas Instruments' Advanced Scientific Computer (ASC), and in Chicago, where it was creating a symbolic assembler for Advanced Scientific Instruments (ASI).[35] In general, new office locations in the 1960s often arose in conjunction with securing a major project in the area. And occasionally it would shut down an office if business did not expand in a city after a major contract concluded, as was the case in Dallas.

Momentum continued into 1962, the year in which the distinguished IBM executive Cuthbert Hurd—a close advisor to Thomas Watson and later to Thomas Watson, Jr.—left IBM to join CUC as an executive and as chairman of the board of directors.[36] Hurd's stature as IBM's former Applied Science Department Director, who helped to lead IBM into the computing business, was widely known throughout the computer industry and further added to CUC's strong reputation.

Computer Usage's revenues grew steadily in its first five years and beyond. The company plowed substantial portions of its earnings into new business areas and new locations. In 1961, CUC had record annual revenue of $1.3 million and net income slightly over $37,000 and was hiring away talented programmers (and sales personnel) not only from IBM but also from other large companies. In 1961, it hired senior programmer Judy Morrison from the New York Telephone Company and senior programmer Yvonne Saleh from Socony-Mobil Oil.[37]

By the mid 1960s the company's mix of business programming services went far beyond traditional back-office data processing applications—accounting, human resources, and payroll—to include specialized programming for sales forecasting, project management, plant management and maintenance, life insurance actuarial analysis, and "business gaming."[38] Its science and engineering applications for the government, defense contractors, the petroleum industry, and various other trades continued to expand rapidly. Among the applications were simulation, command and control systems (including submarine control and air traffic control), training simulators, data extraction and analysis from the US Census, and mathematical modeling of water and petroleum resources. At this time, CUC also entered facilities management (a new segment of the computer services industry), providing personnel to run a customers' data processing operations—often completely.[39]

The company's Computer Usage Facilities Management (CUFM) subsidiary was a natural outgrowth of CUC's computer time brokering business, where it could balance loads at different customer sites to optimize usage of expensive computer resources. The facilities management segment of the computer services industry had been invented a few years earlier by H. Ross Perot with his startup Electronic Data Systems (the focus of chapter 6).

By the mid 1960s, CUC had expanded to twelve offices in or near major US cities. In addition to New York, Washington, Los Angeles, and Chicago, it now had offices in Baltimore, Boston, Houston, Montclair (New Jersey), Mount Kisco (New York), Palo Alto, Philadelphia, and San Francisco. The company had more than 350 employees in 1965 and more than 700 by the end of 1967—the majority of them in the job categories of systems analysts and programmers, with lesser numbers in sales and administration. Likewise, revenue and net profits grew substantially by the mid 1960s and continued to accelerate for several years.[40] In 1965, revenue exceeded $6 million net income was nearly $350,000. And a mere two years later, revenue had more than doubled to $13.26 million annually, and profits exceeded $600,000.[41] Net profits—generally around 3 to 5 percent in the company's first dozen years—were a relatively small share of revenue in part because of CUC's rapid expansion (the expense of obtaining new offices and hiring and training new employees) and adding new computer services segments or businesses, much of which was funded organically. In addition to accelerated expansion, heightened competition—from C-E-I-R, CSC, and bundled services from IBM, and others—was also pressuring CUC's operating margins.

Increasing fierce competition and particularly misguided rapid expansion during the 1970s, left CUC vulnerable. After Elmer Kubie was diagnosed with diabetes and chose to leave the firm in 1968, Cuthbert Hurd, as chairman of the board, recruited the head of IBM's Federal Systems Division, Charles Benton, to become CUC's president (Sheldon had resigned four years earlier to pursue educational and various business interests). Benton wanted to grow the sales organization greatly and create a very large, broad-based information technology company, a plan that included moving into software products and hardware peripherals.[42] In 1969 CUC had its first annual loss. In the 1970s, executing Benton's strategy, CUC expanded into a number of new areas—from disc drives and tape drives to central processing units and installing turnkey minicomputer systems. Eventually nearly all these businesses were shut down as they failed to consistently contribute to the bottom line. Programming services accounted for $10 million in revenue for CUC in 1967 and just $200,000 in 1974. With a few losing years during the 1970s, losses became the norm by the early 1980s. In 1982, 1983, and 1984 the company had annual losses of

$1.5 million, $400,000, and $2.2 million, respectively. In 1985, CUC's last full year in business, it lost $2.4 million on a mere $1.5 million of revenue.[43] Unrestrained expansion—geographic expansion and expansion into many businesses in which it had little experience—resulted in a firm that provided an uneven quality of services and products in an increasingly competitive IT field. In programming services, Computer Sciences Corporation became more focused on federal government business and Computer Usage lost out mightily to CSC in this area. And overall, CUC failed to compete effectively against Perot's formidable EDS in facilities management. In 1986, Computer Usage Company, the enterprise that launched the independent programming services industry, declared bankruptcy and was liquidated.

C-E-I-R, Inc.

The Council of Economic and Industry Research would become a major competitor to Computer Usage Company in the second half of the 1950s, but in 1952, when it was founded, it was a nonprofit organization and did not supply computer services. It probably was not on the radar of John Sheldon or Elmer Kubie until late 1956 or early 1957, about two years after they founded CUC. The Council of Economic and Industry Research's assembly of programming talent, an entrepreneurial-minded leader joining the organization in late 1953, and the economics of managing expensive internal computing resources led the organization to become the first significant enterprise to enter the programming services field after CUC.

A group of accomplished researchers, including Wassily Leontief of Harvard University and Frederick Moore of the RAND Corporation, launched the Council of Economic and Industry Research on November 26, 1952 as an economic and operations research organization to fulfill a two-year $350,000 research project for the US Air Force to begin in the middle of 1953.[44] The project involved conducting complex analysis and modeling to determine optimal bombing targets to economically cripple the Soviet Union in the event of a military conflict. The economic modeling to accomplish this task required calculating simultaneous equations with thousands of coefficients. This Cold War economic research organization was connected to computing only because it would take the advanced speed and accuracy of digital computers to be able handle such complex sets of equations. Ironically—and indicative of the complexities of military contracting—the standout economist founders of the Council of Economic and Industry Research neglected to account for the timing of the money they would need (and hence the interest on funds they would need to borrow), and stood to potentially lose thousands of dollars if they carried out the contract as specified. Digging in their heels, rather than flexibly re-negotiating, the small Council of Economic and Industry Research team refused to start the project and no work on it was completed during 1953.[45]

Herbert Robinson was selected to solve the impasse, to take over as the sole leader of Council of Economic and Industry Research, and to re-launch the organization (in Washington) in March 1954.[46] The organization was still required to carry out the founding project on the original terms. As the Cold War escalated week by week after August 12, 1953, when the Soviet Union tested its first hydrogen bomb, the Air Force wanted bombing target ratings as soon as it could get them.

Robinson, a British economist with a doctorate from the London School of Economics, had immigrated to the US after World War II. Before the war he had been on the staff of Lord Cherwell, Winston Churchill's top science advisor. After the war, he had worked for the US Veterans Administration, for various international relief agencies (among them the World Bank), and for the Defense Production Administration. The DPA, which Robinson served from 1951 to 1953 as Deputy Chief of Foreign Requirements, was an independent US federal government agency that was an outgrowth of the December 16, 1950 Defense Production Act.[47] It had been created to help oversee industrial production in support of the US effort in the Korean War. In late July 1953, after the Korean Armistice Agreement, Robinson was looking for work. A contact at the National Science Foundation who was aware of Robinson's impeccable economics research credentials put Robinson in touch with Frederick Moore and other RAND Corporation researchers. Though Leontief stayed on for a time, most of the other founding researchers left the project as Robinson came on to lead the organization. The government would let the other researchers off the hook for the legally binding contract only if other competent individuals took it over.[48]

Robinson faced considerable financial risk in signing on for this endeavor and essentially taking it over from its founders in early 1954, but he had the optimism and the strong work ethic (though not the business experience) of a true entrepreneur. He estimated that the maximum cost of borrowing money to get the original project completed would be about $30,000 (which he believed gradually could be made up as the project ran its course), and agreed to put up slightly more than half of that amount—his life savings to that point. He got the rest from two former DPA colleagues (Rudolph

Figure 3.1
Herbert W. Robinson, longtime president of C-E-I-R, Inc., circa the 1960s. Courtesy of Charles Babbage Institute, University of Minnesota.

Johnson and Malcolm Catlin) and a consultant. Catlin and Johnson would become vice presidents, and serve with Robinson on the board of directors. Robinson was betting that he and his colleagues would be able to create a longer-lived organization. He hoped that one successful project would lead to other contracts, and that this nonprofit organization might become a stable home for him.[49]

Leontief, of Russian heritage and educated in Russia and in Germany (he had a PhD in economics from the University of Berlin), came to the US in 1931 to work at the National Bureau of Economic Research and later became a researcher and a faculty member at Harvard University. In 1949 he used the Harvard Mark II to calculate various complex equations for economic models, and hence brought considerable experience and expertise in computer-based economics to the founding project for the Council of Economic and Industry Research. Later (in 1973) he was awarded a Nobel Prize for "input-output tables" that demonstrated how inputs from one industry produce outputs for consumption or inputs for a different industry (work analogous to the founding Air Force project at hand). He was a mentor to three future Nobel laureates, including Paul Samuelson.[50]

During 1954 and 1955, Robinson assembled a staff of more than thirty specialists in input-output economics, some of whom were recruited away from the Bureau of Labor Statistics and the Department of Commerce. They produced dozens of industry reports on various Soviet industries, as specified by their contract.[51] Language specialists were offered subcontracts to help with the task. Every hire (and every contractor) had to obtain Top Secret clearance. The data processing/computer time was contracted out as well. The Air Force contract was successfully completed in less than two years, with all staff and sub-contracting entirely devoted to the project. Though the organization had lost between $10,000 and $20,000 completing the project, its success, as Robinson had hoped, set up the Council of Economic and Industry Research for other data-intensive economic modeling projects.[52]

Because the Council of Economic and Industry Research had lost money in each of its first two years, Robinson proposed in 1956 that it re-incorporate as a for-profit entity under the name C-E-I-R, Inc. This would allow greater flexibility in raising capital, offer benefits for writing off losses, and most importantly, provide potentially greater financial rewards for Robinson and others. The organization was able to obtain additional contracts and Robinson, with support from his executive team, decided to acquire an IBM 650. C-E-I-R, Inc. was expanding its staff at this time and needed more space. Rather than stay in Washington, the company moved to larger quarters in Arlington, Virginia.[53]

Between the start of 1955 and late 1956, C-E-I-R had grown from a staff of 32 to 68. With the move to Arlington and the increase in the company's personnel, Robinson had an "epiphany" and said to his leadership team "Let's get a 704."[54] The IBM 704—a powerful scientific computer system priced at $2 million and the first to use FORTRAN—leased for more than $40,000 a month for most installations. Robinson communicated to his staff that IBM was willing to lease a 704 and that C-E-I-R should seize the opportunity. Robinson recalls that "everybody in the company was against it."[55] The entrepreneurial leader reasoned that C-E-I-R could sell computer time as well as programming services during downtime on

Figure 3.2
C-E-I-R facilities in Arlington, Virginia housing an IBM 704, circa 1960. Courtesy of Charles Babbage Institute, University of Minnesota.

the system to cover the cost, and perhaps even more. Despite continuing opposition, he moved forward with acquiring a lease on an IBM 704 in 1957.[56] With that decision, C-E-I-R, Inc. firmly launched itself as a programming services enterprise as well as a service bureau business. Before long, these were C-E-I-R's primary businesses and the businesses that propelled its rapid growth.

One of the keys to C-E-I-R's successful transition to a computer services company was hiring William Orchard-Hays, a talented scientist and programming expert, away from the RAND Corporation in the mid 1950s. "[W]e hired a really outstanding man to head our computer operation," Robinson reminisced.[57] At RAND, just before joining C-E-I-R, Orchard-Hays had teamed up with Harry Markowitz. Orchard-Hays had been the principal programmer for the algorithm for Sparse Matrices—a type of matrices, invented and named by Markowitz, that were widely used in operations research. (They took advantage of the large number of zeros in a matrix for a particular desired operation to enable rapid solutions.) Markowitz's co-developed programming language, Simscript, was optimized for writing simulation programs and quite influential in this specialization. And his development of "portfolio theory," would make Markowitz famous and earn him a Nobel Prize in economics.[58] Orchard-Hays' achievements were less in the spotlight, but he was deeply respected in computing circles and was instrumental to C-E-I-R's becoming the fastest-growing broad-based programming services company in the world in the late 1950s and the early 1960s.[59]

In building a programming services business, Orchard-Hays instituted a programming aptitude test. The company hired about twenty trainees on the basis of their performance on that test. Most of these hires, although they showed aptitude, had no programming experience and took considerable time to train. Some early programming applications were unique to a client. Other, more standard types of applications could be modified or customized for different clients in a particular industry—for example, for the electrical utilities industry C-E-I-R delivered a circuit analysis program and a load flow program. One early challenge Robinson related was that his company sought to capitalize the training as an expense and to capitalize the development costs of a software program. Arthur Andersen accountants, however, would not allow their clients, including C-E-I-R, Inc., to capitalize training or software programs—these had to be written off immediately.[60] Programs, which would later be referred to as software, were something new and by their very nature were "intangible."

By the late 1950s, it had become increasingly possible to hire experienced programmers for more senior assignments. Herbert Grosch—who had had a distinguished career working at the Watson Scientific Laboratory, at IBM (twice), at General Electric's Evendale facility, and in MIT's Whirlwind project—signed on at C-E-I-R at the end of the 1950s. Grosch was widely known in computing circles for his editorials and observations, and especially for a mid-1960s observation on the relationship of scale to efficiency in computing (discussed in chapter 7).

Grosch recalled that Robinson's refined English background and social graces, coupled with his lack of computer knowledge or experience, sometimes helped him sell programming services contracts to clients. Grosch stated that Robinson "a remarkable fellow" who "had two great advantages. First, he didn't know beans about computing or computers, so he could promise anything his clients wanted with a straight face. Second, there was no way [anyone] ... could realistically test the resulting software. The SAGE people knew all about that but were off in another Pentagon universe ... ."[61]

Figure 3.3
Herbert Grosch working on an IBM CPC (Card Programmed Stored Program Calculator). The 1949 announced IBM CPC was an electronic digital calculating system that lacked the stored program capability of the IBM 650 computer. Courtesy of Charles Babbage Institute, University of Minnesota.

One of Grosch's first assignments was helping to oversee an IBM 709 installation shared by C-E-I-R at Union Carbide's New York City headquarters. C-E-I-R, in contrast to CUC, had a mixed business model of sometimes acquiring computers and sometimes using customers' machines and service bureaus or sharing installations. In addition to the facility shared with Union Carbide in New York, C-E-I-R shared a computing facility with Del Monte Foods in San Francisco. Another of Grosch's early assignments was helping to oversee C-E-I-R's launch of an office in London on March 1, 1960.[62]

Around the same time as the London expansion, C-E-I-R was also expanding quickly in the United States, both organically and through acquisitions and mergers. C-E-I-R's Arlington IBM 7090 installation (August 1960) was the world's first non-defense installation of a 7090, and, of course, the first to be available for contract services for customers/clients. By that time

C-E-I-R's Manhattan Research Center had expanded greatly, occupying the entire 16th floor of the Union Carbide Building. In that building C-E-I-R also had office space for personnel on the 35th floor and a new IBM 7090 installation (replacing multiple IBM 709s) on the 36th. By late 1960, C-E-I-R's Manhattan Research Center had forty employees, most of them programmers and systems analysts.[63]

In 1960 C-E-I-R implemented a strategy of expansion through acquisitions. On July 1, C-E-I-R acquired (by stock) General Analysis Corporation of California. General Analysis had been founded in 1955 and had a quite similar profile to C-E-I-R, with businesses in conducting economic surveys, operations research, and economic and business gaming, and with a rapidly growing computer services business to take full advantage of data processing resources. With the merger, the former General Analysis headquarters immediately became the new C-E-I-R Los Angeles Research Center.[64]

Similarly another computer services acquisition, DATA-TECH Corporation, became the new C-E-I-R Hartford Research Center. An engineer who had gained computer and programming experience at United Aircraft, Peter Harris, founded DATA-TECH in 1959 as a service bureau. Though DATA-TECH had been in existence for only a year when C-E-I-R acquired it in late 1960 with another stock-only deal, it was already profitable and focused entirely on the rapidly growing programming services and data processing services industry segments. Harris had a difficult time securing business in DATA-TECH's first six months, but his firm then won a substantial contract to create the first state computerized Motor Vehicle Registration System in the United States. Harris served as technical director for the Hartford Research Center for a couple years, and the Connecticut Motor Vehicle Registration System was a successful high-profile project for C-E-I-R in the early 1960s.[65] Hartford also led to C-E-I-R's first possession of a non-IBM computer system, with this facility operating an RCA 501 mainframe.[66]

In addition to these new facilities gained through mergers and acquisitions, C-E-I-R expanded to Houston in 1960, opening a sizable research center (headed by Orchard-Hays) that would focus on computer services. Whereas C-E-I-R's Arlington Research Center had a broad mixture of economic and operations research in the late 1950s and the early 1960s, its other major facilities, including those in New York, London, Los Angeles, and Hartford, were overwhelmingly computer services operations, and were increasingly focused on programming services.[67]

As C-E-I-R, Inc. entered the 1960s, Robinson's strategy—exemplified both by internal growth and by major mergers and acquisitions—was to grow the company quickly so as meet the needs of the rapidly evolving programming and data processing services needs of its customers.[68] The business of selling computer time, and to a large degree the business of providing programming services (as continually redeploying talent to new cities was a recipe for high turnover), were very much tied to place in this era.

By the end of 1961, C-E-I-R had new centers in Boston, San Francisco, Paris, and Mexico City. Robinson and C-E-I-R were focused on revenue and expansion, not on net profits. After recording its largest profit to date in 1959 ($76,702), C-E-I-R started the new decade with four consecutive years of net losses, totaling more than $4.5 million. From 1959 to 1963 the company had enlarged its workforce from about 200 to just over 1,000. In that period its annual revenue had gone from $3.45 million in 1959 to $16.6 million in 1963.[69]

Robinson believed that if C-E-I-R did not quickly gain a substantial widespread geographic presence, with an office in every major US city and a few strategic offices outside the US, the opportunity might be forever lost. In

Table 3.1
Data on C-E-I-R, Inc. (in 1954 and 1955 a nonprofit corporation, from 1956 to 1966 a for-profit corporation).

	Gross income	Net income	Number of employees
1954 (6 months)	$121,519	($2,274)	32
1955	$388, 123	($1,214)	32
1956	$683,633	$20,650	68
1957	$940,333	$16,083	66
1958	$1,373,820	$48,611	95
1959	$3,459,494	$76,702	205
1960	$9,547,634	($252,293)	664
1961	$10,570,985	($914,724)	883
1962	$15,592,636	($1,909,293)	915
1963	$16,608,243	($1,858,537)	1,012
1964	$16,954,385	$1,601,511	
1965	$20,355,073	$1,845,215	
1966	$22,358,974	$1,450,064	

Source: C-E-I-R, Inc. annual report, 1960; C-E-I-R, Inc. annual report, 1966 (Charles Babbage Institute). Employment numbers were not disclosed in 1964 or in later years.

the computer services business it was far easier to keep customers than to attract new ones. Customers' investment in particular software programs and the familiarity with using certain machines and working with certain programmers further heightened switching costs. If a computer services competitor had gained a stronghold in a particular city, it possibly could be difficult for a newcomer to profitably enter the market in that locale. This was the rationale behind Robinson's decision to expand in the early 1960s at a far faster rate than the company's earnings. It resulted in substantial annual losses.[70]

In 1960 the company had its first loss (about $250,000) since its beginnings in 1954 and 1955. It then lost more than $900,000 in 1961, and approximately $1.9 million in both 1961 and 1962. Securing space, hiring personnel, conducting sales campaigns, and other expenses preceded the inflow of revenue with each new installation, as offices were in what Robinson called their "non-profit-making startup phase." Initial projections by Robinson of a four-month run-up to profitability at new facilities did not materialize, and often the company lost money in a new facility's first year. In the early 1960s, even some longer existing centers were temporarily in the red as new computer systems and additional capacity were added in rapid succession. And acquisitions took time and could be expensive to integrate into the firm.[71]

The early-1960s plans for expansion of computing infrastructure even exceeded the firm's already dangerous pace of growth. In 1961, Robinson announced that C-E-I-R planned to acquire two IBM Stretch supercomputers, one to be installed in Boston and one in Los Angeles. That year C-E-I-R increased its available square footage substantially in its Los Angeles facility to make room for the Stretch system. IBM's Stretch project had begun under contract with Los Alamos National Laboratory (LANL), the only facility to receive delivery of a Stretch machine in 1961.[72] IBM's Stretch, also known as the 7030, was by far the fastest computer in the world, and was, despite significantly missing targeted goals for processing speed, arguably the first supercomputer. Because of lower processing speeds, IBM had to reduce its price from $13.5 million (the figure previously agreed to with LANL) to $7.8 million. Although it was not known at the time, the second Stretch—a one-of-a-kind modified version codenamed Harvest or IBM 7950—went to the highly secretive National Security Agency in 1962. This was followed by several other installations, including at Lawrence Radiation Laboratory, a rival of LANL, and at MITRE Corporation, a major defense contractor that had contributed to the development of the Semi-Automatic Ground

Environment (SAGE). Only eight of the computers were offered, and IBM lost money on producing each one.[73]

C-E-I-R not only contracted to acquire two Stretch machines; it also announced its intentions to do so in its 1961 annual report. Robinson, in his note to shareholders, made the case for accelerated expansion. He explained that C-E-I-R had gone from machines and facilities valued at $467,931 on September 30, 1960 to machines and facilities valued at $6,446,358 only a year later. Robinson's boldness, if not hubris, of the early 1960s, however, would soon be quelled by the primary underwriter of C-E-I-R's stock.

In addition to expanding assets and showing major annual loses in the first years of the 1960s, C-E-I-R also greatly diluted stock ownership through large-scale stock offerings. In 1961 it raised nearly $1.7 million by issuing stock, most of which was underwritten by Lehman Brothers. This precipitated managerial adjustment. In 1962, John Lehman required C-E-I-R to hire George Dick as executive vice president, a newly created position effectively equivalent to chief executive. (Robinson retained the title of president and chairman of the board.) Dick had run RCA's computer division for a short while in the early 1960s, and previously had been at IBM. He immediately hired Robert Holland (who also had an IBM background) as vice president for finance and administration, and together they sought to rein in the firm's excesses.[74]

Robinson had viewed rapid acquisitions and expansion as essential to conquering new geographies for the business. Many talented people joined the firm, but not all of C-E-I-R's technical staff contributed to generating steady revenue to pay for their salaries and benefits and overhead. According to Dick, Robinson did not want to be in the service bureau business, but because the machines had to be paid for he was forced into the business of selling idle computer time. Robinson was most interested in securing contracts to complete major programming services projects and operations research, but securing such contracts often was more sporadic. Dick fired hundreds of people from C-E-I-R during his several years in charge of the company's daily operations, and the corporation stopped providing figures on its labor force, which almost certainly peaked in 1963 at just over 1,000 employees.[75] C-E-I-R canceled the two Stretch machines, and the firm negotiated an undisclosed volume of complimentary programming services to IBM for a time in exchange for the breach of contract. In 1963 C-E-I-R also sold ten of its already installed mainframe computers to bring working capital up to nearly $2 million to position the firm to safely meet its upcoming financial obligations.[76]

With these austerity measures—particularly firing employees and selling some existing computers—business improved to net more than $1.6 million in 1964, its first profitable year since the end of the 1950s. By that year the firm had also reduced the risk of heavy reliance on its largest clients—of its expanded base of 1,800 customers, none represented more than 5 percent of overall revenue, a substantial divergence from the past. In 1964 the company also divested some its international operations by initially selling a 60 percent stake in C-E-I-R, Ltd., its British subsidiary, to British Petroleum Corporation, and the remaining 40 percent over the succeeding two years.[77] The stock recovered somewhat from its early 1960s trough, and John Lehman divested and no longer controlled majority share on the board. While reining in of costs characterized Dick's regime, in the mid 1960s he proposed building a very large signature facility in the Washington area. He was focused on marketing and believed this would further enhance the visibility and reputation of the firm. With Lehman out of ownership control, and many board members disagreeing with Dick on a major new facility, Robinson was able to regain active day-to-day leadership of the company he had built. In 1965 Robinson was able to remove Dick, who immediately signed on as president of American Business Research Bureau, a radio ratings research firm.[78]

Despite strategic cost cutting, C-E-I-R expanded into two important new business areas in the mid 1960s. In 1965 it acquired Automation Institute of America, Inc., one of the largest programming and punch card operation educational institutes in the country, which had facilities in 47 cities.[79] Programming schools had emerged throughout the United States throughout the 1960s to deal with the field's chronic labor shortages. Most of these were small, but a handful of independent chains such as Automation Institute had grown sizable, as had computer manufacturer Control Data Corporation's somewhat similar Control Data Institutes. (See chapter 8.)

In 1966, C-E-I-R acquired a GE time-sharing computer and set up its Multi-Access Computer service, a time-sharing operation in its Arlington facility. Within roughly a year it had 75 time-sharing clients using telecommunication-based remote processing, including the National Bureau of Standards, the Naval Ordnance Laboratories, the National Institutes of Health, Reynolds Metals, and Union Carbide.[80]

With C-E-I-R's programming services (which included many federal government department/agency clients), its new programming school business, and time sharing, it became a particularly attractive target for the Control Data Corporation. In the early 1960s, CDC executives had met with Robinson about a possible acquisition of C-E-I-R, but the two sides could not come to an agreement. With Automation Institute and a rapidly expanding time-sharing business, coupled with C-E-I-R's long proven strength in programming services and service bureaus, the computer manufacturer's executives saw great synergies and CDC agreed to purchase C-E-I-R, Inc. in August 1967 for a stock deal that valued C-E-I-R, the largest independent programming services firm at the time (approximately $36 million).[81] This was just slightly less than 1.8 times annual sales.[82] C-E-I-R's restructuring had pulled it back from the depths of multiple years of annual losses and risky expansion and increased the firm's market value, though it was at best a mildly impressive acquisition price for an IT firm in a fast-growing industry during the "go-go" 1960s.[83]

Computer Sciences Corporation

Computer Usage and C-E-I-R, the first two major independent programming services enterprises started on the East Coast before expanding to the West Coast. Yet aerospace firms in the Los Angeles area were among the earliest adopters of digital computers, and considerable early programming expertise lay with information processing leaders of West Coast aerospace companies. These users were the force that led to the formation of the IBM user group SHARE, Inc. in 1955, an organization for sharing ideas and software programs and routines that grew in succeeding years to have members from all geographies and industries worldwide. Two of the three founders (Fletcher Jones of North American Aviation and Roy Nutt of United Aircraft Corporation) of Computer Sciences Corporation, a company launched in May 1959, were senior data processing managers from Southern California aerospace firms.[84] The third founder, Robert Patrick, also was originally from a Southern California aerospace firm, Convair, though he had learned the programming services business and had risen to become an executive of C-E-I-R immediately before playing a role in helping to launch Computer Sciences Corporation.[85]

By early 1959, Fletcher Jones, who had many contacts in the computer industry, had learned of Honeywell's need for a business language compiler for its Honeywell 800 computer. He also knew of the firm's lack of internal expertise for such a programming project. Jones convinced Nutt and Patrick to join him in forming Computer Sciences Corporation and taking on this compiler project—a proposed "17-person year" effort.[86] Unlike CUC and C-E-I-R, both of which did programming for computer manufacturers down the road, but not as their initial core, CSC started on this basis.

The corporation took the somewhat odd and forward-looking name "Computer Sciences," a term that had first appeared (as "computer science") in a 1956 *New York Times* classified advertisement by Sperry-Univac seeking Livermore Advanced Research Computer (LARC) programmers, but was still obscure terminology at the time of the firm's founding.[87] The term only began to gain a bit of currency (mainly in academic settings) after Louis Fein of the Stanford Research Institute, in an article published in the November 1959 issue of *Communications of the ACM* (five months after CSC opened its office in Los Angeles) advocated "computer sciences" degree programs at universities, a field with distinct applications that drew from other fields, somewhat akin to "managerial sciences." Computer Sciences Corporation preceded the first "Computer Sciences" department (at Purdue University) by three years.[88]

While CUC and C-E-I-R began with modest startup capital, the Honeywell FACT compiler project, as Martin Campbell-Kelly astutely pointed out, nearly eliminated Computer Sciences Corporation need for seed funding as the three founders negotiated a time and materials contract. The project exceeded the specified length, CSC earned roughly $300,000 in its first year, and in the end Honeywell was left with needing to bring the project in house to finish about two thirds of the code—despite the high price paid to their contractor.[89] CSC rode this expertise to do other compiler and programming languages projects (including a LARC compiler and an upgraded variant of FORTRAN II), but simultaneously broadened to become a more diversified programming services supplier, one, in time, focused far more on user organizations than computer manufacturers as customers.

By 1963, CSC had grown to roughly 200 employees (the majority of them programming professionals) and was bringing in revenue in excess of $4 million a year. It had offices in El Segundo (its headquarters for many years), in Houston, and in New York City, and, in addition to its compiler and language specialization, it was doing business applications, operations research, command and control, process control, service bureau/back office, and other systems and application-oriented programming. Early customers after Honeywell included Sperry-Univac, the Jet Propulsion Laboratory/NASA, and several aerospace firms.[90]

Computer Sciences Corporation benefited mightily from C-E-I-R's scaling back in Los Angeles, where by 1963 it had reduced equipment and personnel. CSC, like C-E-I-R and unlike CUC, acquired powerful computers for its facilities rather than relying on customers and service bureaus. Though it had only about a fourth of the revenue of C-E-I-R in 1963, CSC had a greater concentration of programmers in the important Southern California region. C-E-I-R's retrenchment in the early to mid 1960s, particularly within Southern California, opened the way for CSC in that center of the aerospace industry. As CSC president Fletcher Jones commented in 1963, "this [programming talent] is something that is found wanting throughout C-E-I-R … . I understand there were about 15 programmers in Los Angeles supporting the large computer installation [of C-E-I-R's] … . This is very shallow support."[91]

Later, CSC would become a major player in C-E-I-R's (integrated into CDC after 1967) Washington home location, and in 2008 it would relocate its headquarters to Annandale/Falls Church, Virginia. Its first major client was Honeywell, its corporate customer base grew steadily; more than its competitors it developed concentrated expertise in serving professional IT/programming services to the federal government. In addition to contacts and a track record, expertise with the unique attributes of federal contracting proved important to its continuing success. CSC is the only major programming services firm launched in the 1950s or early 1960s that was still thriving in the twenty-first century.

To conclude, programming services companies began in the mid 1950s with Computer Usage and C-E-I-R, followed by Computer Sciences Corporation at decade's end. In the mid 1960s, Sam Wyly's University Computing Corporation, along with a handful of other firms, entered the fray. Wyly, a former IBM and Honeywell salesman who had also spent time working at IBM's Service Bureau Corporation, capitalized on particular circumstances and partnerships to launch his firm.

A major expense for programming services and service bureaus was acquiring computers and facilities for maintaining them. Computer Usage avoided this in its early years by using customer machines and renting time from service bureaus. CSC benefited from C-E-I-R's early downsizing of personnel and machines in Los Angeles, while University Computing arguably may have owed its very existence to C-E-I-R's reckless overextension before its activist-investor-driven retrenchment. C-E-I-R, needing working capital and attempting to return to profitability, was selling its (retail) $1.5 million CDC 1604 in 1963 and Wyly purchased it for the fire-sale price of $650,000. To get the loan, Wyly received a partial advance from a large-scale, future data processing customer, Sun Oil. Rather than finding a facility to house his used mainframe, he creatively partnered with Southern Methodist University to replace its aging UNIVAC 1103-based computer center with his CDC 1604. Students and faculty could use the machine nights and weekends in exchange for SMU's installation and the school's paying the physical set-up, air conditioning, a portion of the electricity, and other surrounding infrastructure needs.[92]

Programming services firms tended to expand into other computer services rather rapidly, including service bureaus (as early as the second half of the 1950s for CUC and C-E-I-R), and time sharing in the 1960s. They also tended to expand to new geographies quickly—particularly before the 1970s, when networked time sharing lessened geographic restraints on the selling of computer time and data processing. Rapid expansion of new offices/locations, new computer infrastructure, and new business segments offered great opportunities as well as substantial risks. It made firms highly vulnerable to strategic missteps or even just the inevitable troughs of economic cycles. Computer Usage's 1970s expansion into many non-services IT businesses proved particularly damaging and led to its downfall in the 1980s. C-E-I-R—for years the largest commercial programming services company—recovered from multiple consecutive years of losses during the early 1960s, and in a position of greater stability though far from true strength, was acquired by CDC. (See chapter 8.)

Computer Sciences Corporation took a more measured approach to expanding its range of services businesses, geographic locations, and machine infrastructure in its early years. At the end of the 1950s and well into the 1960s, it focused on providing systems programming support for mainframe computer firms. While Computer Usage did this to a degree as well, serving IBM on a key effort (systems programming for FAA air traffic control applications), neither CUC nor C-E-I-R made this a decade-long focus in the way CSC did. Computer Sciences' first project, a major time and materials compiler project for Honeywell, allowed it to learn and grow while guaranteeing steady profits. The Honeywell FACT compiler contract forced the customer, Honeywell, to endure CSC's programming schedule and cost overruns. CSC executives took notice to what would have happened if they had not contracted so advantageously, and it taught them the importance of careful costing of projects, which became a key organization capability for CSC.

In the mid 1960s, CSC began to specialize in programming services for the federal government. Its ability to predict the costs of complex IT services projects provided a decided competitive advantage. That ability also benefited CSC in serving corporations and other types of organizations. CSC's early concentration on the aerospace industry in the Los Angeles area, and later in the Washington area, allowed it to have a steady base as it grew in a balanced and cautious way. Some of CSC's government business was for smaller projects, but increasingly as time progressed, many were very large systems. The skills required for such work—making large-scale hardware, software, and networking work in unison, often in "real time" to achieve desired functions—commonly was referred to as "systems integration." In the late 1960s and in subsequent years, CSC became a major "systems integrator." Before then, several other organizations paved the way in this new field—mainframe leader IBM, startup systems integration specialist Informatics, Inc., and (most importantly at the beginning) the nonprofit System Development Corporation.

4 Integrating Systems: System Development and Informatics, 1955–1969

As was discussed in earlier chapters, firms that specialized in consulting, in data processing, and in programming services gave rise to and largely defined the computer services industry in its early years. Nevertheless, another enterprise launched in the mid 1950s, System Development Corporation (SDC), quickly dwarfed the programming labor force of the other computer services companies, and would continue to do so throughout the remainder of the decade. In the late 1950s and the early 1960s, System Development had two or three times as many programmers as IBM.[1]

System Development, though properly a part of the computer services industry, differed from other firms in the industry in nearly every respect. First, it was initially a new division of the nonprofit RAND Corporation, and soon thereafter it was a separate nonprofit corporation.[2] Second, it had a single major project in its first four years—truly its raison d'être: programming and integrating an air defense radar and computer system, and training personnel to use it.[3] Third, to meet the needs of that project, SDC began to hire 700 programmers at a time when there were estimated to be only 1,200 programmers in the United States.[4] Fourth, SDC's "programmer churn" was unusually high—it employed thousands of programmers in its first five years, and former SDC programmers helped populate and propel other computer services companies, and computer and software firms. Fifth, SDC's initial project was not only on a different scale from any of the prior or concurrent projects from programming services companies; it was the first major system focused on real-time networked operations. For that reason, it completely transformed the possibilities for computing. Implementing a large-scale real-time system involved complexity far beyond the capabilities of existing computer services specialist companies, or even IBM. For that reason, what SDC did was referred to by a different name: systems integration.

To understand the origin and the early evolution of SDC, we must first explore the pre-history—technical and organizational—behind its unprecedented systems integration project.

Spinning Whirlwind

In 1944, the four-year-old MIT Servomechanisms Laboratory, which had worked on fire control and gun aiming to aid the war effort, launched a project the goal of which was to come up with a universal flight simulator.[5] That analog-computer-based project—referred to as Project Whirlwind, guided by a 26-year-old engineer named Jay Forrester, and originally estimated to cost $200,000—evolved into a $8 million multi-year effort that was the real-time digital computing design basis for the networked system of radar and computers known as the Semi-Automatic Ground Environment (SAGE).[6]

An effective universal flight simulator required real-time functionality, and Forrester soon recognized—influenced by conversations with MIT graduate student Perry Crawford in 1945—such a capability would require advanced digital computing and could have far wider applicability than flight simulation. Forrester's success in escalating Project Whirlwind's scope was the product of his determination and his skilled rhetoric in presenting the ongoing work—spinning the broad impact it could have for national defense—to skeptical funders from the Office of Naval Research (ONR). At various points in the second half of the 1940s the project was on the verge of termination, but Forrester's focus on securing funding kept it going, as an important Servomechanisms Laboratory colleague, an engineer named Robert Everett, concentrated on the technical details. Late in the project, Forrester also won over a few important allies outside of ONR, none more important than George Valley.[7]

In 1949, Valley, an MIT physicist who served on the Air Force Scientific Advisory Board, conducted a preliminary assessment of the nation's air defenses. He was appalled by outdated equipment at radar stations. This initial assessment led to the formation, in 1950, of an Air Defense System Engineering Committee, which came to be known more commonly as the Valley Committee. Valley recognized the importance of reliable digital computers operating in real time for air defense, and a colleague alerted him to Forrester's Project Whirlwind.[8] Real-time computing, which required unprecedented memory and processing capabilities, was made possible, in part, by Project Whirlwind's invention of magnetic core memory.

The Valley Committee's work, the progress and possibilities of Whirlwind, and a report from an advisory group of 28 leading scientists and engineers on "Project Charles" in the first half of 1951 led to the formation of MIT's Lincoln Laboratory later that year. Lincoln Laboratory was a federally funded MIT center dedicated to air defense research and development. It was located at Laurence G. Hanscom Field in Bedford, Massachusetts. The 1952 Project Lincoln contract funded the Air Force's ongoing work toward a radar and computer air defense system, which was named the Semi-Automatic Ground Environment (SAGE) in 1954.[9]

The SAGE project aimed to develop and deploy a system that could detect, confirm, and communicate to quickly respond to a Soviet bomber attack—the primary threat in the early1950s, though one that was largely obsolete by 1958, when SAGE was first (partially) operational. The advent and testing of Soviet intercontinental ballistic missiles (ICBMs) by May 1957 led to new vulnerabilities that SAGE was not designed to address.[10]

SAGE Computers

Lincoln Laboratory needed a computer contractor to refine the original Whirlwind computer and build dozens more such computers to operate the real-time air defense system at radar centers throughout North America. IBM and Remington Rand beat out three other companies to become finalists. Forrester believed that IBM's technical staff had "a much higher degree of purposefulness, integration, and esprit de corps" than Remington Rand.[11] Lack of integration between two disparate Remington Rand computing operations—one in Philadelphia and one in St. Paul—likely hurt that firm's bid.[12]

In late 1952, IBM received the initial contract to collaborate with Lincoln Laboratory to modify the design of the original Whirlwind computer. This ultimately set it up for the Air Force contract it received roughly a year later to build 46 Whirlwind-derived digital computers that would be installed in duplex (23 systems at directional centers—geographically dispersed North American locations) and would be dubbed the Army-Navy/Fixed Special eQuipment-7 (AN/FSQ-7, often abbreviated to Q-7).[13] To help ensure constant reliability in an era well before meaningful advances in fault-tolerant computing, one computer in each duplex would be in operation and one in standby mode.[14]

The Q-7 was by far the largest computer ever produced. These duplexed systems possessed 50,000 vacuum tubes, weighed more than 250 tons, and covered about half an acre of floor space.[15] The Q-7 drew its memory core and many of its most critical design elements from Whirlwind, and some of its other elements from IBM's Defense Calculator. Teams from Lincoln Laboratory and from IBM's Project High (named for High Street in Poughkeepsie, where the IBM team was headquartered) met frequently to discuss and negotiate the design of the Q-7.[16]

IBM's top leaders were thrilled to receive the lead contract for the computer hardware for SAGE, at a reported $30 million per duplexed system. In addition to the Q-7 computers at the 23 directional sites, IBM contracted to produce the computers (also to be duplexed) for three combat centers. The combat centers, which would have AN/FSQ-8 ("Q-8") computers (modified Q-7s with specialized displays), had the role of coordinating and communicating responses to attacks. IBM received $500 million on this hardware effort by the end of the 1950s, revenues that helped the company to build a base in large-scale computer systems as well as to gain know-how in real-time systems.[17] In the late 1950s, 7,000–8,000 of IBM's technical employees were devoted to SAGE, about one eighth of the firm's entire US-based staff at that time. Though IBM's leaders were grateful for the Q-7/Q-8 contract, they turned down the Air Force's request to supply software/systems integration for SAGE.[18]

The unprecedented SAGE hardware project presented many challenges, and IBM's leaders believed an additional massive software project would have exacerbated financial risk to the firm. After approaching IBM, the Air Force's Air Defense Command (ADC) courted Bell Laboratories and Lincoln Laboratory, but both organizations declined.[19] Lincoln Laboratory's technical leaders, some of whom had programmed the original Whirlwind computer, served as advisors in helping find a solution and played an important early coordinating role in the SAGE project.[20] Ultimately, however, Air Force leaders looked to the RAND Corporation to take on programming and systems integration for SAGE.

RAND Corporation Air Defense Simulations

The RAND Corporation (the name was derived from the phrase "research and development") was launched as a nonprofit corporation in Santa Monica, California, in November 1948. It was the outgrowth of an Air Force–funded project called RAND—initially a $10 million Air Force contract to Douglas Aircraft Company.[21] Project RAND was dedicated to the "study and research on the broad subject of aerospace power with the object of recommending to the United States Air Force preferred methods, techniques, and instrumentalities for the development and employment of aerospace power."[22]

In its early years the RAND Corporation followed the Air Force–focused mission (as had Project RAND), but soon it began to hire economists, sociologists, psychologists, anthropologists, and political scientists engaged in war gaming, evaluating enemies, assessing threats, and developing national defense strategies to complement the work of its physical scientists and engineers. Two reasons for establishing RAND as a separate nonprofit corporation were to assuage the concerns of Douglas' aerospace competitors about the protection of proprietary information and to allay Douglas' fear that it might be penalized in Department of Defense contract competitions to avoid perceptions of favoritism.[23]

In the early 1950s, RAND Corporation scientists and engineers built a computer system they called Johnniac. It was based on, and named in homage to, a computer design developed at the Institute for Advanced Study in Princeton on the basis of a report authored by John von Neumann.[24] In 1951, RAND had established a small Systems Research Laboratory, employing both computer specialists and psychologists, to focus on man-machine systems.[25]

During 1952, RAND's Systems Research Laboratory conducted a simulation study, "Project Casey," on man-machine operations for early warning air defense, using 28 college students. Based on simulating a radar station in Tacoma, Washington, the project was run out of a billiards room in Santa Monica. The following year, the laboratory completed a more advanced simulation study, dubbed "Cowboy," in which fourteen Air Force officers operated a simulated air defense system.[26] In 1953 the latter effort led to the creation of a joint Air Defense Command-RAND study group, which included the artificial intelligence pioneer Allen Newell and two future leaders of RAND's System Development Division: the engineer Melvin Kappler and the psychologist William Biel.[27]

Concurrent with RAND's important simulation and training work, Lincoln Laboratory built, for testing, a modest physical system using a modified Whirlwind computer, a wired network, and a number of radar towers on Cape Cod. The first tests (flights and efforts to detect and communicate data on the computer network) occurred in October 1953 and demonstrated its basic viability. The Cape Cod System included the largest real-time control program in the early to mid 1950s, all of it written in machine language (35,000 instructions).[28]

The partnership that emerged between MIT (Lincoln Laboratory), RAND/System Development Division/System Development Corporation, and IBM continued throughout the succeeding five years (and beyond) on the Air Force–sponsored SAGE's development, deployment, and training. Lincoln Laboratory played a continuing meaningful role. Such a large and long-standing research, development, and implementation effort was without precedent for a university, and MIT leaders decided to spin Lincoln Laboratory off to become a nonprofit research corporation (named MITRE) in 1958, with Robert Everett as the first technical director.[29] The programming and systems integration for SAGE was accomplished largely by System Development Corporation (SDC) during the second half of the 1950s. This massive buildup of a new entity—possessing for a time unparalleled computing systems integration know-how—bestowed both opportunities and challenges as the organization sought to define its proper role after the original core SAGE programming effort.[30]

System Development Corporation

RAND's Systems Training Program was the direct predecessor to the corporation's System Development Division. The program, led by RAND's Melvin Kappler and William Biel, expanded quickly in 1954 with an initial $1.2 million Air Force contract, and soon thereafter, a supplemental $1.5 million. The latter was for training the Air Force's 27th Air Division and for setting up many training sites around the country for the Air Force's Air Defense Command (ADC) personnel. The System Research Laboratory's Systems Training Program partnered with and received support from computer specialists in RAND's Electronics Department and its Mathematics Department, including computer pioneer and future director of RAND Computer Sciences Department Willis Ware.[31] It also partnered with the sub-contractor IBM, which developed cathode ray tube monitors for the training system for use on RAND's IBM 701—the eleventh 701 installed (delivered in October 1953).[32]

The Systems Training Project employed a few dozen psychologists, electrical engineers, and programming specialists. The psychologists were heavily focused on human–machine interfaces. Their interests and expertise lay in an area of study the Advanced Research Projects Agency (ARPA) would draw from for its first Information Processing Techniques Office (IPTO) director less than a decade later (in 1962)—J. C. R. Licklider, an MIT psychologist in the Electrical Engineering Department, who had recently left to join nearby electronics firm Bolt Beranek and Newman before taking the reins at IPTO.[33] Licklider, recruited by MIT physicists Leo Beranek and Richard Bolt (to MIT and later to their company), was the one psychologist among 27 physical scientists and engineers working on Project Charles. He pushed harder than anyone for human factors research at the origin of Lincoln Laboratory. As Licklider later recalled the first years of the lab, "I created a group in Lincoln Laboratory, which was about half experimental psychologists and half electrical engineers ... to work on two aspects of the presentation of [the] information problem: building the stuff and then getting it to work, making a good interface with the user."[34]

This presaged Licklider's leadership of IPTO, setting the office on a course that transformed computing by focusing on human factors—funding key research (creating centers of excellence) in time sharing, artificial intelligence, graphics, and networking.[35] These were all fields to which the SAGE project—Lincoln Laboratory, RAND, and System Development—contributed substantially in the years before the launch of IPTO. While none of the psychologists at RAND in the mid 1950s would reach the stature of Licklider, they shared his approach, and were intimately focused on understanding and improving human–computer interfaces to achieve important goals.[36] The *combination* of RAND's training experience in air defense for the Air Force (largely by human factors/machine-interface inclined psychologists), coupled with the modest sized, but impressive computing and programmer talent at RAND, led to the ADC's interest in approaching the nonprofit corporation for programming SAGE.

On April 2, 1955, ADC leaders met with Kappler and Biel in Colorado Springs and proposed that the RAND Corporation take on the primary programming and systems integration task with SAGE. Kappler immediately replied "Yes, of course," and RAND president Frank Collbohm soon ratified RAND's acceptance.[37] The priority of national defense outweighed any reservations at RAND about a research organization engaging in such a massive and long-term development project. On July 5, 1955, a small group of leading RAND programmers—including Wes Melahn, John Matousek, Mort Berstein, and Pat Haverty—joined with SAGE leaders at Lincoln Laboratory to formalize plans, which included RAND quickly providing 70 programmers—more than three times the number of programmers within the entire organization at the time.[38] In the first couple of years of the project, some SAGE programming was collaborative between Lincoln Laboratory and the RAND Corporation/System Development Division, before a full shift of the responsibility to System Development Corporation in 1957.

In September 1955, in recognition of RAND's taking on the responsibility of programming SAGE in addition to training, Collbohm changed the group's name from the Systems Training Program to the System

Development Division (SDD). By that time, there were already 300 in the new division, which would grow to 500 by year's end.[39] SDD programmers on SAGE worked with a team at Lincoln Laboratory (as responsibilities and personnel, shifted) as well as locally in Santa Monica. SDD continually secured rented facilities anytime acceptable space became available in Santa Monica and other western Los Angeles locations. This chaotic situation—many physically separated System Development programming locales, including most working near Bedford, Massachusetts, at a SAGE facility at Hanscom Field shared by RAND, IBM, and Western Electric/Bell Laboratories—was solved when RAND, at the end of 1955, secured for SDD a long-term lease for several acres of land at 2500 Colorado Avenue, a block north of Wilshire Boulevard and 2½ miles northeast of the Pacific Ocean.[40] Plans were quickly finalized for the erection of three modern buildings—ready and occupied at different points in 1957 and the start of 1958—with a total of 250,000 square feet.[41]

System Development continued to grow at a breakneck pace and the organization had more than a thousand employees by the second half of 1956. RAND leaders recognized that System Development was both too large and too distinct from the rest of the corporation (having one client—the Air Defense Command—and working more on development than research) to be just a division. Collbohm and his board elected to spin the division off as a separate nonprofit corporation, System Development Corporation. On November 23, 1956, the SDC's Articles of Incorporation were filed with the State of California. Collbohm and his vice president J. Richard Goldstein were elected as the Chair and Vice-Chair of the System Development Corporation's board of directors, and Kappler and Biel became SDC's president and vice president, respectively.[42] The $20 million ADC contract—for SAGE programming and training services—was transferred to SDC as were the facilities under construction.[43]

No organization hired programmers as rapidly as System Development, and no programming project approached SAGE programming in complexity in the 1950s. As one author of a recently published history on the Q-7 machines put it, "this might have been the first time in human history that programs had to be developed which were far too complex for a single person to grasp and understand every detail of the code."[44] Both the complexity of the project and need to hire and train programmers quickly presented major managerial challenges.

Two standard tools System Development used to aid with hiring were administering the Thurstone Primary Mental Abilities Test and Thurstone Temperament Schedules. They also used various types of aptitude examinations. Early in the project SDC "validated" this psychological and aptitude assessment testing infrastructure, but despite its perceived usefulness, there was never a perfect predictor for System Development or any other organization of who would become a talented programmer. Nonetheless, the organization found its studies helpful and periodically disseminated results on findings, which included that men and women had no differences in training course performance, college graduates in mathematics and the physical sciences scored slightly higher than other majors, and those without a college degree performed as well as those with one (with the exception of the two aforementioned sets of majors).[45] System Development advertised positions broadly in trade publications, newspapers, radio spots, and elsewhere to boost the pool of applicants—and generally attracted enough applicants to be able to turn down at least two thirds of them. The pay schedules were dictated by Air Force contracting standards and provided solid though far from standout earnings, and there was little flexibility for hiring authorities. In the first years, SDC offered entry-level programmers \$350 a month. Those with applied mathematics education or background received \$400 a month, while hires possessing an advanced degree were offered \$450 a month.[46]

From the start, System Development Corporation's leaders knew there would be challenges to retention. Work was demanding and employees were building highly marketable programming (and training) skills. While no organizations hired programmers more rapidly in the 1950s, IBM, other computer manufacturers, and emerging computer services firms were hiring programmers quickly as well. SDC rarely made counteroffers, so those offered more compensation elsewhere often left. The corporation was careful not to exacerbate problems through false advertising in recruitment. Advertisements did not promote sunny Santa Monica, because many new hires would be deployed to work in Massachusetts for a time, or at some of the emerging directional centers once the first Q-7's were deployed (in 1958).[47]

Working at a SAGE directional center on programming and debugging could be isolating—some of the centers were far from any major city. Turnover weighed heavy on System Development, but in many respects the organization embraced it both by instilling proper management structures to address rapid churn of employees, and down the road, by promoting that SDC largely educated an industry in the art of large-scale systems programming.

By 1960, SDC had 3,500 employees and was more than twice the size of its once parent RAND Corporation. By that year there were 4,000 former

SDC employees, a number that would reach 6,000 within several years. A large percentage of these former employees were programmers and systems analysts—and the thousands of former SDC staff fed into computer companies, corporate and government installations, the computer services industry, and the emerging software products industry by the early 1960s.[48] As SDC's president, Kappler later recalled that "part of the SDC's nonprofit role was to be a university of programmers."[49] Whether this was fully embraced at the time or was merely a leader's later "spin" on the unfortunate and inevitable turnover in view of opportunities elsewhere, is not certain. Regardless, SDC certainly played that role—and through the early 1960s more so than any other organization in the world. Many were trained internally, but in the first years of SAGE programming, with the core activity in Lexington, Massachusetts, and with RAND and later SDD and SDC employees working alongside those from Lincoln Laboratory on some of the programming, hundreds took part in multi-week training courses taught by instructors from IBM in Kingston, New York.[50] IBM, as an early mover in hiring programmers, also helped to populate the broader industry and user organizations with programmers, but often higher compensation and career advancement opportunities at the computer giant (where at the time lifetime employment, if desired, was the norm) resulted in far lesser rates of programmer turnover relative to SDC.

To better handle turnover—and to manage more effectively in general—SDC employed a matrix management structure. With regard to software management, System Development was an early pioneer in adopting modular approaches to the software development process. Turnover also made careful documentation all the more critical and SDC had strict requirements on documentation.[51] In short, while turnover presented challenges it also forced good managerial and computer programming practices—developing careful processes that later would fall under the general rubric of software engineering.

Nevertheless the immense size of the programming effort and the vast number of programmers, made programming SAGE quite difficult. The core real-time operating system for SAGE was more than 100,000 instructions, roughly three times the size of the real-time operating system on Whirlwind. As one of Lincoln Laboratory's Whirlwind programmers, H. D. Benington, later recalled, "we had 20 people who understood in detail the performance of those 35,000 instructions, they knew what each module would do, they understood the interfaces, and they understood the performance requirements."[52] Benington went on to work on SAGE programming and the added size, complexity, and much larger programming workforce

Integrating Systems

made the latter effort far more challenging. In retrospect, he stated that had the many more instructions been transliterated almost directly to the 35,000 on hand at the time, software development time and costs could have been reduced significantly on SAGE, an observation only clear in hindsight.[53]

The size of the SAGE control program in terms of instructions stood out, but a handful of other computer systems programs had been developed by that time for scientific or business applications with "50,000 instructions."[54] A primary difference with SAGE was not merely its being double the size or more in lines of code, but that these other business or scientific applications used decentralized systems where components did not interact to the same degree as required with SAGE. A real-time, integrated system necessitated a high degree of centralization. As Benington highlighted (at a June 1956 "Office of Naval Research Symposium on Advanced Programming Methods for Digital Computers"), "The [SAGE] control program must be *centralized* (his italics). This complicates the design and coding since communication between subprograms must have a high bandwidth. ... Thousands of central table items must be coordinated between 100 or so component subprograms." He went on to emphasize challenges of testing the SAGE control program, "only the most thorough testing of the entire program ensures that system threads have been carefully worked out, that incompatibilities are discovered, and that all contingencies are accounted for."[55]

Ultimately the SAGE control system or operating system ballooned to 230,000 instructions. Support programs for SAGE contained more than another 870,000 instructions.[56] This work was accomplished by expanding to 500 programmers by May of 1957.[57] A core group of programmers situated at Lincoln Laboratory relocated to the SDC Santa Monica facility during that year, as responsibility on SAGE programming fully transferred to SDC. The operating system was completed by the start of 1958.[58] SDC delivery of software was about one year off the original schedule of the project, with one insider estimating the programming cost at $55/per instruction word.[59] Having the system ready on a massive multi-year project one year after the original target date represented a meaningful accomplishment in a field that would be plagued by very long delays and cost overruns on major programming projects for years and decades to come including IBM's monumental OS/360 programming project—a project that began five years later and was famously chronicled and analyzed in Frederick Brooks' book *The Mythical Man-Month*.[60]

On June 27, 1958, McGuire Air Base in New Jersey was formally dedicated. It was the first SAGE directional center. The *New York Times* reported on the dedication: "The $27 million unit here will be to control air defense in New York, Philadelphia, all of New Jersey, and parts of Rhode Island, Connecticut and Delaware [and] is linked to North American Air Defense Command headquarters in Colorado Springs … .The system relies on huge electronic computers that receive data on all aircraft operating in their sector. The system automatically sorts unidentified planes from friendly planes, tracks them, plots the best interception paths and prepares a plan for a counterattack … . The structure is 175 feet high and 150 feet square. It is manned by several hundred Air Force and civilian maintenance personnel … . A BOMARC interceptor missile site is under construction here for completion next year … ."[61]

During 1958, SDC had swelled to 800 programmers, roughly 40 percent of its overall staff. Many of the other staff were engineers and psychologists (focused on Air Force training for SAGE). From the start Kappler had recognized SDC was assembling programming talent to integrate large-scale computer systems, experience that would be unparalleled, and skills in great demand. He knew this would bring high turnover, but also knew that he and his management team would need to think about how to most effective use the fairly large programming staff that had been assembled. As SAGE directional and combat centers came online in the late 1950s and the first years of the 1960s, the long and tedious task of refining and debugging the system remained—both centrally and at each of the individual directional and combat centers. Hundreds of programmers were devoted to this effort, and the isolation, the nature of the work, and ever-expanding opportunities outside only exacerbated the rate of turnover. For practical and business reasons, and also to show staff that elected to remain at SDC that this represented a viable career path, Kappler and his management team looked for new opportunities for SDC. At question was whether SDC would seek out contracts only for the Air Force, only for defense, for defense and other federal government systems, or for all of these as well as private (non-defense) industry work. In the late 1950s and the early 1960s, Kappler and others decided SDC would focus on federal government systems (as a main contractor or subcontractor).[62]

Along with SAGE, one of the Air Force's other major information technology needs was logistics—keeping track of the supply and movement of weaponry, personnel, and equipment as well as weather conditions, and other data critical to command and control in peacetime and wartime. Shortly after the McGuire SAGE directional center went live, International

Telephone and Telegraph (ITT) won a major contract to be the lead contractor for the Air Force Strategic Air Command Control System (SACCS). ITT, in February 1959, selected SDC as the primary software contractor on the project. IBM was among the major hardware contractors and the system used IBM's AN/FSQ-31 computer system. The networked system involved many smaller IBM 709 computers and would soon incorporate two other systems announced later in 1959: the IBM 7090 and the IBM 1401.[63]

SACCS, also known as 465L, would require an operating system composed of more than a million instructions, more than four times the size of the control system for SAGE. SDC's Chief Technologist for Research and Development computer scientist Clark Weissman recalled the challenge of integration of many disparate computer systems in operation for the Air Force. The SACCS had to effectively communicate and integrate SAGE, the 438L Intelligence System, the 433 Weather System, and other systems. Further, a time-sharing (resource-sharing) system required computer security. Before time sharing, computer security had largely been limited to guarding machines and controlling access to computer rooms and installations. With time sharing, personnel at different clearance levels would be networked to the same resource-shared computers. On SAGE, the hundreds of Air Force personnel using the system would all have extensive background checks and Top Secret clearances and be at SAGE centers. A logistics system, like SACCS, that dealt with weather data, personnel, weaponry, and other equipment would have many more users on Air Bases and also in the field.[64]

This recognition led Weissman to head a team to develop one of the first two operating systems with security built into the design, ADEPT-50.[65] This followed an earlier SDC Time-Sharing System (TSS, which was without a focus on security) funded in the early 1960s by ARPA IPTO under Licklider's direction, which furthered SDC's strong competence in the broad field of computer time sharing. A chief advisor to the Air Force, RAND's Willis Ware, who had close ties to SDC, along with the National Security Agency's Bernard Peters, first articulated the "multi-level security problem" for the government and others in a time-shared computing system world.[66] Weissman, Ware, and fellow SDC scientists Marvin Schaefer and Richard Kemmerer, along with important scientists (Peter Denning, Ted Glaser, Peter Neumann, Jerome Saltzer, Fernando Corbató, James Anderson, and Roger Schell) tied to (MIT) Multics (the other major time-shared system with a security design at the time) pioneered the field of computer security.

One challenge evident throughout initial SAGE programming had been the lack of a quality higher-level programming language optimized for real-time command and control. On SACCS, SDC placed a high priority on developing such a language to ease the task of programming for such a large-scale, real-time, computer and communication system. A manager on SACCS programming, SDC's Jules Schwartz, took the lead on a project to create such a language. The International Algorithmic Language (IAL) at its birth in 1958, renamed Algorithmic Language or ALGOL the following year, became the base, along with learning and work that had been accomplished on SAGE programming. Schwartz viewed ALGOL as a preferable foundation to FORTRAN, the other candidate, for this particular purpose. Schwartz led a SDC team of about 20 programmers and began work in New Jersey at an IIT complex. The overall project was called CUSS (Compiler and Utility System for SACCS), but they still needed a name for the language they were developing. Showing respect for its basic foundations with ALGOL, they came up with "Our Version of the International Algorithmic Language (OVIAL)." But in the less "free-thinking" late 1950s, and with some suggesting the name had too close a connotation related to the "birthing process," a team member suggested a name change to Jules Own Version of the International Algorithmic Language, or JOVIAL. A round of laughter ensued without a decision, but when Schwartz returned from a trip he took shortly after the meeting, SDC's subcontract for the language development specified the name "JOVIAL," making it official.[67] A key attribute of JOVIAL's design was flexibility—a language geared to systems with "contributions of many people," "a variety of situations," and "machine independence," all critical for command and control broadly, and SACCS in particular.[68]

Schwartz's retrospective on JOVIAL in 1978 provides a glimpse into how SDC programmers seeded the computer services and software industries, as well as certain demographics. Of the nineteen programmers listed on the original CUSS project from the late 1950s, Schwartz listed 1978 work affiliations for thirteen (the rest were unknown). Of these thirteen, three (including Schwartz) were with CSC, three were with small IT firms (Data Vantage, Century Data, and Litton Systems), two were with IBM, two were independent consultants, two were still at SDC, and one was with an aerospace firm (Northrop). In addition to showing how the expertise of former SDCers spread throughout various organizations and IT industries, it also conveys the youth of the team. All but one of the known thirteen was still working two decades later. (Overall, nearly 90 percent of late 1950s SDC programmers were between the ages of 22 and 29.) And just over a fourth of the members of the CUSS/JOVIAL team were women.[69]

In the early 1960s, SDC's revenue diversified (including ARPA funding) to a degree compared to its first half-decade, but the corporation was still overwhelmingly an Air Force systems integrator and training supplier. SDC benefited from being a sole-source contractor for many of the Air Force's systems integration needs, and as such, for-profit companies did not want it to compete more broadly—even within the DoD. Kappler, however, believed SDC's ability to compete for contracts was critical to the organization's long-term existence and to serving the nation with SDC expertise. In 1963, SDC made a bid of $30 million to the US Army's Electronics Research and Development Agency (ERDA) for a Command Control Information System (CCIS-70). By this time the systems capabilities of commercial computer companies (IBM and Sperry Rand), certain computer services companies (Computer Science Corporation and Planning Research Corporation), and aerospace contractors (TRW), made them potentially viable alternatives. Despite the US Army's ERDA's encouragement to SDC to bid, System Development's submission of a bid angered for-profit competitors and the Air Force, where senior leaders at the latter believed it threatened the whole concept of a nonprofit corporation. Ultimately TRW received the contract.[70]

The CCIS-70 conflict precipitated Kappler's resignation later in 1963 and the SDC's new president, Wesley Melahn, issued a conciliatory message and policy statement at year's end, "SDC has the quality, competence, and uniqueness of capability to justify being selected for tasks on a sole-source basis, and therefore does not enter into competitive bidding." Further, he wrote, SDC would limit itself to "new frontiers of technology" vacating any technical space "when private organizations demonstrate their capability to undertake such tasks."[71] True to the statement, SDC largely abstained from competition for the next three years or longer.

This was somewhat crippling to SDC, and made it critical to get the multi-year renewal on the SAGE/BUIC (Back Up Interceptor Control—a SAGE backup) contract for software system upgrades and training that came up for bid in mid 1967 (for FY 1968). Recent annual revenue was roughly 10 percent lower over the trailing several years than in fiscal years 1963 and 1964, when it was still receiving revenue from contracts it competed to obtain. Even worse, its net income, with the exception of 1963, was a small fraction (generally less than 1/3) of what it had been in 1959 and 1960 when the original SAGE work was still at or near its height. (See table 4.1.)

Table 4.1
SDC's revenue and net income (in thousands).

	Revenue	Net income
1959	$32,717	$1,565
1960	$42,425	$1,451
1961	$43,638	$547
1962	$47,803	$434
1963	$56,526	$963
1964	$56,823	$329
1965	$50,168	$295
1966	$51,298	($608)
1967	$51,215	$588
1968	$53,337	$689

Source: System Development Corporation annual report, 1968.

SAGE programming and air defense training had provided most of the revenue for SDC's history. At $25 million in FY 1967, the SAGE/BUIC contract was fundamental to SDC's survival. With broad turnover, many senior former-SDC managers on the SAGE project were with private companies such as Computer Sciences Corporation. CSC had Stewart Fliege, a true SDC SAGE expert, and the company had developed advanced systems integration and federal contracting capabilities by the mid 1960s, building upon its first major federal contract (in 1961)—for a flight operations facility for NASA's Jet Propulsion Laboratory (JPL).[72] Meanwhile, a specialist in systems integration, Informatics Inc., formed in 1962, had grown formidable. Other competitors for the BUIC project included International Telephone and Telegraph, Hughes, Planning Research Corporation (PRC), and three major mainframe firms with advanced systems capabilities: IBM, Burroughs, and RCA.[73]

System Development Corporation offered a bare-bones set of deliverables, what the RFP called for, at a low price—$14.4 million for the first year, with unpriced options on two additional years. It narrowly won the competition.[74] More broadly, it re-established the firm competing for contracts—and not just ones like this for the Air Force—but throughout government and industry. In 1968 SDC went through a major reorganization around three broad customer categories: Commercial Systems, Military Systems, and Public Systems. With the Commercial Systems Division it opened its time-sharing services to business and industry and created some proprietary software packages for middle to larger firms. It received contacts from banks, manufacturing companies and other private-sector enterprises. With the Public Systems Division, it targeted civilian federal as well as state and local government departments and agencies. It also made strides to build its business with education (computer-assisted instruction) and library information systems.[75]

With SDC's Military Systems Division, in addition to the major SAGE/BUIC, it came out with the ADEPT-50 security designed time-sharing operating systems for the popular IBM System/360 Model 50, installing it for testing at the National Military Command System at the Pentagon. And for the Navy, SDC launched a systems analysis project of a developmental program for the Naval Air Development Research Center for automating Anti-Submarine Warfare aircraft.[76]

In 1969, formally recognizing its new mission and broader focus, SDC became a for-profit private corporation. Struggles in the early 1970s were followed by expansion later in the decade. In 1972, its third full year as a for-profit corporation, it first crossed the $1 million net annual earnings plateau, but did this on revenue of $51 million for only a 2.3 percent margin.[77]

In the 1970s, SDC's Government Systems Division (federal systems for defense) remained its core, though it did diversify projects and revenue to a substantial degree by the mid 1970s, moving into and/or expanding space projects (for NASA), meteorology, energy, traffic signal control, emergency system command and control (for the Los Angeles Police Department), electronic funds transfer (for a California consortium of savings and loan institutions), education, security and other areas.[78] Aiming for diversity in customer base, in 1976, the firm first achieved a longtime goal (since becoming a for-profit seven years earlier), growing its non-DoD revenue to the point where it represented 50 percent of total revenue. It had profits of $1.6 million in 1978 and $3.8 million in 1979.[79] With pressure from shareholders and at a time of strength, in 1979 SDC worked with Goldman Sachs to negotiate a cash deal with Burroughs Corporation—which was seeking to bolster its Federal and Systems Group. The two firms had collaborated on contracts for DoD systems dating back to SAGE (where Burroughs supplied some of the hardware) up to their recent partnership working on an Army Weapons Laboratory information technology system. In August of 1980 Burroughs offered to acquire SDC for $98 million, which was soon accepted by the System Development Corporation's board of directors. SDC had over 3,800 employees at the time and its trailing annual revenue had reached $168 million, with roughly 70 percent from the DoD.[80] SDC officially became a subsidiary of Burroughs on January 5, 1981.[81]

Informatics, Inc.

As was emphasized by the fierce 1967 SAGE/BUIC contract competition, SDC no longer had a near monopoly on systems integration. In fact, by the early to mid 1960s, SDC was just one of several important players in systems integration, including Computer Sciences Corporation, IBM, Planning Resources Corporation, and Informatics.

Computer Sciences Corporation leveraged its lucrative FACT Compiler contract to invest in developing its federal systems capabilities. In addition to its 1961 NASA contract at the Jet Propulsion Laboratory, it took over two divisions of ITT later in the decade that had extensive systems building experience serving the US Defense Communications Agency. And as previously mentioned, CSC recruited some former SDC technical employees and managers. While continuing to serve commercial clients, CSC more than any of the diversified computer services firms became a federal systems specialist.

IBM, benefiting from its work on SAGE, developed the SABRE (Semi-Automatic Business Research Environment) real-time ticket/seat reservation system in partnership with American Airlines.[82] SABRE was by far the largest commercial systems integration project of the late 1950s and the early 1960s (it is discussed in chapter 8). Meanwhile, IBM's SAGE work also gave birth to its substantial Federal Systems Division, which contracted with the Department of Defense, the Department of Energy, the Department of Transportation, the National Security Agency, various national laboratories, the Internal Revenue Service, the Social Security Administration, NASA, and other government departments, agencies, and entities in the 1960s and in subsequent years.

Planning Research Corporation, an early Los Angeles-based operations research firm (founded by three former RAND Corporation employees in 1954) developed into a significant systems integration contractor in the first half of the 1960s. In the mid 1960s PRC moved its headquarters to Washington to better focus on government contracting.[83] As early as 1963 it was advertising in the *New York Times* to recruit "computer systems specialists/programmers," responsibilities included "design[ing] large files in the fields of logistics planning and intelligent systems reflecting at all times the programming requirements of the operational environment."[84]

In the first half of the 1960s it began to receive substantial contracts from NASA and from other federal departments and agencies.

Like PRC, Informatics emerged in the Los Angeles metropolitan area with strong ties to the aerospace industry, but it differed in that it was founded (in 1962) to be primarily a systems integration firm. In the middle to late 1960s, Informatics was considerably larger than any for-profit systems integration specialist company. As with SDC, Informatics' prehistory is important to understanding its development of capabilities and connections for advanced systems integration.

The early Cold War spawned, and in part was defined, not only by a US early warning radar and computer defense system in SAGE, but also by massive efforts to build offensive nuclear capabilities and delivery systems. In 1952 the US tested its first thermonuclear ("hydrogen") bomb; the Soviet Union followed suit the succeeding year. Both countries heightened their efforts toward missile rather than gravity bomb delivery in the early to mid 1950s—missiles were far faster than bombers, more challenging to detect, and more difficult to intercept.[85] A newly formed company, with the two principals having deep experience in aerospace, would become the leading contractor in the mid to late 1950s—the pivotal period of the US ICBM program. That company was Ramo-Wooldridge.

Ramo-Wooldridge, Inc. (R-W) was founded by Simon Ramo and Dean Wooldridge in Inglewood, California in 1953. They were longtime friends and had some interesting parallels. Born weeks apart in 1913, both received doctorates in physics from the California Institute of Technology at age 23. After parting ways for a decade (Ramo going to work for GE and Wooldridge at Bell Laboratories), they both joined Hughes Aircraft in the mid 1940s and became senior leaders—presiding over a large gifted team of engineers pioneering state-of-the-art fighter aircraft, on-board radar, and missile research and development. In 1953, with Howard Hughes unwilling to sell his firm to suitors and failing to address internal managerial problems, Ramo and Wooldridge left to launch their own firm. They received financial backing from Thompson Products, an established supplier of precision components to the automotive and aerospace industries; initially a major shareholder, Thompson Products fully acquired R-W five years later.[86]

By the late 1940s, Ramo and Wooldridge were among the most respected industrial scientists in the aerospace field. In late 1953 both became members of the newly formed Strategic Missiles Evaluation Group, code-named the "Teapot Committee." The classified committee consisted of some of the best aerospace scientists and engineers from industry (Bell Laboratories, Hughes Aircraft, R-W, and Bendix Aviation) and top universities

(Caltech, MIT, and Harvard). Assistant Secretary for Research and Development for the US Air Force Trevor Gardner appointed John von Neumann to chair the "Teapot Committee."[87] Gardner had worked with Ramo at General Electric in Schenectady and closely followed his rise to head missile research for Hughes. Ramo and Wooldridge's connections and firsthand knowledge of the acceleration of US ICBM research and development—if not direct confidence in becoming a major contractor in this lucrative field—probably was an influential factor in their leaving Hughes Aircraft to form R-W.

Central to Ramo-Wooldridge's ICBM and other aerospace efforts would be advanced computer systems, and the two leaders soon hired Walter Bauer to run the firm's computer applications. Bauer, like Ramo and Wooldridge, possessed strong education and experience. He had a doctorate in mathematics from the University of Michigan (1951), joined the Michigan Aeronautical Research Center, programmed on the Standards Eastern Automatic Computer at the National Bureau of Standards, and helped build the first ground control missile information technology flight system for Boeing's BOMARC interceptor missile. Bauer's initial staff of ten grew to approximately 400 people (programmers, systems analysts, and computer engineers) by the late 1950s as he became the leader of the Data Reduction Center of Ramo-Wooldridge's Space Technology Laboratory.[88] In 1958 Thompson Products acquired Ramo-Wooldridge to form TRW (Thompson Ramo-Wooldridge), Inc., with the Ramo-Wooldridge Division an important part of the large corporation. In the late 1950s the new Ramo-Wooldridge Laboratories in Canoga Park, California spanned 90 acres with many buildings around a central mall.[89]

Bauer, in moving to this new facility of the R-W Division of TRW, worked on developing large-scale computer systems.[90] He led a number of important computer systems TRW projects for the Department of Defense, as well as for other departments and agencies of the federal government. These included machine language programming on the AN/UYK-1 computer for the US Navy's Polaris submarine missile guidance system, various real-time command and control computer systems for the Department of Defense Damage Assessment Center (to assess and communicate damage in the event of a nuclear attack), and the first computerized traffic control system (for the city of Los Angeles).[91]

Bauer was immersed in the geographical center of aerospace research and development (the Los Angeles metropolitan area) and all its immense information technology systems' needs. He was only 10 miles away from SDC, the headquarters of SAGE programming and systems integration leadership, and was very active in (IBM) SHARE, Inc. and (Sperry Rand) USE, Inc., the user groups that brought together leaders in aerospace computing (and in other user industries).[92] Bauer believed the major computer manufacturers—IBM, Sperry Rand, and Burroughs—were not rapidly stepping up to focus on systems integration to meet government needs (exemplified by IBM's passing up SAGE programming, and the Air Force looking to RAND/SDC for this work).[93] Also large aerospace firms were not necessarily the best supplier of information systems—as aerospace contracting was a delicate matter that often drew scrutiny and complaints from the competitors of winning contractors. The RAND Corporation was formed in 1948 to separate work from Douglas (and avoid complaints of partiality), SDC was launched in 1957 as a nonprofit corporation to program SAGE; MITRE Corporation was started in 1958 as a nonprofit to serve primarily the Department of Defense with systems engineering; and (perhaps most significant to Bauer) the government-sponsored, El Segundo-based Aerospace Corporation, the fourth of the major nonprofit corporations, was created in 1960 to take on much of the ICBM R&D and systems engineering that had previously gone to Ramo-Wooldridge. Bauer reasoned an independent computer services firm, focused on large-scale systems work, might be able to maneuver to complement the work being done by the four major DoD-sponsored nonprofits and for-profit aerospace contractors—something evident to CSC's and PRC's leaders at the start of the 1960s also.

In 1961, Bauer approached two of his most talented engineers and managers, Werner Frank and Richard Hill, to join him in leaving TRW to launch a systems-oriented (rather than consulting or data processing focused) computer services company. Both recognized the great risk, as well as potential rewards, and signed on to join Bauer in this entrepreneurial effort.[94] "We were going to develop systems," Bauer later recalled. "Primarily, we were going to develop systems for large-scale computer systems, probably of a military nature. That was our first objective."[95] The government represented a very large market in the 1960s. In 1961 the federal government had an estimated 800 installed computers with operating costs of $475 million and 45,000 people in computing-related positions, excluding "classified and military operational applications," which might have boosted total annual expenditures to $1.5 billion.[96]

Werner Frank's computing career began with the US Army. After receiving a master's degree from the University of Illinois in 1955, he worked on the ILLIAC before joining R-W in 1956. Hill had been assistant director of the Western Data Processing Center at UCLA before leaving that appointment and his position on the business administration faculty to come to

TRW's R-W Division in 1960. In 1962, not long after Werner Frank and Richard Hill signed on, the three of them recruited Frank Wagner from North American Aviation, who had been a founder of SHARE, Inc. and a past president of that user group (which at its very start consisted primarily of computer-using aerospace firms in the Los Angeles area). Frank and Hill put up very small amounts of funds, and Bauer a bit more, but in combination their resources fell well short of what they believed was needed. The company began in Frank's empty house while he was on a final work assignment for TRW in Washington in early 1962.[97]

Though Fairchild Semiconductor secured venture funding in the late 1950s, West Coast venture capital enterprises only began to really take off about a decade later. Bauer experienced great difficulty in securing capital in 1961. He approached leading computer manufacturing companies—IBM, Sperry Rand, and Burroughs—and found that they were not interested; neither were any of the many banks he solicited. Erwin Tomash—a friend of Bauer's and a fellow entrepreneur—offered a solution. Tomash, an electrical engineer, had been one of the early employees of Engineering Research Associates in St. Paul, in the late 1940s. After Remington Rand took over this pioneering startup digital computer manufacturer in 1952, Tomash became a Remington Rand sales manager in Los Angeles and had sold systems to Bauer at R-W. Shortly after Sperry merged with Remington in 1955 to form Sperry Rand, Tomash left to join computer memory supplier Telemeter Magnetics, where he soon rose to become president—and then vice president of Ampex after this larger memory company took over Telemeter Magnetics. Wanting greater control over his career and a major equity stake, Tomash left to launch Data Products Corporation, the basis for which involved acquiring a division of Telex Corporation, which had two languishing projects—one in computer memory and the other a prototype computer printer.[98] Tomash and a few secondary founders put up $250,000 for Data Products, and on the basis of his extensive executive experience he secured an addition $1.25 million from banks.[99] He proposed that Data Products provide the needed startup funds, up to $150,000 over the course of two to three years—for Bauer's "Company D" (used in the absence of a real name) and Bauer, Frank, and Hill would receive Data Products shares commensurate with their investment.[100]

As Frank recalls, they received $20,000 up front from Data Products. The business plan called for Informatics becoming profitable by the second year. With the Data Products financing deal, they relocated from Frank's house to office space above Data Products' headquarters in Culver City, California, in April 1962. The entrepreneurs also benefited from mentorship from Tomash in the early days, as he had past experience as company president.[101]

Ultimately they named the company Informatics General (soon thereafter Informatics). It was an independently run (by Bauer) wholly owned subsidiary of Dataproducts (the new spelling shortly after its launch). The name "Informatics" took advantage of a term popular in Europe in computing—"informatique" in French—that had not gained much currency in the US and presented no trademark hurdles, unlike some other proposed names related to data and information that Bauer considered. By registering a trademark in its un-capitalized form, Bauer and his team essentially restricted the use of the term in the US to refer to his corporation. As variations of this terminology continued to accelerate rapidly in Europe (in universities' informatics departments), for two decades the use of the term "informatics" in the US, as a result of trademark protection held by Bauer's company, was stifled.[102]

Informatics struggled to obtain large contracts in its first half year. Early marketing included both editorials written by Bauer and advertisements that profiled him and his words. He was one of the few computer services company presidents with a doctorate and the only one with impeccable computer systems credentials (C-E-I-R's Herbert Robinson was an economist). In September 1962, Bauer wrote a lengthy article in *Datamation* in which he engagingly analyzed trends and the importance of parallel processing, modularity, and online computing.[103]

Writing influential articles and advertisements and editorials from "Dr. Walter F. Bauer" got Informatics' name into the computing community and undoubtedly had some longer-term benefits, but successful major proposals for federal contracts (and subcontracts of aerospace suppliers' federal contracts) was a skill and an art that Bauer and the other three principals had to learn during 1962. That year they took what they could get, which included consulting work for Packard Bell, Astrodata, Mesa Scientific, and RCA—each project for less than $10,000.[104] The one highlight was a $47,000 contract from Bendix (along with the others, this was for time and expenses offering narrow profit margins), but even this could not keep the firm out of the red as deficits per month reached more than $7,300 during September. This prompted Bauer to go on the road and make visits to Washington and elsewhere to have meetings with officials at various federal departments, agencies, and government entities. Late in 1962, Bauer visited the National Security Agency, Oak Ridge National Laboratory, the Air Force Office of Scientific Research, the Office of Naval Research, the Department of Defense's Damage Assessment Center (DoDDAC), the

Navy's Bureau of Ships, and other departments, agencies, and government entities. Bauer and his colleagues wrote an unsuccessful proposal to DoD-DAC in late 1962, but their interaction with contract officers proved a rich learning experience as they learned to write proposals with greater comprehensiveness and clarity. Early the next year they wrote a successful proposal to DoDDAC.[105] This and several others allowed Informatics to move from their small space above parent firm Dataproducts in Culver City, to their own offices in Sherman Oaks.

Several smaller contracts in 1963 got Informatics on the radar with government departments and agencies. In early March 1963, the initial small contract with National Military Command Support Center (NMCSC) expanded; eventually it would bring in more than $220,000 in revenue. They also received contracts from ONR for analysis of software and systems planning for Naval Tactical Systems that ultimately totaled more than $1 million in revenue.[106] With this business Informatics had both a western (Los Angeles metropolitan area) focus with the aerospace and defense contracts/subcontracts and an East Coast focus (with new offices in both the Washington metropolitan area and in the New York–New Jersey area). They also expanded to the Netherlands. Werner Frank was negotiating a major joint venture with Phillips, but unbeknownst to him, others at Phillips were negotiating with CSC, a formidable systems integration competitor by this time. Ultimately CSC won out and Frank chalked it up to a learning experience of needing to know what is going on within the whole company of a potential partner, not just "one individual."[107]

With both recruitment of talent and marketing, Informatics emphasized real-time system expertise. While they rounded out their business in other computer services areas, systems integration was the core. In late 1964, Bauer wrote a full-page editorial/advertisement for *Datamation* that highlighted five "typical" assignments of Informatics from the past two years: an advanced online system for radar, tracking, and display for the Pacific Missile Range; mission control real-time system work collaborating with IBM for NASA Houston; an online system command and control for the National Military Command System; an automated message switching system for GSA Communications for UNIVAC; and the Naval Tactical Systems software and planning for the Office of Naval Research. Bauer structured his company around expertise in real-time systems and in systems integration and conveyed this simultaneously to recruit talent and customer organizations in the same editorial/advertisements. "There is," he wrote, "a considerable shortage of talent in the general field of software ... bad enough ... but when you further qualify this talent to online systems capability, the shortage is acute .... We canvas constantly with the zeal of treasure hunters .... We have to ... our work demands that each and every Informatics staff member be qualified to take on problems of the most modern type, involving computers controlling displays, communication devices, and analog instrumentation." "This," he added, "limits our growth ... the number and nature of the assignments we can accept."[108]

Bauer was not just a technical systems expert, but also a talented entrepreneur. While Informatics expertise in systems integration presented many opportunities and, as Bauer saw it, "limited" the assignments and personnel the firm could "accept," broad-based software products were only limited by the expansion of computer adoption, which was accelerating rapidly in the 1960s. This led Bauer to investigate and eventually acquire a small division of Hughes, in which John Postley was developing and perfecting a file-generator software product in the middle to late 1960s. Ultimately, in 1968, this became Informatics' MARK IV, which was the first software product to achieve revenue of $10 million and the first to reach $100 million (a product creatively financed by Informatics from some important early customers and investors). In view of the high profile MARK IV gained, it may have seemed that the new software product tail was beginning to wag the systems integration dog. Services, however, remained a very substantial part of Informatics' businesses. Historian Martin Campbell-Kelly, while briefly addressing Informatics' origins as a software contractor, wrote extensively on how MARK IV became a standout achievement in the young software products industry.[109] For this reason (and because products are outside the scope of this book on services), it is not recounted here.

Another early computer services company, Applied Data Research, pioneered just such a transition from services to software products by launching the very first software product from an independent supplier (not a computer manufacturer), a system product called AUTOFLOW, in 1964.[110] While this was a (partial) path taken by some services companies, among them ADR and Informatics, many providers of computer services never sought to move into products (or at very least, services remained their core). This was true of Diebold, Electronic Data Systems, Tymshare, Gentry, and C-E-I-R throughout their history, and it is true today of Automatic Data Processing, Andersen/Accenture, and Computer Sciences Corporation.

Informatics continued to do extensive computer services systems contracting in the late 1960s, in the 1970s, and beyond, but its move to add a software products business at a critical juncture and other factors (including some corporate reorganization at the start of the 1970s and its being acquired by a smaller insurance company customer, Equimatics, which renamed itself Informatics) led to faster-growing systems integration competitors relative to Informatics—CSC and IBM's Federal Systems Division—throughout the 1970s and the first half of the 1980s. By the mid 1980s, Informatics' computer services and products businesses were producing nearly $200 million in annual revenue. Walter Bauer, even through the ordeal with the acquisition by Equimatics, remained the president for the company's entire 23-year history. Informatics—a company that stands out for its important systems building expertise and accomplishments in the 1960s, as well as its role in greatly accelerating the software products industry—was acquired by Sterling Software (a firm co-founded in 1981 and run by Sam Wyly) in 1985 for $143 million in cash in (what at least began as) a hostile takeover.[111]

To conclude, system integration entirely changed possibilities for computing from pioneering System Development Corporation's SAGE programming forward. Informatics became a major player in the systems integration field doing many important government and industry projects in the 1960s, 1970s, and 1980s. The field attracted CSC, which became formidable, and despite passing on the original SAGE programming project, IBM—especially through its Federal Systems Division—became a major force in this specialty. And IBM and American Airline's SABRE was unparalleled with regard to commercial system integration work in the 1960s.

II The Industry's Identity

5 Cooperating Competitors: ADAPSO, 1961–1982

Despite Louis Galambos' insightful establishment of the dual themes of competition and cooperation in examining the emergence and growth of a national cotton textile industry organization a half century ago, trade associations, in general, remain significantly understudied organizations.[1] Sometimes this is an oversight on the part of business, institutional, and economic historians. Other times, the lack of available published and archival sources present insurmountable hurdles.[2] Thankfully, in 2002, Burton Grad and Luanne Johnson, two veterans of the software industry who have done much to promote and advance software history, held a three-day workshop on the history of the Association of Data Processing Service Organizations (ADAPSO). Of even greater value than that workshop and its edited transcript, Grad and Johnson (a past executive director of ADAPSO) helped facilitate the transfer of ADAPSO's extensive organizational records to the Charles Babbage Institute.[3]

In addition to trade associations' intrinsic importance, analyzing these organizations helps to better understand constituent firms, to learn the nature and extent of the circulation of knowledge in a trade, and to gain insight into political economy. Further, studying inclusion and participation in these organizations, and the nature and timing of their activities (including what companies and associations they battled), illuminates the industry's identity.

Founding ADAPSO

Competitive dynamics, geography, political and legal circumstances, the real and perceived value of sharing and coordination, and individual initiative often dictate the length of time before an industry association emerges to serve similarly engaged business enterprises in a new field, and the scope of initiatives an association undertakes. In the first four chapters we explored the earliest segments of the emerging computer services industry: consulting, data processing/service bureaus, programming services, and systems integration. The companies that launched these segments generally sought to expand their scope to include other computer services segments. Understandably—as a result of the service bureaus longer history, greater number, and more active participation in exploratory meetings in 1960—ADAPSO, the first trade association in computer services, defined its initial domain as these data processing bureaus.

Service bureaus had pre-computer tabulation machine roots dating back to the early 1930s. This differed markedly from programming services and systems integration—both of which were industry segments originating in the mid 1950s that were inherently based on digital computers.[4] To a greater degree than even programming services, service bureaus of the early 1960s were local enterprises.[5] Service bureau data processing, by definition, was conducted onsite (at bureaus) and results (printouts and storage media of financial accounting, managerial accounting, payroll, inventory, and other data) often were personally delivered to customers. For service bureaus of the early 1960s (nearly a decade before time-sharing networks reduced geographical constraints tied to place), expanding to new geographies meant installing computers and adding personnel in new locations. Many early companies in this field were small and had operations in a just one city. Only Statistical Tabulating Corporation, IBM's wholly owned subsidiary Service Bureau Corporation (SBC), and Sperry Rand's Univac Services Centers (USC) had anything close to a national (and in SBC's case, international) business. The local basis of many early 1960s service bureau companies enhanced opportunities to share and learn from fellow firms operating in different (therefore non-competing) geographies.

Some trade organizations owe their existence to the banding together of company leaders, who know one another from professional or technical organizations, industry battles, legislative efforts, or other activities. Other trade associations are launched by professionals with trade association experience possessing an entrepreneurial mindset. In effect, every emerging industry opened up an opportunity where an association founder, if not gaining equity in an enterprise, could at least significantly boost their job title, management responsibilities, and salary. ADAPSO's origins rested with the latter.

A National Office Management Association staff member, William H. Evans, conceived of forming an "Office Services Institute" consisting (as the name implies) of businesses offering services to offices. Out of this rather unstructured idea (services to offices varied so widely, and common interests, unified representation, and opportunities for learning might be minimal), the organization that became ADAPSO developed. A couple of Evans' emergent co-organizers—one a manager from Sperry Rand's USC, Romuald Slimak, the other SBC's sales and advertising manager, C. W. Graf—at the initial "exploratory" meeting in 1960 pushed the notion of focusing on data processing services centers or service bureaus.[6] Evans was on board with Slimak and Graf's ideas, and the new name, ADAPSO, slightly preceded the formal inauguration of the trade association at the start of 1961.[7] Both pre-computer tabulating machine and digital computer-based data processing were identified as the underlying tools of the trade for member companies. Though the rapid growth of the latter, and the value of sharing information in this new and complex field, resulted in early ADAPSO members being heavily focused on digital computing.

Despite being a "service bureau" trade association at the start, Evans and ADAPSO did not want to exclude other interested segments of the computer services industry—so consultants, programming services firms (without a bureau), and later, emergent facilities management enterprises were allowed to join as associate (non-voting) members.[8] Membership (full or associate) was initially $100 a year. The organization needed revenue for its small office in Abington, Pennsylvania (in the northern Philadelphia metropolitan region) and Evans' annual salary, nominally set at $15,000 (in 1961), was paid only in part (or at least substantially delayed) in ADAPSO's early years. To aid the association's finances, and to tie fees to means, service bureaus with multiple locations were charged the membership fee for each office or facility (enabling their various bureau managers to attend ADAPSO meetings), but only received the standard one vote.[9]

In October 1961, ADAPSO published its first *Directory of Data Processing Services Centers*, what would become an annual publication covering North America (and later the world). It listed the company/center name, address, phone number, and president or manager. At the time, ADAPSO had 40 member companies (34 voting members). As a new organization in a relatively new field, ADAPSO's board of directors believed it important to include in the directory all identifiable qualifying enterprises (possessing tabulator or computing equipment and providing data processing services) since they far outnumbered members—and hoped listing non-members would spark awareness and soon lead them to join. The 1961 directory listed 208 centers in 34 states (plus the District of Columbia), and 15 centers in Canada. Most of the largest companies (including SBC, which had over 70) only listed their headquarters (Sperry Rand's USC being the exception with all 31 of its locations listed). Most major cities had multiple service bureaus, and a substantial number of smaller cities had at least one listing, including Savannah, Georgia; South Bend, Indiana; Keokuk, Iowa; and Lubbock, Texas. By the start of the 1960s data processing services bureaus blanketed larger towns and cities throughout the North American continent.[10]

By 1965 enough enterprises (along with the payments for constituent offices of chain centers) had joined ADAPSO so that the organization's leaders felt justified listing only members (and associate members)—the widely distributed directory became a marketing tool for members and an incentive to join ADAPSO. In the mid 1960s, ADAPSO also enlisted a former employee of Control Data Corporation, Bernard Goldstein, to travel the country and sign up new members. Goldstein made many contacts, signed up many new members, learned from them, and then launched United Data Centers, a chain service bureau (bureaus in multiple cities/locations). "I signed up members for ADAPSO," Goldstein, who became ADAPSO's president in 1970, later recalled; " ... very successful for them but it was also very successful for me because I saw the opportunity to re-enter the business and build what I called then a network of datacenters."[11]

IBM's SBC, a member like all others with just one vote, did list (and pay the membership fee for all of its centers) in 1965, and the total centers in the ADAPSO Directory that year totaled 228 North American member centers (just under a third of which were SBC), with ten "world member" centers (in Brazil, Germany, Great Britain, Ireland, Mexico, and Switzerland) and twelve associate members.[12] Well over half of the North American service bureaus in operation, however, were not ADAPSO members—though many of the bureaus that weren't members were smaller stand-alone centers rather than larger chain centers (like SBC and USC).

In the mid 1960s, its membership having increased to more than 150 companies (more than two thirds of them full voting members), ADAPSO was on firmer financial footing. It moved its home office from Evans' hometown of Abington to the Graybar Building in midtown Manhattan. In 1967, Evans retired and Jerome Dreyer became the executive director; he then served the organization with distinction for two decades. ADAPSO's annual operating budget, which Dreyer oversaw in his first year, had reached $67,000 and would continue to grow steadily along with membership in the late 1960s and later (membership grew to a peak of 700 organizational members in the mid 1980s).[13]

Circulating Knowledge

The overarching benefit of being an ADAPSO member in the organization's first decade lay not in lobbying efforts (a focus in the 1970s, the 1980s, and the 1990s) or in the marketing boost from inclusion in ADAPSO directories, but rather in being part of a knowledge-sharing community.

While technical user groups such as SHARE, Inc. and GUIDE (both for IBM computer users) and USE, Inc. (Sperry Rand computer users) provided opportunities for individuals to exchange technical knowledge and to share libraries of code, ADAPSO offered something different. Eventually technical user groups attracted some representatives of the computer services industry, but in the 1960s SHARE, GUIDE, and USE provided no programmatic content focused directly on service bureaus or on programming services management.[14] In the late 1950s and the 1960s, SHARE—which dwarfed all other IT user groups—had members from corporate customer installations, from government installations, from IBM divisions; it also had among its members some academics associated with university computer centers. Thus, ADAPSO, which concentrated on business data processing services, filled a major void.

On January 20, 1961, in New York, ADAPSO held its inaugural "Management Symposia"—events convened three times a year in the first two years, and twice a year thereafter, that heavily defined the organization's first half-decade. Romauld Slimak, in opening remarks at the first symposium, articulated the vast need businesses had for data processing support and the mission to cooperate and share within ADAPSO: "I doubt that any of us, individually, has the facilities to offer the full range of services and equipment necessary to meet the needs [of data processing and computation] ... but in concert, borrowing from one another's experiences, disseminating information among us, and organizing special efforts we can promote the broader uses of our facilities, conceive of new concepts, create new applications, establish new approaches to the myriad of problems of business and science."[15]

This initial ADAPSO Management Symposium and the ones that followed covered a vast range of topics, but one commonality was openness in giving talks and taking part in post-talk discussions related to member businesses' organizational structures, marketing efforts, best practices, problems, failures, and new initiatives. As we saw in chapter 3, C-E-I-R, Inc.—by the late 1950s a major programming services company with service bureau operations—faced increasing financial challenges due to its unrestrained expansion in the early 1960s. At the first symposium, William Eaton, vice president and director of C-E-I-R, Inc., spoke with great candor of the problems his firm faced: quickly added locations, the difficulty with filling positions with qualified personnel, challenges with billing clients, scheduling to handle uneven loads, keeping operations efficient, and integrating tabulating equipment and computing equipment. In discussions that followed, executives from other firms asked clarifying questions, related common problems, and shared considerable information on their own operations and experiences.[16]

This openness and culture of sharing occurred during the first Management Symposium despite the fact that participating companies had substantial geographical overlap (especially of headquarters for services operations), and hence were direct competitors in some markets. Greater geographic diversity within ADAPSO membership occurred as the trade association grew rapidly in the remainder of 1961 and beyond. Of the sixteen ADAPSO charter companies, all were represented at this first symposium and three fourths were headquartered in only three cities—seven in the New York metropolitan area, three in Toronto, and three in greater Philadelphia. (See table 5.1.) Of the remaining three, one was headquartered in Chicago and one in Arlington, Virginia. Dayton was the sole midsize city having a company representative (from National Cash Register). By June 1961, membership had more than doubled to include 34 companies with headquarters in a greater number of moderate-size cities, including Milwaukee, Tulsa, Sacramento, and Minneapolis. That new members tended to be smaller companies, some with only one location, further reduced the risk of helping competitors by sharing information and best practices.[17]

Talks and discussions at the 1961 ADAPSO management symposia also introduced a topic that ADAPSO would focus on and debate vigorously throughout its history: ethics and proper code of conduct for services enterprises. This had been the basis for originally not allowing consultants to be full members, because they ranged from highly knowledgeable professionals to uninformed charlatans.[18] In January 1961, ADAPSO formed a "Code of Ethics" committee to develop a code "applicable to service center operations."[19] Differing views and the rapid evolution of ADAPSO prevented adoption of a formally ratified code for several years.

At the first symposium, A. M. Lount of Enelco Limited gave a paper briefly outlining religion-based ethics and humanistic perspectives (drawing from Lewis Mumford) on moral practice before examining several practical instances. To paraphrase, these practical dilemmas or questions included the following: To whom does a software program belong (creator

Table 5.1
ADAPSO's charter members (1961).

Bendix Corporation, New York City
Walter Camenisch, Inc., Philadelphia
C-E-I-R, Inc., Arlington, Virginia
Enelco Limited, Toronto
Ferranti-Packard Electric Ltd., Toronto
Machine Accounting Company, Philadelphia
May & Speh Tabulating Services, Inc., Chicago
National Cash Register Company, Dayton
Nationwide Tabulating Corporation, Hempstead, New York
Philco Corporation, Willow Grove, Pennsylvania
Radio Corporation of America, New York City
Recording and Statistical Company, New York City
Sperry Rand/Univac Service Centers, New York City
Service Bureau Corporation, New York City
Statistical Reporting and Tabulating Ltd., Toronto
Ernst and Ernst, New York Services Division, New York City

Source: "Membership Roster" [January 21, 1961], ADAPSO Proceedings Management Symposium, New York, January 20, 1961: 42 (ADAPSO Records, 1960–1999, Charles Babbage Institute).

or client)? Who bears responsibility for improper use of a program? What constitutes reasonable performance standards? What defines the ethical line in raising fees to an existing client once they are locked in? What can be done to prevent large computer firms temporarily selling services at a loss to win business? In engineering applications, what can properly be handled by a computer specialist and what requires supervision by a properly licensed engineer?[20]

Pricing services was both an ethical question (concerning the unethical practice of intentionally pricing below cost to circumvent competition or price gouging locked-in customers) and a persistent practical one frequently discussed at early symposia. Though much services work was on a cost-plus basis, bidding and fixed prices also were common. With fixed-priced contracting, accurately allocating machine time, employee time, and supervisory/managerial time could be complicated. On fixed-priced contracts, the uncertainty of programming and debugging time presented particular challenges.[21] Companies of all sizes could make mistakes or ill-informed bids and commit to work performed at a loss.

Early ADAPSO management symposia also contained significant technical content specific to computer services applications. Unlike SHARE or USE (focused on a particular computer manufacturers' systems), or the Association for Computing Machinery (content more on what came to be called computer science), ADAPSO's technical papers and discussion were not system specific and almost always directly related to technical and managerial practices for services enterprises. An example was a talk at the second ADAPSO Management Symposium (in St. Louis in May 1961), by Malcolm Smith of Sperry Rand USC, titled "The Significance of COBOL to the Service Center." Smith had helped develop the pioneering compiler Flow-Matic, and was a member of the CODASYL System Development Group for specifying and maintaining COBOL, a programming language project initiated by the Pentagon for business data processing in May 1959.[22]

The third ADAPSO Management Symposium, held in Los Angeles in October 1961, highlighted that technical content was not just for routine data processing systems but also for systems integration for business and government. It also demonstrated how geography and local markets influenced the content of such meetings, as Los Angeles held a dominant position in aerospace and aerospace computing. David Green, a former engineer and manager at System Development Corporation in the late 1950s with extensive subsequent DoD computer consulting expertise, spoke on applications of operations research to management problems, TRW's Richard Hill (just before co-founding Informatics) discussed the program evaluation review technique (PERT), and Hughes Aircraft John Pettit outlined a total systems approach in selecting hardware, concentrating on the hardware/ software relationship. All three, unlike most ADAPSO attendees, were engineers possessing doctorates.[23]

The locations of the symposia in the succeeding few years maintained the geographic regional diversity (with Washington, DC and Portland, Oregon meetings), and reflected ADAPSO's growing international scope (with the eighth symposium, in 1964, held in Toronto).[24] These symposia continued to expand the scale and scope of presentations and open exchanges on techniques for running efficient operations, effective hiring and managing staff, accurate costing, client industries, marketing, legal considerations, competitive threats, professionalism, and ethics.[25]

The sessions at the 1964 Toronto meeting introduced and substantially addressed the hitherto little explored theme of protecting client records—which included talks by representatives of large firms, including IBM's SBC, and C-E-I-R, Inc. These papers conveyed strong expertise on safeguards to protect reputation and finances by using fireproof safes, redundant records, extensive liability insurance, and carefully written contracts.[26] On July 2, 1959, IBM lost an extensive amount of computing equipment and storage media in a fire (fought by 300 firefighters from 34 companies and creating an estimated $30 million in losses) in the Air Force's computer room in the Pentagon.[27]

The ADAPSO management symposia did much to circulate knowledge and best practices in the 1960s. Members unable to attend could read the papers and transcripts and always rich discussions that followed. Beyond the symposia proceedings, ADAPSO produced a number of other publications in its first decade, including *ADAPSO Management Guidon* launched shortly after the trade association's founding. In 1969, ADAPSO initiated publication of a glossy trade journal, *Computer Services: The Journal of the Association of Data Processing Service Organizations*, that further disseminated managerial knowledge in computer services (and software).[28]

Banking Changes and Challenges

Many papers and discussions at ADAPSO symposia in the first half of the 1960s addressed the value proposition that data centers offered, their expertise, and articulated that such centers were a great option, even for sizable corporations and organizations that could afford internal data processing. ADAPSO members and the broader data processing industry wanted outsourcing to be a continuing proposition, not a stepping stone for clients to test new digital computing equipment before graduating to possessing internal computing resources to do work in house. Above all, ADAPSO members wanted internal computing facilities, at worst, to represent a (partial or whole) lost business opportunity for service bureaus—not direct competition, and especially not direct competition from industries with immense capital and cash flow stemming from regulated monopolistic advantages.

In the interwar years and the early post–World War II era, many insurance companies, larger banks, and other sizable businesses had internal precomputer punch card equipment and tabulation rooms or departments. The largest banks, like the large insurance companies, became early adopters of digital computers for data processing. The Bank of America and its partner SRI International pioneered in developing the Electronic Recording Machine Accounting (ERMA) computer system in the first half of the 1950s—including the much celebrated magnetic ink character recognition (which evolved to the MICR standard) to aid with the increasing challenge of processing checks.[29] Banks, however, varied in size in the late 1950s, and some service bureaus viewed the industry as a significant opportunity. Even in 1965, *ADAPSO Management Guidon* published an article by a banking industry analyst, Robert H. Long, titled "Data Processing Service Opportunities in Banking." Long emphasized that of the 14,000 US banks only 500 "have or have plans for obtaining their own data processing equipment." "With rising clerical costs and increasing transaction volume," he continued, "it is probable that at least 2,500 of the remaining 13,500 banks would be interested in using data processing facilities if available."[30] His analysis was balanced, emphasizing that his was not "a glowing picture of opportunities for service bureaus in bank processing" and that "manufacturer service bureaus" (such as IBM's SBC, CDC, and Sperry-Univac's USC) probably would continue to be best at securing banking business because of "added assurance that appropriate control, confidentiality and timeliness of processing and reports will be maintained."[31]

As the 1960s progressed, ADAPSO members typically viewed banking as more of a threat than an opportunity. The beneficiaries of outsourced bank data processing tended to be SBC and other very large chain centers. The core of bank data processing was check and other transaction processing that had to be done quickly. Big banks purchased sufficient computing equipment to handle peak transaction loads and often had excess computing capacity during slower periods. The fear, and sometimes the reality, was banks offering data processing services (for free or under costs) to achieve greater computer resource utilization and to better acquire or retain business banking customers (in the process, knocking out data processing services competition). One data processing application that banks targeted was offering payroll to firms doing business with their bank (for free or below market charges). Frank Lautenberg of ADP (president of ADAPSO in 1967) commented that "banks' duplicitousness" was a real threat, but noted that ADP was able to "beat the banks ... on the pavement," by being all the more aggressive with meeting with customers and providing excellent service.[32] According to Bernard Goldstein of United Data Center, banks learned that they couldn't "produce results for free."[33]

As historian Thomas Haigh has astutely related, ADAPSO's board of directors first formally recognized the banking threat in 1962 when the US House of Representatives considered a bill regarding smaller banks providing data processing services. This precipitated an ADAPSO Legislative Committee, and in 1964 a Bank Survey Committee. The latter passed a 1966 resolution to prevent illegal (anti-competitive) bank data processing services. In 1969 Bernard Goldstein (ADAPSO's incoming president for 1970) testified before the House Committee on Banking and Currency and later commented that it was a challenge, in such a short time, to educate this committee on what the data processing services industry was and how it was being injured by anti-competitive bank practices. In the mid 1960s, ADAPSO retained Milton Wessel, an attorney with Kaye, Scholer, Fierman, Hays, and Handler, who served with distinction for many years as ADAPSO's counsel. Though Wessel was a gifted attorney, ADAPSO's small budget provided minimal resources for effective lobbying in the 1960s and the early 1970s, especially against the powerful banking industry. Clarified federal regulatory statements on national banks in 1971 provided a major defeat for ADAPSO.[34] The banking issue led to ADAPSO's greater involvement in lobbying and litigation in the 1970s and later. ADAPSO not only continued to battle banks, but also lobbied and sought legal remedies on many other issues. As was demonstrated by its conflict with independent contractors (discussed in chapter 9), the ADAPSO of the 1980s sometimes was the larger association with greater lobbying infrastructure—in this instance resulting in legislative victory in battling software programmer brokerage firms that spawned a small trade association, the National Association of Computer Consultant Businesses.

Analyzing and Serving a Broader Industry

ADAPSO's symposia, meetings, and early publications provided rich qualitative knowledge that was shared among fellow members. To complement this, in the second half of the 1960s ADAPSO began to administer and commission research to provide members with rich quantitative aggregate industry data. Meanwhile, the identity of computer services and of the broader IT industry was also evolving rapidly, and ADAPSO brought in and began to serve additional segments of the industry—time sharing, facilities management, and (outside the services domain) software products.

Time sharing, or resource shared networked computers—in which slices of time were allocated to users at smaller computers or terminals to give users the experience of their own machine—evolved from university research systems (MIT's Compatible Time Sharing System, the Dartmouth Time Sharing System, and the University of California at Berkeley's Time Sharing System) during the early to mid 1960s to a significant segment of the computer services industry by decade's end. Existing (generally larger) data services centers, large companies like General Electric, and fast-growing startups such as Tymshare (the focus of chapter 7) and Com-share, contributed to the launch and rapid growth of a national, and soon to be international, time-sharing industry in the late 1960s and the 1970s.

Batch processing and time sharing were alternative mechanisms to potentially manage expensive computers efficiently. Companies, universities, and other organizations had the distinct challenge of maintaining adequate computing resources at all times, but not wasting them. Some organizations, including many banks and universities, sold or gave away excess computer processing resources. For many this became a hassle that was far from a core competency, and some avoided it altogether. Astute leaders at programming services companies sometimes partnered to manage the sale of computer time with their customers, while Ross Perot, with Electronic Data System's founding in 1962, insightfully made facilities management (taking over a computing facility and balancing computing resources between many clients) a core business—and a new and important segment of computer services by the late 1960s.

Meanwhile, pioneering software products from services companies of the early 1960s—such as Applied Data Research's Autoflow automated flowcharting software package—broke new ground and proved a concept (the software product) that Informatics furthered with MARK IV, a language and file management package that was the highest revenue generating software product of the late 1960s and the early 1970s.[35] With Informatics, software products rose to join computer services as the firm's dual concentration. Its rapid success with software products served as a model for the young software product industry and helped to inspire software product startups such as the Cullinane Corporation—the first software product company traded on the New York Stock Exchange (in 1982, as Cullinane Database Systems).[36] In 1968, IBM's leaders, seeing the writing on the wall and fearful of the Department of Justice and of anti-trust litigation, decided (and in 1969 publicly announced) that IBM would "unbundle" much of its software from hardware and sell it as products. Simultaneously they announced separate pricing (from hardware) on custom programming and other computer services. This helped to facilitate a quick expanding software products industry, and gave a boost to the services industry.

With these changes in the field of information technology, ADAPSO's leaders saw the opportunity to become a broad-based computer services *and* software products trade association in the late 1960s—one serving all the varieties of computer services (programming services, systems integration, time sharing, facilities management, and consulting), as well as the software products industry.

Concomitant to its expanding scope, ADAPSO sought to insure quality research data on its core industry segments for its member firms to allow these companies to have better metrics to make informed decisions. Producing and publishing such reports informed decision makers, and also helped to recognize, define, and promote computer services as an industry. The organization and categorization of industry segments and domains in market research reports also spoke to how the industry was understood (and possibly misunderstood) at various times.

In 1967, ADAPSO commissioned a business school faculty member—Robert B. DesJardins, at the University of North Carolina—to collect and analyze survey data and publish a report with aggregate data on its members and other firms in the trade. It continued to utilize his services for annual surveys the next three years. Also in 1967, Patrick McGovern began devoting full-time work to leading International Data Corporation (IDC) and launching the new publication *Computerworld*. McGovern, an MIT graduate, had founded IDC three years earlier, but was splitting time between his startup research and publishing firm and serving as editor of the journal *Computers and Automation*.[37] ADAPSO hired IDC to analyze and publish reports for ADAPSO members on the computer services industry in the first half of the 1970s. Subsequently they commissioned increasingly extensive reports from well-established Quantum Sciences and from Peter Cunningham's information technology industry research firm INPUT, which he founded in 1974 and propelled to rapid industry respect and success.[38] To ADAPSO member firms such reports were critically important—aggregate information that benefitted all while protecting individual firms' data on size, financials, problems, and other carefully guarded details.[39]

DesJardins' 1967 ADAPSO report was based on 2,000 questionnaires, with a response rate of 50 percent of those with at least $10 million in annual revenue, and a quarter overall. It provides a rich snapshot of the data processing services industry in 1966. DesJardins argued that, given his survey-based research, he could accurately estimate that in 1966 there were 700 US data processing services firms operating 1,130 centers, serving 80,000 clients. About 70 percent of data processing services firms ran a single center—representing 43 percent of all centers and serving 34 percent of all customers. Nonetheless, there was concentration at the top with the 35 largest data processing center companies (5 percent) operating 30 percent of the centers and serving 40 percent of the customers. Forty percent of the firms were four years old or less, 40 percent had been in business five to eight years, 13 percent had been operating nine to twelve years, and 7 percent, or roughly 50 companies, had been around more than thirteen years (thus operating as tabulation machine data processing providers before entering the computer field). Companies with under $150,000 in annual revenue had average profit margins of 4 percent, while those generating $150,000 or more in revenue had average margins of 8 percent. The greatest short-term problems were identified as competition from equipment manufacturers (computer firms), competition from banks, personnel turnover, and equipment delivery. Partially countering the notion of a heavy degree of lock-in, maintaining old business was seen as a significantly greater hurdle than attracting new business. Industry revenue was estimated at $534 million in 1966.[40]

ADAPSO's second through fourth surveys (for 1967, 1968, and 1969) were also produced by DesJardins. Indicative of ADAPSO's broadening scope, these include breakout data on time sharing and software—as well as comparable data solely aggregating results of data processing services centers or bureaus. (See table 5.2.) In the late 1960s, the reports expanded to include categories of workers, the nature of services, and other measures. Firms' largest employment category was keypunchers at 35 percent in both 1968 and 1969, with programmers and systems analysts totaling 20 and 21 percent respectively in those years. In 1967, 55 percent of data center computers were IBM systems, 18 percent Honeywell, 16 percent Burroughs, while Sperry Rand (Univac) and all other makers combined made up just 11 percent. The survey reminds us of the significant and often overlooked occupation of key punchers in the 1960s and heavy presence of Burroughs systems used by service bureaus relative to the firm's much lower overall market share in the computer industry. Hiring and retaining personnel was seen as the top problem in 1968, and marketing the greatest challenge in

Table 5.2

IDC/ADAPSO estimated size of computer services industry (millions of dollars).

	1966	1967	1968	1969	1970	1971	1972
Number of firms	700	840	1,300	1,300	1,400	1,500	1,600
Batch processing revenue	$410	$480	$600	$740	$930	$1,060	$1,230
Online/time-sharing revenue	$20	$50	$120	$210	$330	$440	$570
Facilities mgt./other revenue	$10	$30	$50	$150	$200	$400	$500
Total revenue	$440	$560	$770	$1,100	$1,560	$1,900	$2,300

Source: IDC. ADAPSO Sixth Annual Survey of the Data Processing Service Industry, 1972 (ADAPSO Records, 1960–1999, Charles Babbage Institute).

1969. In DesJardins' fourth and his final ADAPSO (1969) report, he stated he believed the number of firms had peaked at the end of the 1960s and would stay roughly the same or decline in succeeding years, despite the increasing number of data centers.[41]

In 1971, McGovern's IDC took over ADAPSO's annual industry surveys. True to DesJardins' prediction, growth in number of firms did stabilize, with a substantial growth in bureaus or offices (fewer stand-alone centers as opposed to chains). The computer services industry was evolving rapidly as were its boundaries and the perspectives on useful divisions of sectors to measure. The term "software" was included in sectors—and included both programming services and software products. In the late 1960s, most "software" was programming services; by early 1970s, software product revenue was substantial (though still far less than programming services). Table 5.2 provides data on the computer services industry and its sectors or segments between 1966 and 1972, slightly underestimating the overall amount of revenue because "software" is left out (because it includes products). By leaving in number of firms (which presumably includes at least a small number of independent software products firms), this category is probably overestimated slightly between the late 1960s and the early 1970s.

Peter Cunningham's INPUT, from shortly after its founding, became a long-term leading information technology industry research firm. In addition to its annual reports for ADAPSO in the late 1970s and the 1980s, it produced hundreds of reports on industry segments and customized client research. The structuring of the IT industry with INPUT's data collection and presentation separates software services from products—and divides the services side into data processing services and professional services. The former represents the continuance of data processing specialists (such as ADP—a major data processor to this day), while the latter captures programming services, facilities management, and consulting.

ADAPSO commissioned analyses of the computer services and software industries (including projections) helped companies plan for the future, as well as lent credibility to lobbying efforts that these were important, impressive, fast-growing US industries worthy of consideration—particularly with regard to what many believed were anti-competitive threats from banks and computer industry powerhouse IBM. Evans and Dreyer, along with ADAPSO's board of directors and the volunteer presidents, carefully balanced concerns for small, medium, and large company members in ADAPSO's first decade. No representative of IBM SBC was ever elected president. Nonetheless SBC representatives actively participated in

Table 5.3
INPUT/ADAPSO estimated size of US computer services industry (millions of dollars).

	1976	1977	1978	1979	1980	1981	1982
Processing services	$4,000	$4,700	$5,580	$6,700	$8,810	$11,111	$12,484
Professional services	$700	$1,000	$1,200	$1,550	$3,472	$4,491	$5,329
Total	$4,700	$5,700	$6,780	$8,250	$12,282	$15,602	$17,803

Source: INPUT, Annual Computer Services Industry Report(s), 1976, 1977,1978, 1979, 1980, 1981, 1982 (ADAPSO Records, 1960–1999, Charles Babbage Institute).

symposia and discussions sharing knowledge and best practices. SBC as the largest chain center took some occasional jabs from ADAPSO members, but often appeared outside of the broader anti-IBM sentiment that existed in ADAPSO. Because SBC was a wholly owned subsidiary (as a result of the 1956 consent decree with the Department of Justice), it probably was watched closely by the government. IBM chose to divest itself of SBC, selling it Control Data Corporation on favorable terms in 1973 in partial settlement of Control Data's anti-trust lawsuit against the computer firm. ADAPSO member representatives tended to be more vocally critical of IBM's dominance and leasing practices and prices to data centers than anti-SBC. "In the IBM Corporation," Bernard Goldstein recalled, "SBC was sort of a Siberia … . If you weren't in the hardware side … or research … you were sent to SBC … to Siberia."[42]

By the early 1970s, an increasing number of software product companies were joining ADAPSO, as well as software products divisions of services firms. With the rapid growth of startup time-sharing firms such as Tymshare in the second half of the 1960s, these services companies were also joining. Dreyer, the ADAPSO Board, and various presidents saw this as an opportunity to expand the association's membership size, resources, and power (in lobbying and helping to shape policy). To give adequate voice to, and have a structure conducive to the effective sharing and coordination, ADAPSO in effect became a multi-division organization at the start of the 1970s through the institution of different "sections" such as data processing/data centers, time-sharing services, professional services/consulting, and software products. The elected ADAPSO presidents of the mid to late 1970s and the early 1980s illustrate how this leadership responsibility was spread around over time between members of different industries as all the various major sections were represented: United Data Centers' Bernard Goldstein and Data

Processing of the South's A. S. "Buck" Blankenship (data processing/service bureaus), Tymshare's Thomas O'Rourke and Com-share's Richard Crandall (time sharing), and Management Sciences John Imlay and AGS Computers' Lawrence Schoenberg (software professional services).

For central functions (decision-making and meeting programs), all sections had a voice at the table. The structure allowed for breakout section programs. Another innovative activity of early ADAPSO meetings was "industry roundtables," where CEOs shared information and practices to better run their businesses. The roundtables tended to be set up with small numbers of individuals in the same industry (ADAPSO section or a further subdivision), but not direct competitors (different regions for instance) to encourage greater cooperation. Following the advice of ADAPSO's general counsel, Milt Wessel, they were always careful to have a disinterested party in the room should these sessions ever raise concerns regarding collusion and anti-trust violations.[43]

Lobbying was increasingly important by the mid 1970s, exemplified by the move of ADAPSO's headquarters to the Washington metropolitan area by the late 1970s. ADAPSO's ability to cooperatively bring together a variety of companies (initially data centers varying widely in size) and increasingly different IT industries and industry segments in the late 1960s and the 1970s (programming services, facilities management, time sharing, professional services, and software products) represented its greatest success. Through symposia, publications, hallway discussions, lunches, dinners, and cocktail parties, emerging industry leaders helped educate one another—they taught, they learned, and they networked (many a merger and acquisition had its initial roots in an ADAPSO meeting or contact).

As the IT industry grew in the personal computer and World Wide Web eras, the Information Technology Association of America (the trade association's new name as of 1991) was challenged by many more targeted trade associations. As David Sturtevant, who had served as ADAPSO's vice president for public communications in mid to late 1980s, later recalled, "in 1985, 1986, 1987, the association got a real challenge from the Microcomputer Software Association ... . It was the first time in the association's history that we didn't figure out how to incorporate a new group that was a logical extension of where the industry was headed."[44] Nonetheless, ITAA remained significant (with about 400 members), even if only one of a number of associational players, in the international computer services and other global IT industries.

Conclusion

Although ITAA's role as a services and software trade association was diluted to an extent by competing IT associations in the 1990s and later, its earlier history as ADAPSO truly helped to shape and advance the computer services industry. It did this through promotion, facilitating members' sharing information and best practices, aggregating industry data, lobbying, and other activities. It also helped provide identity to the industry by incorporating new segments and becoming a multi-divisional IT trade association (with all of its divisions representing segments of the computer services industry except the notable exception of software products). Thus, ADAPSO stands as a strong example of how important trade associations are to industry identity and development, particularly in an industry's formative years.

6   Managing Facilities: Electronic Data Systems, 1962–1984

Henry Ross Perot and his 1962 startup company Electronic Data Systems (EDS) launched the field of facilities management, which rapidly became a significant new segment of the computer services industry. Facilities management initially was defined by EDS as on-site work at a client organization's data processing facilities.[1] This industry segment, however, differed from programming services in that it involved taking over a client or customer's entire data processing operation on a long-term, fixed-price basis.[2] Soon, the long-term, fixed-priced, all-inclusive contracts, rather than the site of the work, became EDS's and facilities management's defining characteristic.

Many larger companies in a range of industries (including insurance, retail, banking, chemical, petroleum, food processing, aerospace, automotive, and other manufacturing) quickly jumped on the bandwagon to acquire digital computers in the second half of the 1950s or in the early 1960s. Digital computers potentially offered new possibilities. They were perceived as the wave of the future and companies wanted to use computers either to get ahead or at least not be left behind. Beyond the practical side of what these systems could bring to an organization, possessing and using this technology conveyed success and prestige. Some less scrupulous consultants pushed clients hard to acquire computers without proper evidence of any advantages these systems might offer, and computers were aggressively marketed by skilled salesmen from IBM, from Sperry Rand's Univac Division (trusted firms that often had long supplied punch card tabulation equipment to corporate data processing departments), and from elsewhere.[3] Once computer systems were installed, some corporations and other organizations' data processing departments found it difficult to use these expensive systems in a cost-effective manner. Actual computer use often varied widely from day to day, resulting in substantial, and differing, levels of overcapacity much of the time.[4] And many computer departments either did not want to enter into a side business of selling excess computer time and offering computer services, or they were willing to do so but not good at it.

Overall, most companies and other organizations had substantial unused computer time despite running three shifts a day, six or seven days a week. In the first half of 1962 (just months before EDS's launch), System Development Corporation conducted a major survey of thirty California business and scientific computer installations (in the Los Angeles and San Diego metropolitan areas) about personnel, processes, and equipment utilization. These included aerospace, petroleum, and insurance companies, as well as banks and educational institutions. Both scientific and business installations reported, on average, less than 10 percent supervisory staff. At scientific computer centers the mean was 33 percent programmers and 35 percent data processing operators or key punch operators. At business installations there were on average 24 percent programmers and 51 percent data processing and key punch operators. Despite many types of facilities stating round the clock operations, the mean for utilized computer operating time (combined business and scientific installations—these were not broken out in SDC's study) was 46 percent, with 18 percent for needed scheduled and unscheduled maintenance.[5] In other words, just over one third of computer time went unused (or perhaps closer to one fourth—still a substantial amount—if added processing hours increased needed maintenance time in the same proportion). In view of this waste, the computing field was ripe for Perot's facilities management model.[6]

Facilities management presented a solution where data processing costs frequently could, over time, be trimmed through learning and better system management. EDS sometimes retained a portion or all of the existing staff of a company's internal data processing operation (who became EDS employees) in taking it over to provide facilities management. At other times, EDS set up a completely new data processing organization (in the 1960s through leased space and computer time) for the customer. During the course of a longer-term contract—typically five to seven years in the company's first decade—EDS achieved efficiencies, and especially in the latter years of a contract, earned sizable profit margins. Resources (either computer processing capacity or personnel) from one facility could be used for others to utilize equipment and labor with the utmost efficiency. One thing that was important to companies was not only being able to spend less on data processing, but also being able to predict longer-term data processing costs more accurately.

As a result of its original model, EDS typically was referred as a "facilities management" company for many years, long after it established its own large-scale regional data centers in the early 1970s to offer contractual services work on its own computing equipment.[7] In part, this was because of EDS's start in launching a new field, but also there often were distinctions between EDS large-scale data centers and traditional service bureaus.[8] EDS commonly was taking over entire data processing operations (becoming the data processing department) of large corporate and organizational clients on a long-term basis. Traditional service bureaus tended to have shorter contracts for specified work and fewer corporate giants as customers. That said, the largest chain centers, such as SBC, Control Data, and Sperry Rand's USC, did have many major corporate clients, as did large computer services specialists (for particular applications) like ADP with payroll. And even if contracts were shorter, they tended to get renewed time and again—so while distinctions between service bureau and "facilities management" data centers existed, differences between the two were somewhat blurred by the mid 1970s. EDS, like CSC, did develop into a diversified computer services provider within its first decade and offered programming services, systems integration, and consulting. Likewise, CSC and other integrated services providers, grew significant facilities management businesses. EDS, however, was without equal in focusing on facilities management in the second half of the 1960s and throughout the 1970s.

While System Development Corporation, with its large size and rapid employee turnover, was a major force in populating and educating the early software services and products industries, former IBMers also contributed mightily to the workforce of these industries. Moreover, IBM, to a far greater degree than SDC, seeded the top executive ranks outside of IBM. An impressive number of former IBM sales, engineering, and managerial staff founded and/or became leaders of important IT companies. This included CUC founders (in 1955) Elmer Kubie and John Sheldon, CUC's later chairman of the board Cuthbert Hurd, and Amdahl Corporation (launched in 1970) founder and leader Gene Amdahl.[9] And more recently, the five founders (in 1972) of software products powerhouse SAP (Dietmar Hopp, Klaus Tchira, Hans-Werner Hector, Hasso Plattner, and Claus Wellenreuther), the founders of both human resources software specialist PeopleSoft (started by David Duffield in 1987), and IT advisory and information services standout Gartner Group (started by Gideon Gartner in 1979)—all were former IBM engineers or sales personnel.[10] Even Marc Andreesen, who co-founded Netscape in 1994, had been an IBM intern.[11]

Of these IBM-trained company founders, none would have a higher profile than H. Ross Perot, the first self-made billionaire in the IT industry, whose fortune grew and shrank in volatile fashion with price fluctuations of EDS stock. In the early 1970s, at a tumultuous time for the securities industry, he attempted to bail out a large Wall Street brokerage to learn the brokerage data processing business, and as one prominent journalist put it, respond to personal calls from Attorney General John Mitchell and Secretary of the Treasury John Connally to save the securities industry. He also was a celebrated "rescuer" of imprisoned EDS employees and other "hostages" in 1979 in Iran, was for a time General Motors' largest shareholder (after it bought EDS), engaged in a highly charged and often public battle with GM's leader Roger Smith, and was a 1992 independent candidate for the presidency of the United States, receiving 19 percent of the popular vote.[12] Of primary concern for this book, is that Perot founded and rapidly grew a company that became, and long remained, one of the largest enterprises (by revenue) in the computer services industry.

H. Ross Perot, EDS, and Facilities Management

H. Ross Perot, born in 1930 in Texarkana, Texas, was an average-performing student in elementary school, in junior high, in high school, at Texarkana Junior College, and at the US Naval Academy, but demonstrated drive and leadership skills in other pursuits, becoming an Eagle Scout, getting elected class president at the Naval Academy, and serving as an assistant fire-control officer in the Navy. As he was finishing his four-year naval commitment, Perot had a chance encounter with an IBM executive at a naval event who encouraged him to apply to become an IBM salesman. In 1957 Perot joined IBM to sell computers working out of its Dallas office. Successful from the start, he had little trouble exceeding his quota and becoming part of IBM's esteemed "Hundred Percent Club." Serving on IBM's sales force, he saw how inefficient many customers were at using their computer systems—and how these customers faced uncertain data processing costs in hiring staff or consultants to program and operate the machines.[13]

In 1962, Perot conceived of a business plan to launch a new company to seize this opportunity of mainframe organizational customers' inefficient use of expensive computing resources. Initially, this plan took the form of brokering computer time, while his facilities management concept was first implemented in the following year. Perot took with him IBM's dress code, but in contrast with IBM he rewarded those who joined him with an ownership stake rather than commissions—a necessity, as he had founded EDS with a mere $1,000. IBM's sales representatives Milledge Hart and Tom Marquez followed him in rounding out EDS's initial three-person marketing team. Perot, Hart, and Marquez possessed valuable knowledge of which Dallas-area IBM customers had substantial unused computer time and they approached those firms to be the broker to sell it to others. This was first achieved by selling unused Southwestern Life Insurance Company excess IBM 7070 time to Collins Radio (both Dallas-based companies).[14]

In 1963, Perot implemented his facilities management model, which entailed taking over a company's data processing facilities and operating it on a long-term fixed-price basis, thereby launching an important new segment of the computer services industry. Perot and Marquez promoted this idea to Herman Lay, CEO and chairman of the Dallas-based potato chip maker Frito-Lay. Lay was skeptical at first, but it appears that in his mind IBM's attempt to thwart EDS with a "whispering campaign" lent credibility to the seriousness of Perot and EDS, and led him to take a second look. That year, Frito-Lay, EDS's first facilities management customer, signed on to a five-year contract to have EDS run its data processing department—a deal that generated revenue of $5,500 a month for EDS.[15]

In the succeeding twelve months, EDS got contracts to manage six other organizations' data processing operations. Some had computers already; others had them on order. By mid 1964, EDS had thirty professionals on staff and Perot was spending most of his time recruiting talented programmers, systems analysts, and engineers. The model was attractive to top management at corporations—for once they could know their computing costs in advance rather than just accept uncertainty. Any costs above the EDS quoted contractual amount came out of the pocket of Perot's firm. With success in Dallas, Perot planned expansion to other cities and states. While knowledge of firms, contacts, and spare computer capacity in the Dallas metropolitan area had been helpful, Perot knew the opportunity of his facilities management model was not unique to Dallas. In 1964 the company had its first customer outside the Dallas metropolitan region.[16]

The early success of EDS captured the attention of leading trade publication *Datamation* in 1964. A short article titled "On-site Computer Depts. Making Friends in Dallas" stated: "A fixed-price, packaged approach to EDP (electronic data processing) is making sense to a number of Dallas firms … and giving IBM there a splitting headache."[17] It continued outlining EDS's process, "After a preliminary analysis a bid is submitted … . If accepted, EDS takes on the whole schmear—analysis, design, programming … then runs the programs on computers (IBM 7070, 7074, two 1410s and three 1401s) on which they've purchased blocks of time."[18] In Fiscal 1964, EDS brought in $400,000 in revenue, but only achieved a narrow (1 percent) profit margin.[19] The initial small margins were in part a result of the tail end of early long-term contracts being more profitable (through efficiency gains) and in part a result of EDS's aggressively plowing funds back into the firm to expand.

In the mid 1960s and in subsequent years, EDS continued to bring in more and more facilities management contracts in Texas and elsewhere. This included winning a major contract in 1966 to provide facilities management to PepsiCo's cola operations. Fortunately for EDS, PepsiCo was the product of New York City–based Pepsi-Cola's 1965 merger with Frito-Lay, already a customer of EDS.[20]

One of the keys to EDS's early success was the 1965 Medicare legislation, part of President Lyndon B. Johnson's "Great Society" program. EDS programmed general-purpose software for Medicare processing (payments/reimbursements) that it then customized to meet different states' requirements. During 1966, EDS began to reap the benefits from its investment and gradually added new states. In 1968, one fourth of EDS revenue was from Medicare and (less profitable) Medicaid processing—in roughly a dozen states. This helped to boost the company's once razor-thin net profits to 20 percent.[21]

A Public Company and Perot's Public Presence

Before 1968, EDS had been able to plow profits back into the firm to cover expansion costs. In the late 1960s, ambitious growth plans far outdistanced resources, and Perot investigated an initial public offering of EDS stock. Working with the New York investment bankers R. W. Pressprich and Company, EDS's small initial public offering brought in a much needed $5 million, and for the first time put a market value on Perot's remaining shares (over 80 percent ownership)—worth $154 million. In less than two years, with the stock bubble of the "go-go years," the value of Perot's shares grew to $1.4 billion in March 1970 before an EDS stock collapse of roughly 50 percent on April 22 resulted in Perot's stock losing an unprecedented (in the computer industry at least) $445 million in a day.[22] Weeks later EDS shares dropped further to a typical valuation based on earnings and were about 85 percent off their all-time peak set several months earlier.

Though EDS's stock offering in 1968 was modest, the event caused a media sensation within, and far beyond, the information technology industry. A cover story on Perot in the November 1968 issue of *Fortune* declared him "the fastest, richest Texan ever."[23] Still wealthy, but losing a large fraction of his fortune in 1970 also brought many media headlines. Meanwhile, Perot had become increasingly involved in national and international politics and philanthropy. At the urging of Henry Kissinger (President Richard Nixon's National Security Advisor and future Secretary of State), Perot launched and heavily funded the United We Stand Committee, an effort to improve the conditions of American prisoners of war in North Vietnam. He also donated about $2.5 million to Dallas' public schools to assist inner-city schools that had primarily black and Mexican-American students, and gave generously to the Boy Scouts and the Girl Scouts.[24]

Even before Gene Amdahl caused problems for IBM by offering plug-compatible machines and equipment shortly after the 1970 launch of Amdahl Corporation, Perot and EDS were thorns in the side for Thomas Watson Jr. and IBM. Not only did EDS cut into potential services revenue either bundled with hardware, direct, or outsourced to IBM's SBC, it *disrupted* the relationship between IBM and its customer—as an outsourced computer department, EDS was in the middle. IBM wanted to place IBM mainframes, as well as serve and influence customers on timing and types of upgrades. The presence of EDS changed the equation. EDS focused on long-term contracts and above all reducing waste and adding efficiency. Greater utilization and efficiency, balancing computer-time resources between many IBM customers, could potentially cut into surging demand for IBM systems. Further, unlike other services providers, EDS, following both Perot's example and his rules, had mimicked the conservatively dressed professionals who helped distinguish IBM. Instituting elements of IBM's formal professional culture was all the easier in that many of EDS's early employees, like their founder, were former IBMers. Perot also looked to his Navy roots and hired many clean-cut ex-military men.

Targeting New "Verticals": Wall Street, Banking, and Insurance

In 1970, Perot named one of the original IBM members of the EDS triumvirate, Milledge Hart, president of EDS. Perot remained CEO and chairman of the board and was very active in the day-to-day leadership of EDS into the early 1980s.

Perot, Hart, and a senior vice president, Mort Meyerson, had spotted an opportunity in Medicare processing in the second half of the 1960s. Between 1967 and the middle of 1971, EDS made $72 million processing

Table 6.1
Electronic Data System's revenue and net earnings (in thousands).

	Revenue	Net earnings
1972	$90,955	$13,603
1973	$111,882	$15,200
1974	$118,734	$15,349
1975	$123,896	$14,648
1976	$132,952	$13,602
1977	$164,188	$16,428
1978	$217,837	$19,666
1979	$274,298	$23,702
1980	$374,661	$28,890
1981	$454,614	$37,816
1982	$503,335	$46,967

Sources: Electronic Data Systems annual reports, 1981 and 1982. Data for 1981 and all earlier years are as stated in the 1981 annual report.

Medicare claims, roughly two thirds of its income during that period.[25] Likewise, Perot saw the volatility of the stock market and the challenges many Wall Street firms were having with securities trade processing in the late 1960s and in 1970. Beyond the major problems with fulfilling transactions in a timely and accurate manner, with the bear market of 1970, some were having trouble meeting stock exchange regulatory obligations for capital ratios.

The volume of trades on the New York Stock Exchange increased from 5 million on a typical day in 1965 to about 12 million on a typical day in 1968, overwhelming existing mechanisms to transfer securities and keep records.[26] The required introduction of computers to address challenges was haphazard and painful, and it led to many back-office problems—even well into the 1970s. The problems were severe enough by 1968 that the New York Stock Exchange, the American Stock Exchange, and National Association of Securities Dealers banded together to contract for a team of scientists at the RAND Corporation to evaluate problems and seek technological fixes by using computer simulation techniques—an area in which RAND, through Air Force work, had unparalleled expertise.[27] Despite these efforts, transaction processing and record keeping systems continued to plague

Wall Street for years as a result of the need for more and better computing/software infrastructure and as a result of revenue problems precipitated by the recession in the first half of the 1970s. Ultimately the transaction processing problem was a major factor in one sixth of NYSE-focused brokerages' going out of business during that decade.[28]

One of the largest brokerages (in fact, the second-largest through much of the early 1970s) with substantial problems was F. I. DuPont Glore Forgan and Company. Perot hoped that EDS could gain facilities management business and underlying know-how in back-office brokerage data processing from that firm. An EDS team including Perot, Hart, Meyerson, and a few other executives decided to put up most of the funds necessary to lend $5 million to F. I. DuPont Glore Forgan and Company in order to allow it to meet the New York Stock Exchange's capital requirements. In addition, the EDS team acquired (for about $2 million) Wall Street Leasing, a data processing concern servicing F. I. DuPont Glore Forgan and Company, thereby transferring to EDS a long-term contract to manage the brokerage's data processing facility. Perot and his small team then financed a new company, PHMFG, Inc., to further assist the brokerage with funds as needed.[29]

Part of the rationale behind Perot and his team's endeavor in dealing with the F. I. DuPont Glore Forgan and Company was seeing an important business opportunity, and part of it appeared to be Perot's patriotism. Perot had received strong encouragement to help the struggling brokerage company and to aid his country and the fledgling financial industry from Attorney General John Mitchell and Secretary of the Treasury John Connally. Ultimately, Perot and PHMFG, Inc. took over control of F. I. DuPont Glore Forgan and Company (with an 80 percent ownership stake), and on May 14, 1971 a new concern—DuPont Glore Forgan, Inc.—was launched.[30] Without intention, Perot had become the brokerage industry's largest individual investor. Morton Meyerson, a lesser financial partner in this effort funded through Perot's selling 900,000 shares of EDS for roughly $57 million, became president of DuPont Glore Forgan. Meyerson and Perot tried military veterans to be brokers.[31] This clashed with the liberal culture of New York City stockbrokers in the early 1970s, when marijuana and cocaine use, informal dress, and an ethos of individualism were common.[32] The brokerage's problems, however, went beyond cultural differences, and the financial challenges proved too great for Meyerson to right the sinking ship. In a last-ditch effort, Perot had acquired another brokerage—Walston, Inc.—for $10 million in order to merge it with the troubled DuPont brokerage to form DuPont Walston, Inc. Perot lost an estimated $60 million dollars on the DuPont and Walston investments, and EDS lost the roughly

15 percent of its profits from defunct data processing contracts with the merged brokerage.[33] Meyerson returned to EDS in 1975, and two years later he replaced Hart (who had resigned) as the firm's president.

Wall Street was only one of the areas where EDS was seeking to diversify its facilities management and data processing services in the early 1970s. Unlike horizontally oriented data processing specialists like ADP with payroll, or service bureaus with basic back-office processing largely common to many different businesses (accounts payable, receivable, human resources, etc.), EDS's strategy at the start of the 1970s was to offer services to "verticals" (industries having specialized needs and requiring specialized expertise)—to serve the complete information technology needs of companies in targeted large industries. The company also focused on economies of scale. Rather than relying solely on a constant balancing act of efficiently utilizing human and computer system resources from small facilities (the centers or data processing departments EDS managed on behalf of clients), EDS adopted a hybrid strategy that included creating several large regional data centers.[34]

In 1972, EDS revamped and expanded its Dallas area center and added two centers, one in Camp Hill, Pennsylvania and one in San Francisco. The Dallas center—a campus that included the corporate headquarters—was fully owned by EDS—building, land, and equipment. In view of the large amount of life insurance business, with its clients predominantly on the East Coast, the Camp Hill center was officially named the EDS Life Insurance Industry Center. This offered online systems to allow insurance firms to improve client access to up-to-date information, an essential element to improve decision making, operations, and overall customer service. EDS, always looking to do more with less and to make the fixed-price contracts result in greater margins, was able to trim its Life Insurance Industry Center staff by 15 percent and maintain all service during 1972. While EDS sought to gain efficiency through strategic reductions of staff working on particular contracts, the growth of data processing and opportunities resulted in EDS's workforce expansion in the first half of the 1970s despite tough economic times. Some talent was brought on by taking over a company's data processing department; other talent was recruited. EDS substantially expanded its training infrastructure for educating programmers, systems analysts, and marketing specialists in 1972, training many hundreds of new and existing employees.[35]

By the mid 1970s, EDS was operating 54 mainframe computers. That number included 17 mainframe systems EDS owned, seven of them at the Dallas services center and most of the others at the other regional centers. The EDS-owned computers included fourteen IBM System/360s, one IBM System/370, one Burroughs mainframe, and one Honeywell mainframe. In addition to the aforementioned centers, EDS had 23 smaller data centers across the country, for which it leased equipment and space from customers under contracts of five to ten years. Most of these facilities—including those in Seattle, Miami, San Diego, Boston, Grand Rapids, Burlington (Vermont), and Madison—had IBM systems. Although EDS experienced decreases in net earnings in 1975 and 1976, it was still able to increase annual revenue growth for both years by 4.3 percent and 7.3 respectively despite the sluggish economy of the mid 1970s, and to return to rapid top-line and bottom-line growth in 1977.[36]

EDS's early facilities management contracts typically were for five to seven years, but by the mid 1970s EDS was sometimes contracting for ten years to, essentially, serve as a company's IT department. Many economists, including Ronald Coase and Oliver Williamson, have pointed out the potential challenges (though not necessarily insurmountable ones) associated with long-term contracting.[37] While it is costly to continually write out and negotiate short-term contracts, and changing a supplier of services might be disruptive and damaging, lesser durations offered flexibility. Meanwhile it is difficult to foresee all contingencies and write them into a long-term contract. EDS's contracts specified fixed monthly payments as well as variable amounts at predetermined rates based on volume of transaction processing. Nonetheless the needs of a customer could change dramatically within a five-year, a seven-year, and (especially) a ten-year period. EDS and customers would renegotiate as was necessary, and the relative lack of reported problems (such as lawsuits for breach of contract and other reasons), coupled with long-term repeat customers, suggest that EDS and its customers acted in good faith and fairness and to a degree defied the logic of economic (transaction cost) theory regarding markets versus hierarchies and the challenges of long-term contracts. One reason this may have been the case was that many of EDS' customers had begun with hierarchies, operating their own data processing department both before and during the early days of digital computing. For many organizations, digital computing was a rude awakening to technical difficulties (hardware failures, software bugs, etc.), fluctuations in costs (especially with programming software), and, most significant, the heavy inefficiencies of idle computer time and the hassles and risks associated with entering the business of computer services or the business of brokering computer time. The advantage of fixed costs (generally a set monthly amount plus additions at an established rate based on increased data processing transaction volume) was a true benefit, coupled with the expertise and learning EDS staff brought from one client to the next. For EDS large insurance companies, banks, utilities, and others were critical to its future—and if it sought to hold up a client by refusing to negotiate unforeseen contingencies or new services in a fair manner, word could quickly spread. For customers, although EDS was not the only game in town for facilities management, it had invented this field and had a strong reputation for success, hardworking professionals, and creating efficiencies. Fixed-price contracting meant EDS generated profits only to the degree it performed well—and it could only be highly successful if it continually reduced cost per specified unit of work (generally a data transaction). Absolutely critical was effective appraisal of jobs before initial contracting. EDS teams would descend on a potential client for major site evaluations characterized by extremely hard work and long hours to accurately evaluate costs to write an effective contract. This was a defining organizational capability that EDS cultivated, with much success, from the 1963 Frito-Lay contract on.

EDS had problems with certain customers from time to time, but generally maintained excellent relations. Perot's problems with F. I. DuPont Glore Forgan and Company were the exception not the rule. In terms of processing stock transactions for the brokerage, things had improved dramatically once EDS took over that function. The problems with the brokerage stemmed from resentment over the bailout that Perot felt that he in part was pushed into by high-ranking leaders in the Nixon administration—a bailout resulting in a transfer of ownership that F. I. DuPont Glore Forgan and Company leaders saw as opportunistic. Perot believed he was helping out his country, but the leaders of F. I. DuPont Glore Forgan and Company believed it was an excuse to seize control of a legendary brokerage during a moment of temporary vulnerability.[38] Perot's conflict with the brokerage's former leaders were rooted with the change of ownership (the brokerage's executives resenting losing control), not contracting. This, along with a clash in personalities between EDS leaders and the brokerage's executives caused bitterness—a situation that would, with some variation, play out again a decade later, with Perot and EDS on the other side of an acquisition.

Expanding Scale and Scope and Becoming a GM Division

Although the failure of F. I. DuPont Glore Forgan/DuPont Walston resulted in a substantial financial loss to Perot, he was still a wealthy individual, and he and his team had positioned EDS to prosper by serving numerous firms in the banking, life insurance, utilities, and other industries in the second half of the 1970s. One source of early success, health care processing, had been faltering, but Meyerson revived it through the savvy appointment of a talented system operations manager, Ken Riedlinger, to head health care efforts for the Eastern region—winning a North Carolina Medicaid contract in 1975 and a Medicaid contract in Tennessee in 1976. By 1977, EDS employed more than 6,300 people, nearly double the total several years earlier.[39]

Many of these employees came from clients' data processing departments that EDS took over. For existing data processing managers this could be a trying experience, but most successfully survived the transition to what was often the more formal ways of EDS. Becoming an EDS employee often opened up opportunities greater than being stuck in a corporate data processing department—seldom a path to the executive ranks. Skilled technical and managerial leaders could move up the ladder at EDS, perhaps to one of the larger data centers or the Dallas headquarters. And if the timing was right with EDS's volatile but generally advancing stock (after the 1970 plummet), they could meaningfully advance their overall compensation through stock incentives. Some, however, did lose out by borrowing funds to buy options that became worthless as a result of the DuPont Walston collapse, the recession, and the overall weak stock market of the early to mid 1970s.

For years EDS had a small center in San Juan, Puerto Rico, and had added a bit of business in Europe, but well into the 1970s it was still overwhelmingly a US-focused company. At the middle of the 1970s this appeared to be changing as it first received a contract from King Abdulazziz University in Saudi Arabia, and then, in 1976, a $41 million contract for data processing for Iran's social security system. The glowing discussion of the latter operation in EDS's 1978 annual report coincided with the beginning of revolutionary change in Iran. Suddenly payments to EDS ceased, and the company, which issued all the health care and social security checks for Iran, stopped providing services and ordered all but two of its 80 employees in Iran and their families to go home. The two EDS executives who remained in Iran were imprisoned, and Perot traveled to Tehran and financed a commando group to try to rescue them. Ultimately the efforts were successful, and the two EDS executives and others escaped and joined their rescuers to come home. To publicize the episode, Ross Perot's wife, Margot, suggested hiring the novelist Ken Follett, who agreed to write a book if he could have a free hand. The deal struck involved a million-dollar advance to Follett from the publisher, with both Perot and

Follett empowered to cancel the project at any time and with a guarantee that Perot would reimburse the publisher if the book project were to be canceled or not completed for any reason.[40] This resulted in the popular book *On Wings of Eagles*, which was also made into a television mini-series with David Niven portraying Ross Perot. The book and the TV series added to Perot's fame and mystique, which were unrivaled among leaders of computer services companies.[41]

EDS's business continued to expand rapidly in the late 1970s and the early 1980s. Like Computer Usage, EDS tried its hand with businesses outside of services—in EDS's case, minicomputers and consumer electronics distribution—but by the time EDS did this the company had far deeper pockets than CUC and these unsuccessful ventures were merely small hiccups in a highly successful run for EDS. Back in 1976, EDS had won a $2 billion Texas Medicaid contract, one of a handful of massive contracts—providing roughly 14 percent of EDS's net profit over a half-decade. When the Texas Medicaid contract came up for renewal, Bradford National underbid EDS and was apparently in line to get the business; however, through lobbying and legal effort, Perot and EDS's lawyers prevailed and EDS won the highly lucrative renewal.[42]

In 1982, EDS brought in more than $500 million for the first time and had net profits of more than $45 million. Perot and Meyerson knew that, in recent years, smaller to mid-size contracts had no longer moved the revenue and net profit dial to help perpetuate EDS's impressive growth in net earnings (and its even faster growth in revenue).[43] EDS needed very large contracts, such as the $656 million contract with the US Army that was signed in 1982. That ten-year contract, up to that point the largest ever signed by the Army for data processing, was for connecting 42 army posts to create a nationwide data processing network. EDS beat out bids from CSC and IBM for this contract, which was expected to result in the employment of 700 EDS workers.[44] Back in 1976, EDS had roughly one sixth of the $800 million US facilities management market by revenue, a lower figure because it was not yet a significant player in the DoD facilities management market dominated by IBM's Federal Systems Division, CSC, Control Data, and others.[45] By the early to mid 1980s, EDS's primary targets were the Department of Defense, other departments and agencies of the federal government, and *Fortune* 100 companies—among them General Motors. Its market share gradually expanded in a fast-growing field in which outsourced facilities became increasingly common.

In late 1983 and early 1984, a Salomon Brothers investment banker who was evaluating potential acquisition targets in consultation with General

Motors' leaders, floated the idea of acquiring EDS to the GM leaders and floated the idea of GM as a potential buyer to EDS's top executives. At the time, GM had more than 700,000 employees in 39 countries and sold more than $60 billion in products and services annually.[46]

Automobiles are composed of thousands of parts, produced by many outside suppliers, and computers are critical to logistics—including the electronic data interchange GM and other automakers utilized in coordination with their supply chain in the 1970s. Beyond this—and the standard financial accounting and managerial accounting critical to any large corporation—computer-aided design and computer-aided manufacturing were critical to GM, whose recent reorganization of its brands had been disruptive to its data processing department. Further, the 1980s were a time of heightened international competition (especially from Toyota and Honda) and of heightened interest in bold organizational and manufacturing process innovation. This was evident in the early-1980s partnership of GM and Toyota at the NUMMI plant in Fremont, California, where Chevrolet Novas (and, later, Geos and Chevrolet Prisms) and Toyota Corollas were built, with essentially the same platforms and the same general designs, on the same assembly lines. For GM, much of the purpose of this was to better learn Toyota's successful methods of "lean manufacturing," which reduced "waste" in the production process.[47] Streamlined computer services could also aid with timely and efficient coordination of supply chains and implementation of just-in-time delivery (of components from suppliers) and other elements of Toyota production processes.

Roger Smith, General Motors' board chairman and CEO, became interested in EDS in early 1984. He met with Perot, Meyerson, and others. Meyerson asked him how much business the data processing department at GM did in revenue in a year (resulting from transfer pricing, as the department essentially served GM, not outside customers); Smith responded about $2 billion, but said that GM would like to achieve annual revenue of $3 billion in the future.[48] GM had a huge data processing organization, but did not have the competency needed to succeed well in selling computer services to external markets. The prestige of GM and the auto corporation leaders' strong interest in the broader impact innovative EDS could have on the automaker appealed to Perot and his executive team. GM's internal needs alone ensured near-term growth for EDS, and GM's leaders conveyed their desire to see EDS grow supplying externally as well. Moreover, the substantial offer, which was a considerable premium to the existing share price, proved financially attractive to Perot, Meyerson, and others. More than anything, Perot wanted an ever-growing, innovative enterprise and thought becoming part of GM might accelerate this. In the middle of 1984 GM, acquired EDS for $2.5 billion in a stock deal.[49]

Often the classic economic case for choosing vertical integration as opposed to long-term contracting (selecting managerial hierarchies over markets to coordinate resources) is the potential problems of a supplier (once entrenched) holding up their customer for remediated (higher) compensation or making decisions that are adverse to the customer.[50] Economists have looked to GM's acquisition of ownership control of Fisher Body in the second half of the 1920s perhaps more than any other business relationship and deal concerning the hierarchies versus markets (or make versus buy) question—where scholarship for decades highlighted GM's acquisition of Fisher Body as a textbook acquisition to avoid hold up and ensure necessary asset specific investments from the supplier. Ronald Coase, however, through careful historical analysis (correcting both some misstated facts and interpretations), convincingly argued (in 2000) that GM acquired Fisher Body for reasons other than potential future holdup (which did not apply)—and even though GM leaders wanted to and did acquire Fisher Body, there is not a compelling, supportive economic argument that demonstrates they needed to do this for efficacy or efficiency, instead of signing a long-term contract.[51]

There is likewise not a compelling economic argument for why GM needed EDS as a captive department rather than a long-term contractual facilities management supplier. Unique to information technology, the efficient utilization of human and computer resources might seem to provide a partial justification. However, EDS had managed and balanced resources (owned by clients) quite efficiently throughout the history of the company. More than anything it may have been the timing seemed right for two men—Smith and Perot—to do a bold, high-profile deal symbolic of growth and innovation for both firms.[52]

The relationship between Smith and Perot, however, deteriorated quickly after the acquisition. Perot and EDS staff found GM regimented, bureaucratic, and unwilling to adopt the innovative culture EDS promoted. Perot believed Smith was feared and not approachable to many managers within GM, including those at the new/revamped EDS computer services division. Smith saw Perot, GM's largest shareholder, as a disruptive GM board member. The battles between Smith and Perot became increasingly public during the first half of 1986.[53] Overall, the many conflicts between Smith and Perot are thoroughly detailed in Doron Levin's book *Irreconcilable Differences: Ross Perot versus General Motors*.[54]

Late in 1986, Perot, Smith, and GM's board of directors agreed for Perot to sell his shares of GM and vacate his post on GM's board for $700 million.[55] Meyerson had been the day-to-day manager of GM's EDS services division and continued in this role. Two years later Perot teamed with investors and founded computer services company Perot Systems. In addition to serving GM's massive computer service's needs, the EDS division did succeed in its strategy of focusing on large contracts, especially for the federal government. In 1987, GM's EDS division secured a guaranteed $542 million 12-year US Navy contract to supply a 58 city computer network logistics system that financial analysts believed could ultimately bring in $1 billion.[56] Between serving GM and outside customers EDS revenue within two years more than tripled to $3.5 billion annually and its employment nearly tripled from just under 14,000 to around 40,000. In 1996 when GM divested its EDS division, it had 100,000 employees.[57]

Conclusion

During the 1980s, facilities management increasingly become a commodity service for many corporate and organizational customers. CSC had grown its facilities management business, IBM had entered into the corporate facilities management business at the end of the 1980s (it had been doing this for federal government installations for many years through its Federal Systems Division), and Unisys (the product of a combination of Burroughs and Sperry Rand's Sperry-Univac Division) and other computer manufacturers had entered facilities management as well. Increasingly, very large contracts with a few select corporate giants and with government agencies (such as the Department of Defense and the Central Intelligence Agency) presented great opportunities and great risks. One GM-EDS employee who was with the firm/division from 1989 to 2005 commented: "We signed bad contracts, riskier contracts, which we might not have done in the past because you have to have the growth."[58] Bidding on sizable fixed-price contracts involved great risk, and EDS had some missteps once it became independent again—which helped precipitate its reduced valuation and it becoming an acquisition target. It was acquired by Hewlett-Packard for $13.9 billion at a time of relative weakness and reduced margins (attributable in part to the growth of overseas challengers, particularly in India).

7 Sharing Time: Tymshare, 1965–1984

The May 1964 issue of *The Atlantic Monthly* carried an article, titled "The Computers of Tomorrow," in which MIT industrial engineering professor Martin Greenberger stated "Computing services and establishments will begin to spread throughout every sector of American life, reaching into homes, offices, classrooms, laboratories, factories, and businesses of all kinds."[1] Most of the article analogized delivery of computer time sharing—inexpensive terminals or small computers connected to mainframes (possessing specialized operating systems to quickly rotate processing between users)—with delivery of electrical utilities. Greenberger wrote of the "information utility" (the term "computer utility" took hold more in the succeeding several years).[2] He speculated on what this service might look like (assuming it private rather than public): "[T]he industry will probably be dominated by one or two firms of giant proportions. Logical candidates among existing companies include not only the large communication and computer enterprises, but also the computer users."[3] Among these major industrial users, he identified insurance companies, banks, brokerages, and scientific and engineering firms. He concluded his article articulating that "an information utility may be as commonplace by 2000 AD as telephone service is today."[4]

Greenberger's article was in part an outgrowth of his organizing a conference several years earlier on computing's future as part of celebratory events for MIT's 100th anniversary—participants at this conference included many of the foremost figures in computing: Norbert Wiener, Vannevar Bush, Marvin Minsky, Jay Forrester, Robert Fano, Herbert Simon, Grace Hopper, Alan Perlis, Claude Shannon, John Kemeny, and Gene Amdahl.[5] Greenberger's views undoubtedly were also shaped by his witnessing pioneering digital computer, programming, and networking accomplishments at MIT for the past decade—Lincoln Laboratory, SAGE, Compatible Time-Sharing System (CTSS—operational in 1961, the first significant time-sharing system), and the 1963 ARPA IPTO-supported Project MAC (Multiple Access Computer/ Machine-Aided Cognition—which funded development of CTSS's ambitious follow-on Multics time-sharing system as well as artificial intelligence research).[6]

Greenberger's "information utility," for individual home users did not materialize during the time-sharing era (from the mid 1960s to the mid 1980s). The computer utility concept for home users arguably has been realized to a degree in the World Wide Web era with cloud services, but this has fallen far short of complete standardization and uniformity we associate with utilities. Nonetheless, on certain individual predictions, he was quite prescient. He forecast time-sharing providers would serve many businesses and industries and offer custom programming and software products (the latter a precursor to recent Software as a Service, or SaaS). With regard to suppliers of time sharing, Greenberger was correct that the time-sharing industry would be dominated by a few firms.[7] What he did not anticipate was that this would only be true after a shakeout of roughly 100 startups, and that one of these survivors, Tymshare, Inc., would become the second largest time-sharing company in the world. The largest, General Electric (GE), sold its mainframe computer business in 1970 to Honeywell, but retained and expanded GE Information Services, or GEIS, which for two decades focused on time-sharing services.

Time sharing was a disruptive technology that sparked many debates in trade and academic publications in the second half of the 1960s regarding the relative efficiencies of batch processing and time sharing. Ultimately, underlying goals for applications and how computers were employed made the most difference as to whether users succeeded at optimizing processing resources. As was mentioned briefly in chapter 3, in the mid 1960s the computer industry consultant Herbert Grosch made an observation famously labeled as Grosch's Law—that stated the costs of computer systems increase at a rate equivalent to the square root of their power. This statement on economies of scale—the most powerful computers being more efficient—certainly highlighted the importance of supercomputing and utilization. It also lent some weight to the time-sharing model and computer utility idea. At times, computer use was purely a question of efficiency, but there were also qualitative differences in expanding experiences as true "users." And while batch was efficient for processing data, time sharing offered programmers greater efficiency in perfecting and debugging software. Time sharing of the 1960s facilitated a level of user interaction with machines not previously possible (as computer operators tended to run machines in batch environments)—and not equaled or exceeded before the era of the personal computer.

Time sharing was adopted by most of the largest service bureaus during the middle to late 1960s. This included IBM's subsidiary Service Bureau Corporation, Control Data Corporation's Data Services Division/Professional Services Division, and programming services specialist C-E-I-R (which CDC acquired in 1967). These enterprises, which appealed particularly to scientific and engineering computing users, typically would offer time sharing along with traditional batch processing, and networking, remote batch processing (batch with input-output delivery through telephone line-based networks rather than mail or courier). Over time, the time-sharing industry redefined and to a significant degree subsumed the service bureau industry. Non time-sharing service bureaus continued to exist in the 1970s and in subsequent years, but tended to be small stand-alone centers. And a meaningful portion of "time-sharing" firms' revenue generally was derived from services that did not necessarily utilize time-sharing applications.

IBM offered limited forms of time sharing in the mid 1960s, but these efforts had minimal impact. The company only entered strongly into the field with its "CALL 360" service in 1968, which it quickly transferred to subsidiary SBC because of scrutiny by the US Department of Justice.[8] Time sharing had both offensive and defensive strategic elements for IBM. A thriving time-sharing industry could reduce the demand for customer computer installations—some firms opting for purchasing computer time on their premises rather than installing a purchased or leased computer, or using time sharing for periods of peak demand to avoid obtaining additional computers. Restraining the development of an independent time-sharing industry was in IBM's interest. At the same time, time sharing was an opportunity to tie firms not yet able to buy or lease IBM equipment to IBM's family of compatible computers. IBM's lack of foresight and planning with time sharing for its planned System/360 was a major factor why Fernando Corbató and others at MIT (who had used an IBM 7094 on CTSS) rejected IBM in favor of GE for hardware on Project MAC's Multics. IBM's Time-Sharing System for System/360 Model 67 was not released until October 1967, shortly before CALL 360 was introduced as a time-shared service for the IBM System/360 (models 40 and 50).[9]

As historian Martin Campbell-Kelly and economist Daniel Garcia-Swartz argue in an important article on the economic history of the time-sharing industry, this segment has been marginalized by scholars and was highly significant to the overall computer industry.[10] This chapter seeks to complement their work with a business history-focused case study of Tymshare,

Inc., a remarkable firm launched in 1965 very shortly after industry leader GE essentially kicked off the time-sharing industry (GEIS).[11] It briefly discusses the academic (and nonprofit research corporation) origins of early time sharing as well as GEIS. It then focuses attention on Tymshare before concluding with highlights on a broader industry context.

Computer Time Sharing's Academic Origins

The concept of time sharing emerged in the late 1950s as a new and redefined instance of certain elements that had occurred years earlier—using a computer at a distance (a lone individual using a George Stibitz early 1940s relay computer more than 100 miles away) and multiple simultaneous system users using a mainframe for a single purpose (terminals with cathode ray monitors networked to SAGE computers in the late 1950s). Time sharing differed in that it involved multiple users for different applications simultaneously sharing a computer. Conceptually, two early articulations of time sharing stand out. In 1959 National Research Development Corporation's (Great Britain) Christopher Strachey presented a paper (at a UNESCO conference) on debugging a computer while it was used by another individual in normal application mode. Independently, earlier that year, MIT's John McCarthy circulated an influential MIT internal memorandum with multiple debuggers simultaneously utilizing a modified IBM 709 system.[12]

Many important figures contributed to MIT's pioneering work with time sharing. Herbert Teager and subsequently John McCarthy led MIT computer planning committees that first discussed time sharing and launched an early time-sharing project. Robert Fano, the first leader of Project MAC, was a major force with MIT's time-sharing research and development.[13] MIT Computation Center scientist Fernando Corbató, however, stands out in becoming the principal leader of both CTSS and Multics, the school's influential time-sharing systems. The latter effort began as a partnership between MIT, Bell Laboratories, and the electronics conglomerate and computer manufacturer General Electric.[14]

CTSS had no true rival in late 1961 when it was first successfully tested (at least in the open community outside the confines of the National Security Agency and the classified world). Bolt Beranek and Newman, Inc. had a complete DEC PDP-1–based time-sharing system (that both John McCarthy and J. C. R. Licklider helped develop) in September 1962.[15] And System Development Corporation (SDC) developed an important time-sharing system for AN/FSQ-32 (Q-32) computers operational in the middle of 1963, which received an enhancement contract from ARPA IPTO Director J. C. R. Licklider. Licklider and his successors made time sharing a major focus of IPTO funding in the early to mid 1960s and hoped Project MAC could influence developments on the East Coast and SDC and the Q-32 time-sharing system could do likewise for the West Coast.[16] As part of this effort, Licklider also funded Project Genie at the University of California at Berkeley to develop a time-sharing system for a Scientific Data Systems (SDS) 930.

In 1964, as IPTO funded Project Genie, the National Science Foundation funded John Kemeny and Thomas Kurtz of Dartmouth College to further an effort they had conceptualized a couple years earlier: the Dartmouth Time-Sharing System (DTSS). The Dartmouth System, operable by mid 1964, initially used a donated GE 225 and GE Datanet-30 message switching machines—the latter, the corporation had developed for Chrysler and could accommodate 40 simultaneous users.[17] DTSS, which by 1966

Figure 7.1
A Q-32 time-sharing installation in the Santa Monica headquarters of System Development Corporation. Courtesy of Charles Babbage Institute, University of Minnesota.

accommodated about two hundred simultaneous users, also benefited from the corresponding programming language that Kemeny and Kurtz created for the system, Beginners All-purpose Symbolic Instruction Code (BASIC). Among the computer manufacturers, GE, SDS, and DEC played early leadership roles with time-sharing hardware, GE standing out in becoming the first major time-sharing services provider.

In the second half of 1965, working closely with Kemeny, GE designed a parallel system to DTSS. GE Corporate Headquarters in New York City and selected GE computer customers were trial users late that year. General Electric's Information Processing Center (IPC) was the services side of the GE Computer Department, and initially focused on augmenting hardware contracts—it was losing money as a batch processing services business. IPC's new manager in 1965, Warner Sinback, however, succeeded in launching a time-sharing enterprise during his first year at the helm, and importantly, gained approval to separate it from GE's Computer Department (significant for Simback to have free reign as a manger and also because GE several years later would sell its Computer Department to Honeywell). In January 1966, with the formal split with the Computer Department, the business unit was renamed GE Information Services (GEIS) and moved to Bethesda, Maryland. Early customers included Hughes Aircraft, Chrysler, Boeing Aircraft, Bechtel Corporation, and AT&T.[18]

As Warner Sinback recalled, IPC/GEIS ran an inexpensive full-page advertisement in a small-circulation accounting trade publication with the heading "Would you like access to a $1 million Computer?" According to Sinback, IBM executives were "so perturbed that they ran [an ad] in the *Wall Street Journal* within a month that said 'Would you like access to a $5 million dollar computer?'"[19] Just as Ross Perot's facilities management model threatened IBM's relationship with customers and offered more efficient usage of expensive processing time, so too did the new time-sharing business. As United Data Centers founder and future Tymshare executive Bernard Goldstein put it, time-sharing companies were "offering raw machine time for firms to avoid installing their own computers."[20]

In the second half of the 1960s, GEIS aggressively entered most major metropolitan areas of the US. By late 1968, GEIS had secured 40 percent of the $70 million time-sharing market. It developed a nationwide network with a few regional hubs to effectively utilize computer processing resources.[21] Although GEIS stands out, oral histories, archival documents, and other resources provide a unique opportunity for a richer analysis of a highly innovative Silicon Valley startup—Tymshare, Inc.—that created an equally influential network (TYMNET), grew rapidly, and became the longtime and quite formidable number-two firm in an industry that evolved into an oligopoly.

Tymshare's Origins

Tymshare originated at the hands of Thomas O'Rourke and David Schmidt, who left General Electric to launch the startup time-sharing company in 1965. O'Rourke, the more senior of the two, had an electrical engineering degree from University of Washington and had been a district manager in Los Angeles for GE's Computer Department. He rose to become a regional manager for the entire West Coast (working out of the Phoenix Computer Department headquarters and later Sunnyvale, California) before losing the position to the son-in-law of one of GE's vice presidents. As a concession, O'Rourke was offered a position running GE's military electronics group in Washington. In May 1965, having moved his family many times during his four years in the Air Force and his thirteen years with GE, and wanting to stay on the West Coast (and near Sunnyvale), O'Rourke left GE. He recruited David Schmidt, a top programmer, to join him in forming Tymshare Associates.[22]

O'Rourke had been an advocate of donating computers to Dartmouth eighteen months earlier and had carefully followed the DTSS project. He believed there was a viable business opportunity in time sharing. Both O'Rourke and Schmidt withdrew their retirement holdings and put up their life savings—$15,000 and $10,000 respectively. O'Rourke, having worked on ERMA in Los Angeles in the late 1950s, contacted Bank of America's George Quist (later a founder of influential venture capital firm Hambrecht and Quist, subsequent underwriters for Apple Computer, Netscape, and Amazon.com). Bank of America and Quist provided half a million dollars in venture capital for a 50 percent ownership stake.[23]

O'Rourke and Schmidt ordered a GE 225, which was to be delivered in the fall of 1965, but the order was canceled and the down payment returned when GE informed them that their credit was insufficient. O'Rourke believed that the true reason for the cancellation was that GE had decided to enter the time-sharing business and didn't want to sell a machine to a future competitor.[24] Warner Sinback of GE offered O'Rourke a senior post in helping to run GE's time-sharing business, but O'Rourke had already launched Tymshare, partnered with Schmidt, and taken venture funding; there was no turning back.[25]

In January 1966, with the new name Tymshare, Inc., the company was formally incorporated in California, setting up shop in dilapidated rented office space on Distel Drive in Los Altos. The lone bathroom did not have a roof.[26] O'Rourke was president and chairman of the board; Schmidt was executive vice president and became the general manager for technology.

O'Rourke asked the distinguished venture capitalist Arthur Rock to provide additional financing to secure a Scientific Data Systems (SDS) computer. Though Rock did not invest directly, he helped get an initially reluctant Max Pavlesky, SDS's president, to offer a generous lease for a new model SDS 940 (then under development); it was to be delivered in the second half of 1966.[27] The SDS 940 was derivative of pioneering work done by Mel Pirtle, Butler Lampson, Peter Deutsches, and others at the UC Berkeley to modify an SDS 930.[28] As part of the agreement with SDS, Tymshare promised to help line up four other SDS 940 sales. O'Rourke delivered on that promise by arranging for the acquisition of one SDS 940 each by two other time-sharing startups (Comshare of Ann Arbor, Michigan, and Dial-Data of Newton, Massachusetts) and one each by SRI and Shell Oil.[29] Initially, owing to expensive long-distance telecommunications/data charges, time sharing was thought of as a local business—an office serving a metropolitan area. Thus, at the start Tymshare and Comshare were not direct competitors, and they partnered on early efforts in systems programming.

Table 7.1
Established firms and other enterprises in the time-sharing industry, 1963–1966.

		Original headquarters	Original system
1963	Bolt Beranek and Newman	Cambridge, MA	DEC PDP-1
1965	IBM (later SBC).	White Plains, NY	IBM 1460
	General Electric (later GEIS)	Phoenix, AZ	GE 225
	Tymshare, Inc.	Los Altos, CA	SDS 940
	Keydata Corp.	Watertown, MA	Univac 491
1966	Call-A-Computer	Raleigh, NC	GE 225
	Comshare, Inc.	Ann Arbor, MI	SDS 940
	Dial-Data	Newton, MA	SDS 940
	VIP Systems	Washington, DC	IBM 1460
	Allen-Babcock Computing	Los Angeles, CA	IBM/360–50
	Applied Logic Corporation	Princeton, NJ	DEC PDP-6

Source: Auerbach Corporation, *A Jointly Sponsored Study of Commercial Time-Sharing Services*, volumes I and II, *Description of Companies Interviewed, Final Report*, 1968 (Charles Babbage Institute).

Before Tymshare got its SDS 940, both Tymshare and Comshare accessed the University of California's SDS 930 between 2 and 6 a.m.—the only time the machine was available to them. Comshare's leader, Robert Guise, sent Rick Crandall and his wife, both Comshare programmers, to team up with the Tymshare programmers Ann Hardy and Verne Van Vlear, the latter a former systems analyst from GE.[30] Along with O'Rourke, Schmidt, Neil Sullivan (a marketing specialist), and a secretary, this was Tymshare's entire workforce for months. As they designed and built a system with commercial applications in mind, the small two-company software development team benefited from the time-sharing operating system designed and programmed by two University of California doctoral students, Peter Deutsches and Butler Lampson: the Project Genie Berkeley Time-Sharing System.[31]

Ann Hardy had senior technical and managerial responsibilities virtually from the start at Tymshare. She later became the first female vice president of the company, a position she held for roughly a decade, and was one of the earliest female vice presidents of a major IT firm. Hardy was a Chicago native who had moved to New York to take education courses at Columbia University after completing her bachelor's degree at Pomona College. On the advice of a friend, she took IBM's Programmer Aptitude Test in 1956 and was hired as a programmer to work on the STRETCH supercomputer development project in Poughkeepsie. Lawrence Radiation Laboratory (in Livermore, California), after acquiring a STRETCH in 1961, soon hired Hardy. After seeing a Tymshare advertisement in *Datamation* at the start of 1966, she phoned the office of Tymshare, Inc. to inquire into the possibility of leasing a terminal for the Hardys' house (her husband, Norman, also worked at the lab) to enable them to work remotely. Tymshare's leaders had never thought of a home connection, and the conversation quickly took a different path.[32] Ann Hardy soon visited Tymshare, and was hired in February 1966. She had the task of programming (along with Verne Van Vlear) a time-sharing operating system with a "Monitor" (an effort managed by Hardy) and an "Exec" (managed by Van Vlear).[33] This work (aided by the Crandalls from Comshare) was done first on UC Berkeley's modified SDS 930, and later in the year on a prototype at SDS in Santa Monica (to enable time-sharing operability for the forthcoming SDS 940).[34]

When the SDS 940 was delivered in fall of 1966, Tymshare moved to larger space, on East Meadow Drive in Palo Alto, that accommodated the mainframe. The firm's rapidly growing staff soon included Norman Hardy (who joined in 1967) and several influential sales managers.[35]

At the start, O'Rourke concentrated on marketing to engineering firms on the peninsula and Bay area. This limited geographical focus had the advantage of avoiding or lessening long-distance charges. Tymshare—like GE—initially provided free time on an experimental basis, but switched customers to pay service by the last months of 1966. In doing so, O'Rourke could then show his investors that checks were being sent and that time sharing, in fact, was a viable business.[36]

GE had been aggressive in making Berkeley their second time-sharing location (after Phoenix) in order to compete directly with Tymshare. In concentrating on engineering applications, Tymshare was able to secure a small group of customers. Owing to Tymshare's initial focus on engineering, Los Angeles aerospace companies, such as Lockheed, became a major base for the company. GE's discounting or eliminating phone charges to Southern California firms pushed Tymshare to quickly open their second office in the Los Angeles suburb of Inglewood in early 1967.[37] Later that year they expanded to the East Coast with an office in Englewood Cliffs, New Jersey—initially served via phone lines, but by early 1968 possessing a SDS 940.[38]

In 1967 Tymshare was the first vendor for a time-sharing operating system for the SDS 940 that could pass stringent "acceptance tests" of Harvard University—they did this in a matter of several days after numerous others over the past five months had failed. As Ann Hardy remembers, the SDS salesman thought he could "never make the sale" to Harvard until the Tymshare team arrived and aced the tests.[39] Before the end of 1967, Tymshare was also aggressively targeting business data processing applications.[40]

In its early years. Tymshare benefited from having as sales managers John Jerrehian, Ray Wakeman and Ron Braniff, who each came to manage one of its first three sales regions, Northern California, Southern California, and the Northeast. O'Rourke worked closely with the sale managers.[41] Braniff, like others, emphasized O'Rourke's personable nature, straightforward style, and high expectations for his employees. Braniff described O'Rourke as tough, ethical, informal, and fun to be around. "[T]he culture of our company," he said, "grew up around his personality."[42] This continued even as the company expanded, the combined staff growing to more than 100 for Tymshare's three locations (with a total of five SDS 940s) during 1968.[43]

While O'Rourke focused on sales, Schmidt concentrated on overseeing the technology. In 1969, Schmidt became enamored with what he believed to be a great opportunity for Tymshare: building time-sharing computers (and thus competing with SDS, GE, and DEC). In 1969, unable to get the board's support for adding that business, he left Tymshare to launch Multi-Access Systems Corporation (Mascor) to fulfill his vision. Schmidt's timing could not have been worse. The 1970 recession hit IT particularly hard. The $20 million Schmidt was promised was pulled away by financiers, and his vision was not realized, though he later had a successful career with several IT businesses.[44]

Tymshare's fortunes played out far differently during the recession of 1970. Tymshare reached the $10 million plateau, having grown by 60 percent since the previous year and achieving its first of many continuous years of positive net earnings.[45] With strong results in the first half of 1970 the company had a track record to complete a highly successful initial public offering of stock in September of that year that quickly sold out then rose to a premium above the $6/share issue price. Tymshare IPO raised $3.6 million, which it used to acquire additional equipment and facilities, as well as to retire debt.[46]

Tymshare's Tymnet, Acquisitions, and Applications Software

In Tymshare's first several years, expanding to target new areas—Los Angeles, Englewood Cliffs, Dallas, or Washington (the latter two added in 1968)—involved securing offices and computing equipment in a new city. One way to speed this process was to acquire existing time-sharing companies—of which Tymshare initially looked to other enterprises that were based on the SDS 940. In 1968, Tymshare acquired Dial-Data—and, with it, offices in Newton, Massachusetts and five SDS 940 computers. This also brought new managerial talent in Dial-Data president Lou Clapp, who became a Tymshare executive vice president with Schmidt's departure.[47]

That same year, Tymshare hired LaRoy Tymes from Lawrence Radiation Laboratory. He was recruited by his friends Ann Hardy and Norman Hardy.[48] Tymes possessed not only a fitting last name for working at the company, but also considerable programming and network design acumen. The pioneering work of Tymes and Norman Hardy in developing a network, TYMNET, freed Tymshare of the yoke of largely limiting geographical markets to metropolitan areas in which Tymshare had offices and mainframes.

Norman Hardy, a UC-Berkeley-trained mathematician, joined Lawrence Radiation Laboratory in the mid 1950s. There he found a "beautiful combination" of mathematics, physics, and computing. He joined IBM for a time to work on the development of STRETCH, but rejoined Lawrence Radiation Laboratory once STRETCH was delivered. In the mid 1960s, he had also had an opportunity to visit MIT and interact with the Project MAC

Multics team, after which he convinced Sidney Fernback, the director of Lawrence Radiation Laboratory's Computation Department, to fund a project to design a time-sharing system for the newly acquired CDC 6600—a project Hardy then worked on with Tymes. After a brief return to IBM to work on the Advanced Computer System, Hardy joined his wife at Tymshare in 1967. What drew Norman Hardy to Tymshare was the exciting new technology of time sharing. He saw O'Rourke as a "businessman," and had "never ... had a lot of respect for businessmen," but he believed Schmidt to be "computer savvy ... with a fundamental talent for knowing what was [technologically] feasible."[49]

Hardy and Tymes conceptualized and created TYMNET. One key to making it work, as Norman Hardy reflected, was the idea of statistical multiplexing. The first time-division multiplexing was not optimized for efficiency; it did not take advantage of the fact that "most teletypes were idle most of the time." Statistical multiplexing allowed there to be variable amounts of buffering "in the machines, and in the nodes."[50]

In 1968 Hardy and Tymes, through many conversations, convinced O'Rourke to back the networking project that became TYMNET. Tymes, who had both a bachelor's degree and a master's degree in mathematics from California State College at Hayward, firmly believed that the existing business model of opening new offices to target metropolitan areas was not economically viable—it was inefficient and limited the potential customer base. Ultimately this insight proved true for Tymshare, GE, and other companies that built networks, as well as for faltering competitors that didn't build networks.[51]

Although Tymes had a vision for the full network, he sold the idea to O'Rourke in pieces, solving intermediate problems. He began with the terminals; the existing SDS 940 computers had relatively expensive Customer Terminal Equipment (CTE). Tymes had the idea of a minicomputer at a fraction of the cost of CTE, using one to be programmable to accommodate different terminal types and baud rates (information transfer in communication channel). Thus the hardware could be far cheaper and utilization optimized. They initially used inexpensive General Automation, Inc.'s SPC-12 minicomputers—a "12-bit, very anemic, special purpose processor," as Tymes recalled. After quickly demonstrating the concept, they graduated to more powerful 16-bit Varian 620i minicomputers serving as both the base (to interface the network to the host, or time-shared computer) and remote (to drive the terminal). The nodes of the TYMNET, as it became fully operational in 1970, were interconnected by full duplex synchronous leased private lines at 2,400 and 4,800 bits per second.[52]

Figure 7.2
The Cupertino time-sharing facility of Tymshare Inc., with Varian 620/i minicomputers used as "remotes" to enable TYMNET system (circa early 1970s). Courtesy of Charles Babbage Institute, University of Minnesota.

The most critical element of TYMNET was the "supervisor," which built circuits, performed diagnostics, kept statistics, as well as other functions. Multiple supervisors were run for redundancy with one in active mode. Tymes programmed the supervisor in late 1969. Every node on TYMNET contained a "leprechaun," which issued high-priority diagnostics and passed log-in information to the supervisor in active mode. A master user directory (MUD) was programmed by Van Vlear. This contained all passwords and access control information. It was one-way encrypted, offering significant security.[53]

TYMNET facilitated the delivery of computer time sharing far more broadly and economically. By 1972 it connected more than 40 cities in the United States, a number that continued to grow rapidly domestically and internationally. TYMNET was developed contemporaneously to the ARPANET, but as Tymes later reflected, different in important respects. First, TYMNET was privately funded and had to be "profitable right from the very beginning." Second, the ARPANET "originally did not have terminal interfaces, it was entirely computer to computer." Third, ARPANET could not have a centrally controlled point like TYMNET, it was distributed. It needed to be distributed because, as Tymes articulated, a central control point for ARPANET was "politically not viable." Fourth, Tymes emphasized, the ARPANET, like today's Internet, is vulnerable to "logjams, thrashings." "Nobody," he claimed, "ever successfully hacked into TYMNET." Finally, Tymes added, TYMNET had no use for TCP, it had "flow control, node to node, on every section of every circuit, so it was not possible to flood a particular node with excess data. ... There is no such mechanism in the internet." While TYMNET never getting hacked is probably an exaggeration, it may have been true through Tymes' involvement with it from the late 1960s to 1981. Regardless, TYMNET's creators, Tymes, Norman Hardy, and others, considered security concerns far more seriously than protocol groups for the more interoperability-focused Internet.[54]

TYMNET opened up entirely new possibilities to serve customers effectively in new geographies as well as to build new applications and provide networking infrastructure for companies. In the early 1970s the National Institutes of Health approached Tymshare about providing network access to a large database they had created on poison antidotes. This turned into a major application, operable in 1972.[55] Later in the decade, they sold underlying software and services to assist TRW with establishing a large internal computer network—TRWNET. In 1976, joint-ownership with Taylorix-Tymshare in Frankfurt and Marubeni in Tokyo, furthered TYMNET's geographic reach to include 90 percent of Western Europe and much of Japan.[56]

As TYMNET grew rapidly in the first half of the 1970s, the Federal Communications Commission suspected that Tymshare was evolving to become a "communications company."[57] This led Tymshare to proactively spin off TYMNET as Tymnet, Inc. a wholly owned subsidiary in 1976. That year Tymnet, Inc. applied to become a "common carrier," which the FCC granted. With this status, Tymshare's TYMNET had no restriction on being a "message communication" service, which was previously limited under its "shared use" designation.[58] This was the overwhelming upside, the limitation (which proved not to be onerous) was, as a common carrier, Tymnet, Inc. was subject to regulation on rates.[59] Throughout the 1970s, TYMNET, along with software applications and various acquisitions, propelled the firm's rapid growth.

With the development of TYMNET under way in the late 1960s, Tymshare secured a location on Budd Road in Cupertino, California that soon became the computer center to house the SDS 940 computers. The company rapidly expanded at this location and the emergent Cupertino campus became Tymshare's corporate headquarters at the start of the 1970s.

In its first years, Tymshare was engaged in software development applications internally, which it complemented with software packages and

Table 7.2
Tymshare's revenue and net income (in thousands).

	Revenue	Net income
1972	$27,175	$1,617
1973	$39,513	$2,973
1974	$52,682	$3,638
1975	$64,412	$5,064
1976	$81,837	$6,713
1977	$101,174	$8,008
1978	$149,559	$10,594
1979	$193,042	$14,644
1980	$235,854	$18,743
1981	$289,687	$15,670
1982	$297,025	$8,809

Source: Tymshare, Inc. annual reports, 1976–1982 (University Libraries, University of Minnesota).

programming capabilities acquired through acquisitions. Initially (and internally) it developed engineering application packages. By the end of 1968 it had three time-shared enabled applications packages completed: COGO, CAPT, and Conversational ECAP. COGO was a coordinate geometry system to solve civil engineering problems such as highway design, right-of-way survey, and bridge geometry. CAPT was a conversational language that enabled programming and operating numerical control in manufacturing. Conversational ECAP was for electronic circuit analysis. Like business data processing applications later, these early systems helped differentiate Tymshare's time-sharing services to customers.[60]

Tymshare used acquisitions to expand rapidly during the 1970s and the early 1980s. By 1968, more than forty companies had entered the time-sharing industry.[61] A few quickly faltered and disbanded, but most new entrants survived until the nearly year-long recession of 1970, which led to an initial shakeout in the industry. Much of 1969 saw unparalleled exuberance in software, computing, and time-sharing companies and their stocks. There were more new time-sharing companies formed that year (26) than any other in the industry's history. The number of new firms dropped to 16 in 1970, 5 in 1973, and only several others were launched between 1974 and 1978.[62] Tymshare was well established by 1970 and faired far better than most time-sharing companies. In fact, it expanded through joint-ownership agreements into both Canada and France that year—taking minority ownership in Tymshare, Canada and CEGOS-Informatique and Credit Lyonnaise.[63]

Tymshare's acquisitions accelerated rapidly throughout the remainder of the decade. Some acquisitions were to reach new geographies (even with TYMNET, geographical coverage was important to both costs and serving customers). Of equal and sometimes greater importance, acquisitions allowed the firm to gain software packages targeted at specific industries, as well as to secure talented applications programmers. Some acquisitions achieved both geographic expansion and software/industry capability goals—especially the 1974 acquisition of United Data Centers. It brought fifteen new offices, primarily in the eastern United States and one in Montreal, as well as important software applications.[64] Like most Tymshare acquisitions, it was an all-stock deal. It represented its largest acquisition up to that point, with a value of approximately $6.1 million.[65]

United Data Centers' founder and leader, Bernard Goldstein, who had previously worked at Control Data, launched UDC to become a chain of batch-focused data centers. He had grown it through acquisitions with an eye toward business applications—including UDC's acquisition of

Dynafacts (in Wichita, Kansas), which had an automated tax return software package. Goldstein later recalled that Tymshare wanted UDC for its business software applications, its capabilities to offer "solutions for the marketplaces."[66] After the acquisition of UDC, Goldstein became Tymshare's mergers and acquisitions specialist for several years.

In 1976, Tymshare reorganized into three marketing subdivisions: Information Services, Industry Services, and Marketing Services. This formalized strategic planning and organization in targeting particular industries for customers. O'Rourke emphasized in the 1976 annual report that this would be achieved both through internal resources and through acquisitions to gain certain industry specific capabilities.[67] Three targeted industries for Tymshare acquisitions (by O'Rourke and Goldstein) in the mid to late 1970s were tax preparation, travel, and medical data processing. This strategy was implemented through acquisitions of Unitax (1976), Autotax (1978), Western Twenty-Nine (1976), Medical Data Systems (1976), and Medical Information, Inc. (1978).[68]

In the early 1980s, bank card processing was targeted by Tymshare through acquisitions of Bancard of Rhode Island and TeleCheck (both in 1980). Unlike Dial-Data, new acquisitions also furthered the expansion of Tymshare to a wider set of hardware systems including models from minicomputer specialist Digital Equipment Corporation and IBM. Tymshare's 1975 acquisition of Alan-Babcock Computing provided IBM mainframes. Tymshare's organizational customers also developed applications software that ran on the network. Some businesses made deals with Tymshare to serve as a selling agent for their applications. Norman Hardy emphasized the flexibility of Tymshare in its early years and O'Rourke's willingness to consider all types of arrangements that had revenue and profit potential.[69]

Tymshare's varied acquisitions brought new employees from different corporate cultures, which sometimes proved challenging. This also was true with its international joint ownership arrangements. With the latter, in the mid 1970s, Tymes and Norman Hardy were meeting with CEGOS, Tymshare's jointly owned company in France. According to Tymes' recollection of the corporate planning meeting, "The French people were aghast at our proposal that the names of all the French customers had to be in the MUD that was to be stored in the United States ... . And Norman Hardy said, 'why not?' A supervisor is a supervisor; its geographic location really isn't material. If it's controlling the net it does have to have all the information of all the users to function." Reflecting on the situation, Tymes said "[T]o us it was a pure technology problem, but to the French it was a matter of pride."[70]

Acquisitions and subsequent integration could be even more difficult. Many of Tymshare's 1,500 employees by the end of 1976 had joined the company after their firm or organization was acquired by Tymshare. Lynn Sanden, who joined as a secretary in 1969, worked her way up to the position of human resources manager. She recalled the difficulties after Tymshare, in 1977, acquired a center of about 20 research scientists at SRI, Douglas Engelbart's famed oN-Line System Center (NLS), which he founded (as the Augmentation Research Center) in the mid 1960s. For Tymshare executives, the SRI-NLS acquisition was completed to advance efforts in office automation and to develop software for the office of the future.[71] Though Engelbart adjusted well to the new environment and became Tymshare's chief scientist, many of the former SRI-NLS computer scientists did not want to be acquired, and that led to "hostile" attitudes.[72] A number of the former SRI-NLS computer scientists resigned and joined Xerox PARC or other businesses or organizations.

In other cases disagreements materialized with regard to responsibilities, job titles, compensation, or deployment of technology. In acquiring Western Twenty-Nine, the agreement was for the newly organized unit, and its head Pat Brent (former leader of Western Twenty-Nine), to receive a portion of compensation in profit sharing. However, Brent's unit was administratively set up in a way that it was exceedingly difficult to show a profit (internal or transfer pricing on what the unit was charged for TYMNET was high).[73] And with the 1978 acquisition of Medical Information, Inc., to bring time sharing and applications software to hospitals and nursing stations, what Tymshare could actually offer was never carefully considered by O'Rourke. Ann Hardy had written a memo opposing the acquisition on this basis, but Tymshare nevertheless went through with the acquisition.[74] It was a bust, and Tymshare completely divested of medical information processing several years later in 1982.

The advent of personal computers in the mid 1970s and the acceleration of the personal computer industry in the second half of the 1970s did not go unnoticed by Tymshare's leaders. Tymshare's subsidiary Tymnet was not only providing the network infrastructure for Tymshare, but selling networking service to customers that were not purchasing Tymshare's time sharing. In the early 1980s, O'Rourke estimated, half the business was coming from the TYMNET communication services-only side, and of the two that was the growth business. O'Rourke had pivoted strategically to move more of the business "to communications," to the

Table 7.3

Tymshare's acquisitions and mergers, 1968–1982.

	Acquired or merged division or company
1968	Dial-Data
1970	CEGOS-Informatique and Credit Lyonnaise, France (partial ownership)
	Tymshare, Canada (partial ownership)
1971	Computer Systems Division, Graphic Control Corporation
1972	Computer Complex TS Ops
1973	Valley Computer
1974	United Data Centers
	Tronics, Inc. (merger)
1975	Alan Babcock, Inc.
	Quelex Data Systems
	Leasco Response, Inc.
1976	Unitax, Inc.
	Simplified Data Processing (merger)
	Taylorix-Tymshare, Germany (partial ownership)
	Marubeni Corporation, Japan (partial ownership)
1977	Western Twenty Nine, Inc.
1978	Auto Tax
	Medical Information, Inc.
1979	Valdata Division, TRW
1980	Medistat
	Bancard of Rhode Island
	TeleCheck
1981	Electronic Travel Services, Inc.
	Payment and Transfer Services, Inc.
	Microband Corporation of America
1982	FTC Communications

Sources: Corporate History Collection, Computer History Museum; http://corphist.computerhistory.org/corphist/view.php?s=themes&id=5 (accessed July 24, 2015).

Figure 7.3
An operator at a Tymshare 430 monitor station, circa 1981. Courtesy of Charles Babbage Institute, University of Minnesota.

Tymnet subsidiary. O'Rourke later commented that outsiders missed what Tymshare was doing in the early 1980s: "They felt we were riding a dead horse with time sharing, we were frantically trying to change horses but not doing it abruptly because these guys [the time-sharing business] were cash cows ... ."[75]

Tymshare had created a valuable network in TYMNET, but Tymshare had declining net income in both 1981 and 1982. In 1983, when TYMNET linked 400 cities in 42 countries, Tymshare became an acquisition target of McDonnell Douglas' Automation (i.e., information technology) division (commonly known as "McAuto"). McDonnell Douglas' initial offer in 1983— $378 million—was withdrawn late in the year amid third-quarter reports showing Tymshare had lost $139,000 in the first nine months of the year. Early in 1984, McDonnell Douglas came in with an offer roughly 20 percent lower—$307.5 million—and it was accepted by Tymshare's board.[76] McDonnell Douglas broke the non-network portion of Tymshare into many divisions, some of which were sold; it kept TYMNET for five years before selling it to British Telecom in 1989.[77]

The Evolution and Fate of the Broader Time-Sharing Industry

GE Information Services (GEIS) took the early lead in the industry and never looked back. Like Tymshare, it developed a major computer network in the late 1960s.[78] GEIS's Network I had remote hubs in Los Angeles, Kansas City, Atlanta, Teaneck, Schenectady, and Washington that linked its suburban Cleveland (Brook Park) central computer complex to buffer units (multiplexers) in more than forty major metropolitan areas of the US by late 1969. Similar to Tymshare, GEIS developed specialized software applications targeting particular industries—in part through acquisitions in the banking and energy data processing fields.[79] In the mid 1980s it became a major supplier of Electronic Data Interchange (EDI), offering business-to-business networked communication systems that linked inter-firm supply chains IT systems in the motor vehicle, aerospace, and other industries. By 2000, the former GEIS, having been renamed Global Exchange Services (GXS) a few years earlier, had revenue of about $640 million and was focused on supply management EDI operations for about 100,000 business units of roughly 30,000 multinational firms. In 2002, GXS sold a 90 percent stake to the venture firm Francisco Partners for $800 million.[80]

IBM had officially entered time sharing in mid 1960s with QUIKTRAN and DATATEXT, but those were limited specialized offerings for "interpretive program execution" and "text editing" (respectively) for certain pre-IBM System/360 machines, including the IBM 1460.[81] IBM participated in the field in a substantial way in 1968 with "CALL 360" services that in a matter of months served 34 US cities to sell time on IBM System/360 Model 40 and Model 50 mainframes located in Los Angeles, San Francisco, Chicago, Cleveland, Philadelphia, and New York. Virtually from the start of CALL 360, this effort caught the attention of the US Department of Justice, which, under the 1956 consent decree, had prohibited IBM from service bureau operations in forcing it to spin off its Service Bureau Division as Service Bureau Corporation (SBC), a wholly owned subsidiary that had a separate workforce and got no breaks on purchases of IBM computer systems.[82] Early speculation in the computer trade press in 1966 about whether time sharing was another form of leasing (allowed with restrictions under the 1956 consent decree) or a new form of service bureau (not allowed) appeared to tilt toward the latter (at least in the US Department of Justice's interpretation) by late 1968, since many service bureaus were moving to add time sharing.[83]

While many historians have emphasized software products with IBM's unbundling, Department of Justice officials were also heavily focused on

IBM's impact on the fast-growing computer services industry and its relationship to the hardware business, including time sharing (potential anticompetitive advantages either in hardware or services by the computer leader). In early December 1968 an IBM spokesperson stated, that for several months the corporation had been "reexamining its methods of doing business in the US to determine what support services should be separately offered and priced."[84]

Figure 7.4
Lead operator Bob Warno (left) and systems programmer Carlos Navarro (right) in the Chicago office of Comshare, Inc., 1979. Courtesy of Charles Babbage Institute, University of Minnesota.

More than a month earlier, on October 23, 1968, IBM's leaders, in fact, had already made a major decision on services by shifting "commercial time-sharing" services away from IBM's Data Processing Division to its subsidiary SBC; by then SBC was a slower-growing, low-margin operation.[85] In 1972, SBC had profits of only $1.5 million. (SBC's margins in those years tended be about 2 percent.) In 1973, at the time of the transfer of SBC to Control Data Corporation as part of the settlement of its lawsuit against IBM, two thirds of SBC's business was still from batch processing.[86] Thus, in 1973, SBC probably had less revenue and smaller profits from time sharing than Tymshare, and far less than GEIS. Adding SBC to CDC's time-sharing business probably propelled it beyond Tymshare in revenue to second place in the industry, far distant from GEIS (though Tymshare probably regained the number-two position in the industry later in the decade as CDC's time-sharing services business grew more slowly).

Conclusion

By the late 1970s, time sharing, an industry that had begun in the mid to late 1960s with a flurry of startups and divisions of a few computer manufacturers, had become an oligopoly dominated by GEIS, Tymshare, and CDC. By the early 1980s, the impact of inexpensive personal computers sent the industry into decline and the network infrastructure of these firms was far more valuable than the waning time-sharing business. In 1982, industry revenue peaked to $1.75 billion, dropping $100 million the following year and more rapidly thereafter.[87] Time sharing had offered a new interactive model that greatly expanded the number of direct users of computers, allowed smaller businesses and organizations computer access, and offered efficiencies so that organizations with computers could purchase added time as needed rather than acquire additional machines. Time-sharing companies and facilities had changed the service bureau model—by the late 1970s revenue from time sharing was nearly double that of batch processing.[88] While the notion of a computer utility had played out to a degree for many organizational users in work settings, exceedingly few individuals used time sharing in their home. Arguably the information or computer utility vision would reappear more broadly in the late 1990s and the 2000s—in the form of the Internet, the World Wide Web, and cloud computing—but this has fallen short of fully realizing idyllic utility rhetoric and goals expressed by Martin Greenberger, Robert Fano, and others in the late 1960s.

8 Expanding Capabilities: IBM and Control Data, 1960–1988

The more than three dozen books published on IBM's history (a few by scholars, many more by journalists) have a decided focus on computer hardware. This is understandable in view of the organization of IBM's business divisions, its marketing, and how it defined its profit centers. In the 1960s, the 1970s, and the 1980s, most of IBM's revenue and earnings came from computer hardware. The iconic mainframes—IBM 1401, System/360, and System/370—have been given center stage in this literature, with occasional bit appearances by software. There has been a little coverage of programming languages (FORTRAN), operating systems (OS/360), transaction processing systems (CICS), and databases (DB2). Absent are IBM's immense and varied services, sometimes bundled within hardware contracts and less often separately priced (especially before 1970). Discussion and analysis of IBM's maintenance services, like the maintenance of many technologies in different eras of the past, is entirely absent from the secondary historical literature.[1] And all too often, custom programming for business and scientific applications also is ignored, or minimally addressed. An exception is SABRE, the real-time airline reservation system that IBM and American Airlines created in the 1960s, but even in that case the secondary literature tends to highlight the fact that SABRE was a transformative application without exploring how the project helped change IBM as a company.[2]

While IBM's substantial work in services has been obscured, so too has Control Data Corporation, IBM's competitor in mainframes. Control Data bundled (included within the price of hardware) services less, and established and grew its data services and programming services businesses (as profit centers) from the 1960s forward, but the services activities and businesses of the corporation have been overshadowed by its supercomputer enterprise. This chapter briefly surveys the services work of these two mainframe computer companies. Between the 1960s and the late 1980s, no information technology firm in the world approached IBM in the number of workers devoted to computer services; and no computer manufacturer was more committed early on to making services a recognized business (or businesses—data processing, time sharing, programming services, and systems integration) than CDC.[3]

Continuity and Change in IBM's Field Services

IBM's organizational capabilities with maintenance and applications dates back virtually to the company's C-T-R origins, and to a degree even to its late-nineteenth-century and early-twentieth-century prehistory with Herman Hollerith assisting at major punch card tabulating machine installations. Shortly after Thomas J. Watson established field services education at Endicott in 1916, class after class of field services workers were taught to help customers with punch card tabulation machines (and other office machines) and their applications. Their positions began with a heavy training regimen, but that was merely the start. As equipment and applications evolved, C-T-R/IBM field services workers returned to Endicott, Poughkeepsie, Kingston, and other IBM locations time and again for continuing education. In the late teens, these workers were designated by C-T-R as "Customer Engineers" (CEs), and were part of the firm's Customer Engineering Departments for its various product divisions. In the 1950s, the largest group of IBM's CEs served the Electronic Accounting Machine (EAM) Division, which graduated its 10,000th CE trainee in 1956.[4] By that time, schools had been set up for IBM's rapidly expanding Data Processing Division (DPD), which trained CEs for its early computers—including the IBM 650 and IBM 1401. With the rapid proliferation of IBM computer installations in the late 1950s and the early 1960s, DPD CEs soon exceeded those of EAM, and had no rival in providing services support in the computer industry. These CE departments recruited from accredited technical schools, colleges, and universities and established an extensive infrastructure, where a portion of CEs would later become instructors to other field services personnel in IBM's equipment-specific training courses, while a smaller cadre became authors of "Field Engineering Manuals" for the company's many systems.[5]

Though IBM hired its first female salesperson in 1936, women only made slow inroads into the male-dominated IBM sales culture. Not until ten years later did the first woman become a member of the "100 Percent Club" for exceeding her annual sales quota, and not until the middle years of the century were there multiple female members of the club (three by 1949).[6] In contrast, in 1935, IBM already had held its first "System Service" training school for women college graduates. That year, Virginia Linkenhoker became IBM's first system services employee, soon followed by the other two dozen graduates of the training school.[7] The second class, in 1936, had 31 female graduates, and the numbers continued to grow from there.[8] This was the first wave of women to have both technical positions at IBM *and* customer contact in the field. CEs, as a group, were predominately male, though jobs in CE were more open to women than jobs in sales. Meanwhile, graduates of the 1935 all-women System Service School, and subsequent annual classes, became part of what was referred to as the "System Service Women Corps" of field services technical and marketing employees. In the pre–World War II era this corps was focused exclusively on pre-computer electronic accounting, punch card tabulation, and other office machines. The System Service Women Corps flourished during the war, and through IBM's hiring only select college graduates to become System Service trainees (a higher requirement than for CE trainees), the group earned a reputation for excellence.[9] They helped customers solve data processing problems, and educated both customers and other IBM staff on equipment and programming.[10]

Customer engineers and the System Service Women Corps were critical to setting up and servicing IBM 701s, IBM 650s, IBM 704s, and other machines in the 1950s. On complex, larger installations IBM often placed support staff on site for extended periods of time. In addition to installation, maintenance, and hardware optimization work (typical of CEs), IBM also commonly supplied professional engineers and skilled programmers. A short biographical account by IBM's Glenn Meyers provides rare insight into certain aspects of early IBM field engineering in computing. He received extensive instruction in both pre-computing equipment and mainframe computers and his account hints at how IBM's know-how and organizational capabilities in maintenance and field services were important to the early computer age.

Meyers, a merchant marine radio operator from late in World War II to 1948, "came ashore" that year and became a RCA television technician. In August 1956, at age 29, he was hired as an IBM field service engineer. Immediately designated for servicing a Westinghouse IBM 704 installation, he underwent training for six months. This included training on key punch machines, card readers, line printers, and other machines that predated the IBM 704 but were essential components of the overall Westinghouse IBM system installation in East Pittsburgh. Meyers' training, of course, included training on the IBM 704 mainframe and instruction in assembly language—the language of IBM 704 diagnostics.[11]

For Meyers, and for all of his field service colleagues, training was continual. Looking back at his first ten years at IBM, Meyers recalled he spent a full three years of combined time in Endicott, Poughkeepsie, and Kingston receiving training on the IBM 704, IBM 709, the IBM System/360 series, and various peripheral machines and software. Reflecting on the early years, he reminisced that operational problems on the IBM 704 often consisted of burned-out tube filaments, or as the system aged, "silver migration" (the coated silver shifting between pins with potential differences caused outages). Other times the diagnosis and the fixes were more complex.[12]

Meyers traveled frequently on service calls, serving installations such as NASA Glenn near Cleveland, Redstone Arsenal in Huntsville, and the headquarters of the Chrysler Corporation in Detroit. This travel included driving through snowstorms to reach customer locations quickly and get their system up and running again. In addition to fixes, Meyers spent considerable time at installations on preventative maintenance, as well as implementing techniques to ensure that systems ran at their rated efficiencies. Hurricanes, tornados, fires, and other disasters also required Meyers to rapidly mobilize to assist customers.[13]

As maintenance increasingly involved debugging, and hence a knowledge of programming, the need for more field personnel with a wider range of skills became apparent to IBM's leaders. In 1963 IBM announced a new division to start the following year, the Field Engineering (FE) Division, to replace the former "Customer Engineering Departments" for its main product divisions. With this, IBM's Data Processing (DPD) and General Systems (GSD) Divisions came together to establish a symmetrical five employment class structure for field engineering (in order of least to greatest responsibility): CE Trainee, Associate CE, CE, Senior CE, and FE Specialist.[14] It also established centralized management, regions, and offices, as well as the latest data management and deployment communication technologies.[15] In the second half of the 1960s, many FE staff members in the highest three classes (CE and above) took courses, built skills, and assisted with systems and applications programming.[16] In 1969 alone—after the Field Engineering Division ramped up CE programming courses—more than a thousand CEs received training on OS/360 and a range of associated IBM programs to become "program support" CEs (to assist customers with software). These program support CEs, given unbundling (separate, or broken out, pricing for software and services), suddenly had responsibility for determining an entirely new and separate category of billing for certain "programming maintenance services."[17]

By the late 1960s, IBM even applied its considerable know-how in real-time systems to effectively and efficiently deploy its large base of CEs. Using data from its Brooklyn, New York and Washington Field Engineering Division branch offices, it developed PACE (Programmed Automatic Customer Engineer), an optimizing mathematically programmed dispatching system with an outer adaptive feedback loop adjusting parameters on the basis of a performance index.[18] For IBM, effective logistical responses—deploying appropriate people and replacements parts quickly—was critical. IBM had numerous storerooms at branch sales offices, as well as specialized support centers. In the late 1960s, to further speed the process in major cities, IBM began adding "Satellite Centers" in the downtowns of major cities to cut the messenger delivery time of parts to CEs by ten minutes to an hour when serving downtown installations. In 1969 five such downtown satellites were operating—in San Francisco, Seattle, Minneapolis, Washington, and New York—and more were planned.[19]

IBM CEs were fundamental to IBM's success and strong reputation, despite the fact that only design engineers (and to a lesser extent software engineers) have made their way into the secondary historical literature on IBM. Like CEs, "System Engineers" (a new classification for IBM as of 1960) also have remained in the shadows of those designing and developing hardware, or the engineers leading the highest profile operating systems development projects, such as OS/360. System engineers were particularly important on larger-scale commercial applications and systems integration projects, of which none in the first half of the 1960s rivaled SABRE.

IBM and the Development of SABRE

IBM's work on SAGE computer systems (and earlier collaboration with MIT on precursor's Whirlwind and IBM's modified XD-1) helped the company to prepare for its work on SABRE. No previous computer contract came close to the one secured in the mid 1950s for IBM to supply the Air Force AN/FSQ-7 mainframes to operate in parallel at 23 Semi-Automatic Ground Environment (SAGE) air defense radar communication sites. While IBM had declined on seeking the SAGE programming contract, which went to newly created System Development Corporation (spun out of the RAND Corporation), IBM engineers, in serving these important mainframe installations, worked in concert with SDC on SAGE and systems integration efforts. This set a precedent for a heavy services component on major Department of Defense IBM computer installations, a precedent that carried over to other federal government sites as well as corporate installations.

In 1959, IBM's leaders launched a major reorganization of the company, re-designating its Military Products Division as the IBM Federal Systems Division, and creating an Advanced Systems Development Division to "explore new markets."[20] As part of the reorganization, IBM divided its Data Processing Division into the General Products Division (focused on product development and manufacturing) and Data Processing Division (which focused on applications and marketing). In servicing installations with maintenance, IBM was building on long-established service capabilities and work routines (that carried back to pre-computer punch card tabulation days). In working with applications programming and systems integration, it was seeking to develop new capabilities—a challenging but increasingly fruitful endeavor.

IBM benefited from early learning with SAGE (and its MIT precursors) when its leaders solidified plans to partner with American Airlines to build the Semi-Automatic Business Research Environment (SABRE), a real-time airline reservation system. Back in 1952, IBM hired Perry Crawford, an insider in the Office of Naval Research and in Project Whirlwind, to bolster its know-how in real-time systems. In the mid 1950s, IBM Kingston trained many of the newly hired RAND/System Development programmers for SAGE.[21] And IBM provided teams of engineers to handle "testing, installation, and maintenance" for all SAGE computer centers.[22] These IBM engineers were important to setting up and keeping SAGE computers running, working together with SDC in the mid 1950s to ensure that hardware and software came together to create a reliable real-time system. This was during SABRE's formative stages, when key elements of the system were being defined and established.

Discussion for what became SABRE had begun in late 1953 with a chance encounter between American Airlines' founder and chairman C. R. Smith and IBM's sales manager R. Blair Smith when they were seated next to one another on an American Airlines flight. This conversation resulted in Perry Crawford's being assigned to work with a small group at American Airlines in the mid 1950s. Crawford would go on to lead the design and development project established in 1957, after coordinating a research effort on it in 1956. In 1960, Crawford (and IBM) would name it SABRE—the "SA" standing for "semi-automatic," recognizing and paying homage to SAGE.[23] By late spring of 1958, much of the equipment necessary to operate SABRE had been specified. In 1959, IBM and American Airlines personnel began to program the system. Early estimates placed the total cost of IBM 7090 computers (as in SAGE, duplexed with one operating online and one offline and as a backup), disk storage units, and typewriter terminals with special networking equipment for agents at $30 million.²⁴ The programming was bundled into the hardware cost. For instance, the modified IBM Selectric typewriters that were to be used as reservation terminals were priced at $16,000 each, and about a thousand of them would be needed. IBM developed and continually updated estimates for programming in "man-months." This problematic measure (or rather problematic underlying rationale that boosting human resources in programming could reduce delivery time) would later be famous after OS/360 leader Frederick Brooks reflected on the OS/360 programming project in his book on software engineering, *The Mythical Man-Month*.²⁵

The huge expenditure for SABRE was justified by American Airlines in light of the rapid expansion of its fleet, to ensure strong customer service, and to either fill planes to capacity or to full demand. Back in the 1930s, use of radio telegraphy at American Airlines began to give way to the 60-word-per-minute teletype, in use by all major US airlines by 1938. From the late 1930s on, huge rooms in select airports contained teletypes, many rows of desks, and hundreds of workers with eyes glued to frequently updated reservation boards up front. Some of these rooms were so large that employees near the back would use field glasses to see the boards showing the number of seats available for sale. American, working with the Teleregister Corporation, had pioneered the electromechanical and electronic "Reservisor" in Boston in 1946. The Reservisor, and various later iterations (Magnetronic Reservisor I, with arithmetic capability and memory to hold two weeks worth of reservations, was in place at LaGuardia Airport in New York in 1952), used a plug-board system and cards, but still suffered from delays and tickets were issued manually. This could risk either oversold situations, or the most "perishable product," an empty seat, when demand existed, going unsold.²⁶

For IBM this was a commercial programming effort of unprecedented proportion. As Robert V. Head (one of the SABRE software developers) recalled, "the traditional IBM business model, whereby IBM sold [or leased] the hardware and the customer did the needed applications programming with 'assistance' from IBM, would not see SABRE through to completion." As Head put it, "even though [IBM] had essentially sold hardware to the customer, IBM would have to foot the bill for a large number of onsite programmers."²⁷ These programmers were part of IBM's newly formed Advanced Systems Development Division (ASDD). "The programming task," Head added, "was far more complex and sizable than the sales types and engineers realized … . Our initial estimate was 100,000 instructions."²⁸ Once fully operational, the system, and associated simulation programs developed in testing, included far more than half a million lines of code. Ultimately, between IBM and American Airlines, more than 200 people participated in SABRE's development.[29]

Although IBM's leaders unquestionably underestimated the programming and system integration effort required to create SABRE, internal documents indicate that IBM team leaders were not caught completely off guard. In 1959 they recognized some of the risks, including "each programmer's work affecting and [being] affected by, every other programmers," that "later changes [would] probably occur on a frequent basis … over years," and that "a trail must be left behind the programming." In addition to the needs for interdependent iterative efforts and for quality documentation, IBM managers foresaw that many of the programmers would be "completely inexperienced," and that there would be the great challenge of "division of responsibility" between IBM and American Airlines.[30] These and other challenges all played out in the SABRE project, and in many of IBM's major programming and systems integration projects in the coming years and decades, and also in the projects of other computer companies and computer services enterprises.

Inevitably with integrating large-scale systems, software modifications and hardware re-engineering occurred frequently. In May 1961, IBM and American Airlines presented (to great fanfare) an operable system using redesigned IBM 7090s. For several years thereafter, SABRE was run in tandem with a manual system and many bugs had to be identified and fixed. The SABRE system was not fully operational until late 1965. By that time it could handle 65,000 phone calls, 40,000 reservations, and 20,000 ticket sales a day.[31] IBM leveraged its learning on SABRE to develop Programmed Airline Reservations Systems (PARS), a software system that worked on various System/360 models that IBM delivered to Delta and other airlines in the mid 1960s.

SABRE, which involved what was by far the largest software and systems integration contract for a commercial customer in the computer industry up to that time, was an eye opener for IBM's management. The company gained unique large-scale programming experience that it could use to create transformative mainframe computing applications. The project contributed to IBM's know-how for future software and systems integration projects. But all such efforts presented unique challenges, and, as IBM found with OS/360, coordinating very large programming efforts would always be extremely difficult.

IBM's Systems Engineers

SAGE and the start of the SABRE project, without question, alerted Thomas Watson Jr. and his leadership team to the need for systems specialists—engineers who could do advanced programming, manage major programming projects, and integrate hardware, software, and computer networking to design important new applications and realize new possibilities (including, and especially, real-time systems). However, the origin of the idea of "System Engineers" as a formal class of IBM employees originated with William Lawless, an IBM engineer and executive assistant to vice president Al Williams, who not only looked to present and future needs but also considered the more distant past.[32]

At the end of the 1950s, Lawless was seeking to address the problem of attracting more technical people to the marketing side of the company—to work in the field at customer locations. This was critical to creating demand for IBM computer systems and maintaining and extending its competitive advantage over other mainframe computer firms. At the time, virtually anyone wanting a technical career at IBM would opt to work within either one of the product divisions (Data Processing Division or the Federal Systems Division—for computers), IBM Research (if they had an advanced science or engineering degree), or the Advanced Systems Development Division. The technically inclined already at or joining IBM generally were not drawn in meaningful numbers to data processing field work, in which sales personnel and Customer Engineers marketed and maintained machines. The main opportunity for advancement was to become either an instructor or a marketing manager. Lawless recognized that some people were better suited for technical rather than managerial responsibilities, but of course they still wanted to advance in their careers.[33]

Lawless envisioned a new job family to attract top technical talent to the field services side of IBM. He conveyed his general goal to Al Williams, who encouraged him to put some research together and prepare a proposal. Lawless proceeded to survey the existing technical staff in field services and identified three groups. First, he "found the System Services Women Corps," which had grown steadily since its origin in 1935. By the end of the 1950s, the System Services Women Corps consisted of more than 500 field service women.[34] It was composed entirely of college-educated women who increasingly became involved with computer electronics and computer programming. As the male-dominated sales force adhered to IBM's dress code (dark suit, white shirt, dark tie), the "System Service" women wore "hats and white gloves." Ellen Kerksieck Schaefer of System Services recalled: "Mr.

Watson [Sr.] believed we had a certain image to maintain … . We did a lot of marketing as well as installation … . [We] wired the boards … . We'd take off our gloves do the work and put our gloves back on."[35] Schaefer later became IBM's first female "consulting systems engineer." Another alumna of that first class, Loraine McLennan, a graduate of the University of Iowa with post-graduate training in business at Columbia University, had been with IBM for more than 25 years in technical and managerial positions, working out of Detroit, Chicago, New York, and Los Angeles.[36] Alongside this corps of women, IBM had an all-male Customer Technical Representative group that did many of the same jobs as the System Service Women Corps. Lawless also identified a third group of technical employees with experience in field interaction: Applied Science Representatives, individuals with advanced degrees who helped with installations for scientific applications at universities and laboratories.[37]

Lawless thought there should be a cohesive group that recognized the talent of computer-oriented engineers in the field, rewarded them with appropriate salary and rank, and created various forms of infrastructure for the exchange of ideas and techniques. Drawing on the name of the corps of women, he came up with a formal proposal for a new employee class to be called "System Engineers." He outlined that this was "a profession for people of superior intelligence, solid technical education, a demonstrated flair for problem solving, and the desire to work in a field of unlimited possibilities to make significant contributions."[38] Lawless successfully presented his proposal to Williams and to IBM's president, Thomas J. Watson Jr., and IBM officially established the System Engineer employment classification in December 1960.[39]

Lawless, who was named Director of Systems Engineering, established five levels of SEs: Trainee, Associate SE, SE, Advisory SE, and Senior SE. Later, Consulting SE became the sixth and highest level. With this reorganization, the gender separation of Customer Technical Representative in the System Service was eliminated and coeducational SE training courses were launched. As an employment class, "System Engineer" was "open to system services women."[40] There were trainee courses, but also many advanced ones to allow SEs to further their career within the Field Engineering Division hierarchy, or seek promotions within other parts of the company. In 1961 SE Maggie Wilcox attended advanced training, as did SE Alan Seelenfreund. Both found the course useful to advance their career within IBM. Seelenfreund, who had a background in mechanical engineering and operations research, soon applied knowledge from the course to helping customers by educating them in inventory management simulation.[41]

Lawless and his staff established institutions to help engineers grow intellectually and to learn from one another. This included establishing the Systems Research Institute (where employees took "graduate level" engineering courses), setting up a major annual System Engineering Symposium, and launching the *IBM Systems Journal*. The first SE Symposium was held in October 1961, and the first issue of *IBM Systems Journal* was published the following year.[42] The competitive, symposium program included more than seventy technical papers. Thomas J. Watson Jr. spoke at the first SE Symposium's banquet, stating "the fact that we are beginning to recognize the tremendous need for systems engineering is an indication that IBM is still able to improve and change."[43] That Ellen Kerksieck Schaefer reached the rare and esteemed status of Consulting SE is suggestive of the relative greater opportunity for women to advance in the field engineering path as opposed to sales in the 1960s.

By the end of the first year of launching the SE job class, IBM had more than 1,000 of them; by 1965 there were more than 5,000.[44] Rapid growth continued in succeeding years. In late 1968, IBM's leaders made and soon announced the firm's "unbundling" decision—that it would separately price much of its software and programming services. One challenge to implementing the SE program was that SEs bundled countless hours of work, in serving major IBM hardware customers and their systems needs. SEs supported the marketing mission, but initially were not formally part of the marketing division. Before unbundling, they represented IBM, but they did not have to sell themselves as consultants—they were seen by customers as a helpful and a "free" benefit of buying or leasing hardware. The fear of a newfound need to sell themselves as consultants led some to leave the SE ranks after the unbundling announcement, and others to resign. Many, however, stayed with IBM. Less than two years after announcing unbundling, IBM's leaders made SEs full members of the marketing team (which had an associated technical managerial hierarchy). The number of SEs then continued to grow. By 1985 there were more than 20,000, outnumbering all non-SEs in marketing. Unlike CEs, who were around for initial installation and for shorter spans, SEs became true fixtures at customers' facilities.[45] As Joy Greenwood, General Dynamics' manager of data processing, put it in the first half of the 1980s, "the best SEs think of themselves as an extension of the customer [and] give you that extra effort to help you solve problems."[46] With SEs, IBM advanced its stature in becoming a (or *the*) "solutions" company, which would later become an important component of its marketing.

IBM's leaders made the unbundling decision in the context of escalating scrutiny by anti-trust lawyers in the US Department of Justice. In the early 1960s an independent software products industry had emerged; by the late 1960s it was rapidly expanding. Meanwhile, the computer services industry grew quickly. Many companies in both industries believed that bundling provided an anti-competitive advantage to IBM and thwarted their future growth. At the same time, IBM's hardware competitors believed that the ties of bundling (offering "free" software and services, which was popular with customers), as well as IBM's pre-announcements, made it more difficult to sell hardware (or software or services) to compete with the computer giant. These factors inspired the filing of lawsuits against IBM in the late 1960s by Applied Data Research (largely at the hands of this services firm's relatively new Software Products Division), by Control Data Corporation (highlighting that a pre-announcement hurt its supercomputer business), and by the US Department of Justice (very shortly after unbundling was announced). Ultimately, IBM settled with CDC in 1973, one major component of the settlement being IBM's sale of its Service Bureau Corporation (data processing services centers) subsidiary to CDC at an attractive price. The Department of Justice's case was dismissed—in a changed political and computer industry environment—in 1986.[47]

The pressured sale of SBC to CDC did not mean that IBM was exiting from computer services in 1973; it was merely exiting a particular segment—that of service bureaus or data centers. And SBC had already been separated as a wholly owned subsidiary, as a result of the 1956 consent decree that focused on pre-computer tabulating machines. By the early 1970s, IBM's SEs generated far more value for the company than SBC, when both what was bundled and what was priced out are considered. At the time CDC acquired SBC, that subsidiary was creating substantially less than 0.7 percent of IBM's overall revenue and even a lesser share of IBM's profits. SEs generally had far more education, greater technical ability, and better programming and systems skills than the vast majority of employees within SBC. By 1985, with more than 20,000 in their ranks, SEs made up roughly 5 percent of IBM's workforce, and were growing in number much faster than the company's overall employees.[48]

In addition to serving at many corporate computer installations, IBM Systems Engineers worked at many government facilities. The IBM Federal Systems Division, which initially was largely an extension of SAGE, began in the 1960s to serve (or to provide extended service to) many different government departments and agencies, including the Department of Defense, the Department of Energy, the Department of Transportation, the

Department of Justice, the Federal Aviation Administration, the National Security Agency, the Federal Bureau of Investigation, Los Alamos National Laboratory, other federal labs, and NASA. IBM Systems Engineers ran the control systems that helped to put men on the moon. To a certain degree, the Federal Systems Division was set apart, as it served entities that did sensitive and sometimes highly classified work. In 1992, IBM's Federal Systems Division had revenue of more than $2 billion and a net income of more than $70 million, and employed more than 11,000 workers.[49] Though some of its business was supplying hardware, most often the division was focused on designing and developing large-scale complex systems and applications that also required extensive advanced programming, networking, and systems integration services work—and, of course, considerable maintenance support. Such work was usually done by a combination of federal employees, government contractors, and on-site IBM technical personnel.

In the 1980s, with CEs, Federal Systems Division engineers, the Advanced Systems Development Division, and especially SEs, IBM was very much a services company (with tens of thousands focused on services work) before it focused on services as a (and in time *the*) primary profit center for the corporation. In contrast, services became a major profit center for one of IBM's core mainframe competitors Control Data by the early 1970s and remained so through the 1980s.

Control Data Corporation's Origins

By acquiring two pioneering computer startups—the Philadelphia-based Eckert-Mauchly Computer Corporation in 1950 and the St. Paul–based Engineering Research Associates (ERA) in 1952—Remington Rand gained early leadership in the computer industry. Its lead, however, was soon eclipsed with IBM's entrance into the field. In 1955, the Sperry Corporation merged with Remington Rand; the resulting computer division was named Sperry-Univac after Remington Rand's early and influential computer. Mismanagement, tight research and development budgets, and rivalry between the Philadelphia and Minnesota computer groups, left many engineers disgruntled, especially those in Minnesota.[50] Among the most talented and prominent of these engineers were William Norris and Seymour Cray. Sperry-Univac's Willis Drake (an engineer) and Arnold Ryden (ERA's former treasurer) developed a plan in the first half of 1957 to siphon away some of Sperry-Univac's best technical and managerial talent—with Norris and Cray at the head of the list—for a new startup computer firm that was to be based in Minneapolis. Norris' recent demotion from Sperry-Univac vice president and division manager aided the opportunity to successfully recruit Norris. Drake and Ryden teamed with three Minneapolis businessmen to found the Control Data Corporation (CDC) on July 8, 1957, the incorporation papers listing Norris as president.[51] In September, the small team of roughly a dozen set up shop in a warehouse building at 501 Park Avenue in Minneapolis. Since the Department of Defense would be a primary focus for CDC, Cray was unwilling to leave an existing Sperry-Univac military project—the Naval Tactical Data System—in the lurch (the department and agencies of the DoD would be important customers for CDC from the start); he came aboard several months later, becoming CDC's chief engineer and its leading computer designer.[52]

Control Data's founders, through their own investments and the issuance of stock, raised more than half a million dollars, but the young firm soon developed financial difficulties. Facilities, the purchasing of Cedar Engineering (a supplier of aircraft components), and personnel costs all strained the company's resources as it sought military computer contracts in its first years. Control Data's early financial challenges eased greatly in 1959 when it was awarded several Department of Defense contracts, the most important of them a $2.5 million contract for the first CDC 1604 mainframe computer, which was to go to the Naval Postgraduate School in Monterey, California to be used principally for weather predictions for the Pacific Fleet. With a number of even larger contracts from the Department

Table 8.1
Control Data's first decade.

Fiscal year (ending June 30)	Revenue (thousands)	Net income (thousands)	Number of employees
1958	$626	($115)	260
1959	$4,588	$283	380
1960	$9,443	$552	690
1961	$18,062	$843	1,350
1962	$32,129	$1,543	2,273
1963	$63,111	$3,065	3,495
1964	$95,821	$6,073	6,861
1965	$127,820	$7,913	9,744
1966	$109,222	($1,678)	11,048
1967	$147,512	$8,406	14,881

Source: Control Data Corporation, "Statistical Summary Ten Years," in 1967 annual report (Charles Babbage Institute).

of Defense and the Department of Energy in the following two years, and a growing number of CDC 1604 installations, the company was growing rapidly and moving beyond its early financial struggles.[53] The CDC 1604, designed by Seymour Cray, was the first successful transistorized computer, and at the time of its delivery, early in 1960, was one of the most powerful computers in the world. The company followed the CDC 1604 with the Cray-designed CDC 6600, arguably the world's first "supercomputer," which was delivered in 1964. While CDC was producing the most powerful computers, it was also successfully developing other important computing businesses. By the early 1960s, it had a thriving business in computer peripherals, selling disk drives and other components and sub-assemblies to original equipment manufacturers. It also had fast-growing businesses in computer services.[54]

CDC's Early Data Centers, Programming Services, Engineering Services, and CYBERNET

One of CDC's first six 1604 mainframes was installed at its downtown Minneapolis headquarters. It was used by CDC engineering staff for design work and for the company's internal data processing needs (financial and managerial accounting), but also to launch a new type of business. William Norris was committed to starting and rapidly expanding computer services as a business from early in his tenure leading Control Data. This initial center aided customers purchasing a 1604—they could come to "perfect" planned computer applications programs before receiving their mainframe. The center, from its origin, also served as a traditional service bureau, where a range of scientific and business data processing customers purchased computer processing time and programming services.[55]

The center's first client (in 1960) was Klink Realty, located in northeast Minneapolis. The fixed-price contract of $400 was for developing a program to produce amortization tables. After the program had been delivered and the bill was 60 days past due, the center's manager, making a personal collection effort, "found the office abandoned and no forwarding address."[56] Despite its inauspicious start, the center soon secured business for data processing with three large companies based in Minnesota's Twin Cities: Northern States Power (NSP), General Mills, and the Honeywell Corporation. The fact that Honeywell's Aerospace and Ordnance divisions became frequent customers of CDC caused a rift with Honeywell's top management, which wanted to nurture Honeywell's emerging computer division on the East Coast. Thanks to business from Honeywell, General Mills,

Northern States Power, and other firms, the data center's monthly revenue exceeded $50,000 by the summer of 1961.[57]

In 1961, CDC built a major new facility, which became its future headquarters, in Bloomington, Minnesota. The company also moved forward with plans for a large computing center in the heart of the region soon to be known as Silicon Valley—Palo Alto, California. The latter, in Stanford Industrial Park, was acquired under a lease-back arrangement and was more than 12,000 square feet in area; it housed a CDC 1604 that was installed in December 1961.[58]

CDC had entered the region a year earlier with a sales office in Sunnyvale, California that virtually from the start included an Applications Services Group to aid CDC's computer hardware customers.[59] This group grew organically in serving programming and systems services needs of major customers such as Lockheed, and quickly expanded in lockstep with this rapidly growing technology region in the 1960s and the early 1970s. As R. C. Gunderson of CDC later articulated, "this strategy was born of necessity … . In a move to help the customer gain earlier use, we decided that each [CDC] system should be accompanied by a trained analyst, or applications

Figure 8.1
CDC Headquarters Data Center, Bloomington, Minnesota, 1962. Courtesy of Charles Babbage Institute, University of Minnesota.

programmer, as well as a customer engineer, for a period of six months at no additional cost to the customer." While this initial six months was bundled, as Gunderson explained, "some of these people stayed on at the site, on analyst contracts [billed by CDC] or [leaving CDC] on customer's payroll; for many were worth their weight in gold—to us and our customer."[60] Even if they joined the customer organization, it enhanced a relationship between a customer and CDC—and its growing ecosystem of mainframes, peripherals, and (billed) services. Similar to and largely concurrent with a major CDC Applications Services Group emerging in Northern California, one also was born out of serving aerospace CDC hardware systems customers in Southern California (from CDC's Los Angeles sales office). More generally, CDC's sales offices, especially fast-growing ones, continually added additional programming staff to assist with custom software or with consulting, or to develop software products. (A CDC Software Development Group, which created standard products, also soon was launched in Silicon Valley). The Sunnyvale/Palo Alto locations truly stood out in the early 1960s and in subsequent years as many transplants from Minneapolis and elsewhere (as well as some nearby Californians) went to CDC's facilities on the peninsula south of San Francisco as programmers, systems analysts, and managers.[61]

This Northern California CDC operation, however, also had its share of early organizational challenges. CDC essentially had the same problem throughout the 1960s that IBM had gone a long way to address by adding the job classification of five (later six) levels of Systems Engineers (SEs) in 1960. CDC's "Application Analyst" organization (a designation that in the early years was used interchangeably with "Application Services Group") faced considerable hurdles to upward mobility and had less functional reporting lines in comparison with programming staff at CDC data centers. Securing and retaining top technical talent was difficult insofar as application analysts (systems and applications programmers doing customer field work) were under the sales and marketing division and these analysts reported to sales managers who often had limited technical understanding.

In 1961, Robert M. Price joined CDC as a manager of the Application Analysts/Application Programming Group in Northern California. Price was an engineer with extensive programming experience. He had worked at Livermore National Laboratory (on a Univac), at the Georgia Institute of Technology (on an ERA 1101), at Convair (on an ERA 1103), and at Standard Oil.[62] Because the Sunnyvale/Palo Alto operation had major contracts with Lockheed and other large firms, in just a few years it quickly grew to more than "several hundred" Application Analyst personnel. Price's position created a technical/managerial layer that helped make the Application Analyst position more attractive; however, because of the overarching sales and marketing hierarchy, some awkwardness remained—for example, members of the technical staff reported to sales and marketing leaders. After several years in California, Price—who one day would succeed Norris as CDC's president and CEO—became Sales Manager for International Operations, and soon thereafter, in 1966, General Manager for US Sales. His origins on the applications programming support side raised that side's stature in the corporation and helped facilitate important incremental reporting changes. However, major organizational changes for Application Analysts, facilitating meaningful growth of the field services programming business within the company, required lifting of the artificial barrier within programming services between those in the sales and marketing organization (Application Analysts) and in the data centers (Analytical Services) to create the CDC Programming Services Division, or PSD, in the early 1970s.[63]

Figure 8.2
A software programming group in CDC's Palo Alto facility, circa late 1960s. Courtesy of Charles Babbage Institute, University of Minnesota.

From the first shipments of CDC 1604 systems in 1960 forward, another type of services—Engineering Services—also played a fundamental role. As several in this area recalled, the "product maintenance people received their training by actually helping to build the machines or engaging in 'systems check-out.'" They essentially were sent in tandem with the shipped equipment to become site maintenance personnel. While known as "Engineering Services" for many years, the original name was the "Product Maintenance Group," which soon evolved to the "Customer Engineer Group." In 1962, R. F. Buelow took over as manager of the Product Maintenance Group. In 1963, the company established its first Customer Engineering Parts Warehouse, in Minneapolis, which was expanded multiple times in the 1960s before an even larger CDC World Distribution Center was built in neighboring St. Paul the following decade. In acquiring Bendix's computer division in 1963, CDC's Customer Engineering staff grew from 150 to 300.[64]

Even before CDC's 1970 "unbundling," Customer Engineering was partially set up as a profit center for the firm. Two challenges for Customer Engineering during the 1960s were (lack of) "standardization," and "staffing strategy" in association with the sales and marketing organization. This resulted from the fact that "no two of" the early CDC 6600 computers from Chippewa Falls, Wisconsin (a CDC computer development and manufacturing location Seymour Cray argued for, set up, and ran), "were built alike."[65] This made all aspects of maintenance more difficult. With the latter, sales and marketing leaders wanted Customer Engineers (CEs) on hand for all major sales efforts, but CEs were responsible for keeping existing systems up and running 24/7. Sales and marketing invariably wanted CEs deployed to regions on the basis of sales forecasts, which often were substantially off the mark. This resulted in deployments of CEs (often from the US) to worldwide locations where sales failed to materialize at predicted levels, and shortages in areas (usually in certain regions of the US) where sales exceeded expectations. In 1968, R. C. Hall became vice president of customer engineering. He brought rationalized processes to Customer Engineering and to its integration and its interactions with the rest of the company. In 1968 and 1969, he established protocols that were still in place more than ten years later, working with design, production, sales, and marketing to improve engineering, logistics, finance, quality assurance, and operational support. One part of this effort was convincing top leaders throughout the corporation to be dedicated to "reliability, availability, and maintainability in the design of its products."[66] Under Hall's leadership, a Unified Field Reporting System was established so that

Maintenance Services, the new name in 1969 (in taking on additional roles of "facilities engineering and construction" for new research and development and manufacturing sites), could have the best possible data readily at hand to serve installations. In 1969, under Hall's guidance, Maintenance Services also took on the role of software maintenance or debugging, to enable Application Analysts to concentrate more fully on programming new applications.[67]

On the data center side, by the end of 1963, Control Data was providing data processing services not only out of its large Bloomington, Minnesota, and Palo Alto, California facilities, but also dedicated data services centers in Washington, San Francisco, and Los Angeles. The following year CDC added centers in New York City and Houston.[68] This formed the basis of its newly established Data Services Division (DSD). This division, as one early DSD veteran B. T. von Schmidt-Pauli recalled, was able to "corral between 10 [and] 15 experienced [IBM] Service Bureau Corporation salesmen to form the nucleus of the DSD marketing force" in 1963 and 1964.[69] They had a difficult proposition, as these ex-IBMers had been focused on selling back-office business data processing (unit record equipment and IBM 1401s), and all of a sudden had to transition to marketing the powerful CDC 1604, a computer system that better fit advanced number-crunching, operations research, and simulation needs of the government and scientific computing market. After a steep learning curve, the DSD's marketing staff adjusted to a very different clientele and set of needs. In 1964 the company reported that typical applications for its data centers were operations research, traffic surveying and planning, medical and hospital information processing, school scheduling and grading, and seismic record conversion.[70]

From the launch of the Minneapolis data center forward, Control Data provided programming services out of its data centers. Its service bureau geographical reach and especially its programming services resources grew considerably with the 1967 acquisition of C-E-I-R, Inc., an all-stock deal of roughly $36 million in CDC shares.[71] As it integrated C-E-I-R's personnel and facilities, Control Data, by late 1968, had more than 33 data centers, including centers in Huntsville, Omaha, Detroit, Mexico City, Ottawa, Frankfurt, and Melbourne. Some of these centers utilized CDC 3600 and CDC 3800 computers, and some 1604s were redeployed or retired. Meanwhile, seven CDC 6600 supercomputers were installed in the second half of the 1960s at CDC data centers in New York, Boston, Washington, Bloomington (Minnesota), Houston, Los Angeles, and Palo Alto. With CDC's extensive data center capacity, making efficient use of these resources became all the more important.[72]

Expanding Capabilities

By the end of 1968, Control Data had multiple time-sharing centers in some major cities, with six in Washington, four in New York, three in Boston, and three in San Francisco. In these centers, users could utilize small computers or terminals to access processing resources and software on CDC 3000 series machines or its CDC 6600 supercomputers. Control Data set up high-speed transmission lines to connect centers on the West Coast (San Francisco, Los Angeles, and San Diego), in the Midwest (Minneapolis, Chicago, Cincinnati, and Cleveland), in the Middle South (Houston and Dallas), and on the East Coast (Washington, New York, and Boston). These regional networks also were connected by high-speed transmission lines. Meanwhile, lower-bandwidth telecommunication lines connected the major regional networks to centers in a number of other cities, including Omaha and Detroit.[73] In the late 1960s, CDC named this overall network infrastructure—which was similar in many respects to computer networks of time-sharing services providers GEIS and Tymshare, Inc.—CYBERNET.[74]

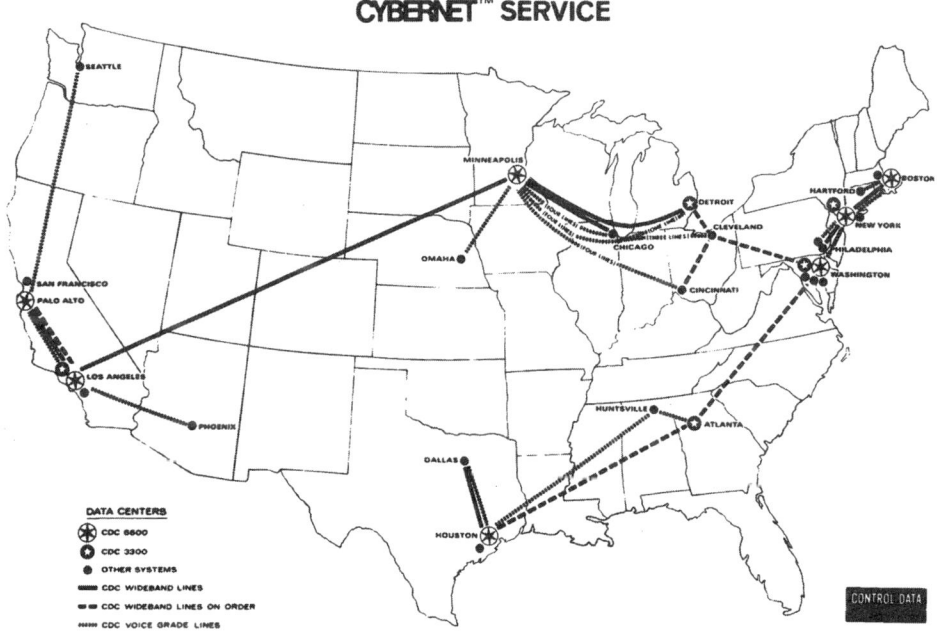

Figure 8.3
A map representing CDC's CYBERNET service, circa early 1970s. Courtesy of Charles Babbage Institute, University of Minnesota.

Organizational users often were purchasing more than just computer processing time. They benefited from libraries of code that CDC maintained in such areas as "operations research, transportation and urban planning, structural engineering, automatic machine tool control, market research, and seismic data reduction and other petroleum applications," and from customized programming services that the data centers provided.[75] Control Data's many sophisticated scientific and engineering users also advanced the software applications field and—following the lead of IBM's users with SHARE, Inc.—exchanged insights and code through CDC's user group CO-OP. The user group had been formed in 1959 shortly before the first deliveries of CDC 1604 systems to customers. As customer drew on both the CDC data centers' resources and participated with CO-OP, these organizations and others also benefited from the Application Analysts who worked on site at their locations. This could be for small projects as well as large ones—an example of the latter was CDC's consulting work in operations research to simulate Panama Canal traffic routing to offer greater efficiency and account for "a wide range of operating conditions, including such remote possibilities as a ship sinking in the canal."[76]

CDC's Educational Services

Historian Nathan Ensmenger richly explored the "software crisis," a labor crisis in the United States (and elsewhere) defined in part by the near perpetual shortage of programmers throughout the 1960s and much of the 1970s (and notions that software was not keeping up with hardware and the trajectory of circuitry advances—Moore's Law).[77] The software crisis, frequently written about in trade publications such as *Datamation* in the 1960s and the 1970s, co-existed with a broader labor crisis that included many new and evolving occupations involving computing. Shortages of computer engineers, software engineers, systems analysts, and programmers often were prevalent, but alongside this there were also substantial shortages in computer technicians, an important occupational category vastly understudied by computer historians.

Norris and Control Data witnessed firsthand the impact of the shortage of computer technicians (a broad category including certain types of computer operators, computer maintenance personnel, and computer factory laborers), and in 1964 they followed through on plans for a proprietary vocational school to be called the Control Data Institute. Existing vocational schools for training in electronics—including the RCA Institute, Brown Institute, and the DeVry Technical Institute—trained electronic

Figure 8.4
Students being instructed in computer maintenance at the Control Data Institute, 1968. Courtesy of Charles Babbage Institute, University of Minnesota.

technicians, but not computer technicians. CDC, other mainframe firms, and computer user organizations needed to retrain or extend electronics education into computing to meet their need for computer technicians. Control Data Institute, from the start, was focused on training computer technicians to work in "manufacturing, development, and field maintenance." It was created to address the company's workforce shortage in manufacturing, and maintenance services (internally, at customer sites, and data centers), but would soon evolve to train technicians, and later programmers, for the broader industry.[78]

Control Data Institute (CDI), which recruited staff members in the spring of 1965, began instruction in Minneapolis in the fall of that year. Swen A. Larsen, a former employee of CDC-acquired DATATROL, was named General Manager of CDI. Taking possession of the facility of the former Gale Institute at 3255 Hennepin Avenue in Minneapolis in the summer of 1965 (the end of Gale's summer session), very little time existed to improve the dilapidated building before the first CDI term. In the early months of CDI, it had problems of roof leaks, air conditioner malfunctions, and falling light fixtures. Nevertheless, the term continued uninterrupted as maintenance work and essential renovations were completed.

The course of study entailed 4 hours of instruction per day for a year—a thousand hours in all—at a tuition rate of $1,795. The price was at the high end of the vocational school market, but, as Larsen saw it, a substantial price was associated with high-quality education and was necessary to support such education. In the initial term, which began in September 1965, there were 72 students (half of them instructed in the morning and half in the afternoon). New cohorts were added at six-week intervals. CDI received considerable attention, including a reported congratulatory letter to William Norris from Vice President Hubert Humphrey. To the company's surprise, its justification (having been involved in internal corporate education for several years) to qualify for GI Bill funding from the start (obviating the two years in existence requirement) was accepted, and students who were military veterans qualified for support under the GI Bill in CDI's first year.[79]

CDI in Minneapolis was a success from its start, and William Norris and the CDC Board authorized plans to launch two more Control Data Institutes in the fall of 1966, one in Los Angeles and one in Arlington, Virginia. The schools in Los Angeles and Arlington required more marketing than the one in Minneapolis and were in the red for a couple of years. Control Data Institutes were created in Dallas and Detroit in 1967, and in Atlanta, Miami, New York, St. Louis, and Rockville, Maryland the following year. Many graduates were hired by Control Data, though a substantial number were hired by CDC's computer industry competitors, as well as computer user organizations in industry and government. From the beginning, coursework prepared computer manufacturing technicians, computer operators, and maintenance personnel.[80]

Most new hires by CDC in programming, unlike the majority of CDI students, were college graduates. Before the end of the 1960s, however, CDI added coursework for "programmer" tracks in response to demand by existing and prospective students. Many small programmer schools had begun in the 1960s, with some having poor reputations for quality and placement of graduates. CDI graduates of the late 1960s had little trouble finding positions, with some receiving multiple offers. Over time, the various CDI locations trained many thousands of programmers. In 1969 there were sixteen CDI schools spread throughout the nation and annual enrollment approached 4,000 students. By the early 1970s, CDI had expanded to offer training in additional areas such as "computer age secretary," "credit specialist," "bio-medical technician," and "radio/television repair," but these courses suffered from uneven or poor instruction and such tracks ultimately faltered. However, CDI's courses for computer technicians and computer programmers were successful in the late 1960s. The Control Data brand, and the focus on quality instructors for these specialties, mattered. The success of CDI spawned imitators within the computer industry (at Honeywell and at Sperry Rand's Sperry-Univac Division) and outside of it (at Bell and Howell and at Sylvania).[81]

The recessionary years of the early to mid 1970s proved difficult for the CDI Division and the school's graduates. In contrast to the favorable job market of the late 1960s, in the 1970s CDI graduates were having difficulties securing job offers. Somewhat similar to the "great financial recession" of 2007–2010, government officials and media responded harshly to CDI and other for-profit educational institutions that were heavily dependent on government loan programs when graduates could not find jobs and defaulted on loans.[82] Despite challenges and critique, CDI expanded nationally and internationally with more than sixty locations by the late 1970s. This highlighted another challenge for CDC, one outside its core capabilities: managing the real estate assets to provide these educational services. But with Norris at the helm, CDI pressed onward. Norris expected it to become a major test bed for an equally favored initiative of his: computer-based education.

William Norris was dedicated to, and was in many ways a pioneering figure in, what today is commonly termed corporate social responsibility. He saw the poverty of North Minneapolis neighborhoods, and of neighborhoods in other inner cities where CDC had facilities, and wanted to help. CDC invested in building a major manufacturing operation in North Minneapolis, and hired people who often had been excluded from the workforce, especially in IT—individuals with little formal education, members of racial and ethnic minorities, and former convicts. Norris saw this as the right thing to do and also as good business. His aim was largely the same with CDI, and he and fellow CDC executives saw PLATO (a computer/software educational and instructional system) as part of a continuing saga of CDC educational services. CDC's goal with CDIs was educating and placing students who did not have the means or inclination to attend four-year colleges and universities in careers—and, in the process, helping CDC and the computer field. Norris also wanted to broaden education and make it more affordable, to use computing to enhance efficiency in education, and to offer new opportunities and techniques for learning.[83]

In the early 1960s, pioneers at the University of Illinois developed a version of PLATO, a high-end graphics-based computer educational/instructional system that went through a number of iterations in succeeding years. Norris and CDC secured rights to market PLATO technology in the early 1970s and sought to do this through CDC's CDIs and by forming partnerships with clients in existing educational institutions—K–12 schools, colleges, and universities—throughout the US and beyond. But PLATO was not a cost-effective, suitable solution for education in the mid 1970s to the mid 1980s, a time when CDC devoted considerable resources to developing, marketing, and deploying versions of PLATO and associated courseware.[84] Similarly, the Control Data Institutes, despite their great start in Minneapolis in the late 1960s, produced inconsistent contributions to the corporation's bottom line. Nonetheless, these institutes did educate many current and future CDC employees, including many who went on to perform maintenance, applications programming, and other services for Control Data and its customers.

Establishing and Rapidly Growing CDC's Professional Services Division

Control Data's educational services had some modest successes (a subset of CDI tracks and locations that thrived for shorter periods of time) coupled with failed long-term business initiatives (PLATO). CDC's overall services operation, however, grew substantially in the late 1960s and the 1970s—both as an activity and as a core business. At the start of the 1970s, CDC rationalized its hitherto disparate services activities and businesses by unbundling services from hardware, as well as through organizational changes.

Throughout the 1960s, CDC typically (at new installations) bundled some maintenance and applications programming services into customers' computer hardware contracts. IBM's 1968 unbundling decision led CDC's management to revisit its own policies and practices with software and services. Norris appointed a "Software Task Force" that met on September 12, 1969 and recommended unbundling, or separate pricing for equipment, software, programming services, and maintenance. The task force believed the transition for customers would be eased because "IBM has set the stage … which should make the concept … easier for our sales force to explain in the market place."[85] Control Data moved quickly with unbundling. After a small CDC team toured the sales regions in the US to inform field personnel about it, CDC publicly announced its unbundling policy on October 1, 1969. On December 19, 1969, Norris formed a Pricing Committee Unbundling Subcommittee; it began implementation of unbundling software and services in 1970. The customer base readily accepted the new pricing for "engineering services," "education services," and "training." There was more resistance from customers regarding systems software and "analyst support" services (applications programming).[86] Nonetheless, the company moved forward with unbundling all these types of services. Initially, CDC set forth a policy of maintaining ownership of all programming services resulting in a software "product" (a customer merely had a license to use what it purchased). Negative response from customers and in the trade press led to a revision of this policy in January 1970 so that all paid-for application services programs and packages (work product) belonged to the customer. CDC still learned from and benefited from applying one programming services effort to the next of a similar type.[87]

Unbundling had many advantages and positive results for Control Data. First, IBM's unbundling helped lessen the advantage that IBM had gained through bundling and its unparalleled support services. Second, IBM's having unbundled first allowed CDC to raise prices (by not reducing hardware prices commensurate to the newly priced services) without losing many customers as a consequence. Third, IBM's unbundling provided momentum to the independent software products and services industries. Overwhelmingly, companies in these industries targeted IBM's base and its market share of more than 60 percent. Also, since CDC was more concentrated on science and engineering customers, it was more immune to the core of the software and services industries, which focused on business applications. And fourth, it forced CDC management to give serious thought to the organization and rationalization of its services (and its much smaller software products) businesses.[88] Beginning in the early 1970s, this led to a much more coherent and integrated services strategy.

From the time of CDC's launching of programming and data services at the beginning of the 1960s, these services were provided to customers by two very different organizations within Control Data: the data centers and the sales and marketing organization. The Data Services Division ran the data centers that grew steadily in number and size—especially with the acquisition of C-E-I-R in 1967. Integrating C-E-I-R doubled the size of the DSD and lent momentum to rapidly expanding time-sharing services operations and optimization of resources with CYBERNET. In the early 1970s, CDC ran more than forty data centers, about three fourths of them in the US and one fourth overseas. Within the data centers, there were advancement opportunities for both technical and managerial staff—upward mobility within and between data centers.[89]

Nothing comparable existed for the "Application Analyst" programmers working out of sales and marketing in the field, who overlapped in function with some DSD staff in providing applications programming for customers. In response, in January 1971, Norris and the board reorganized

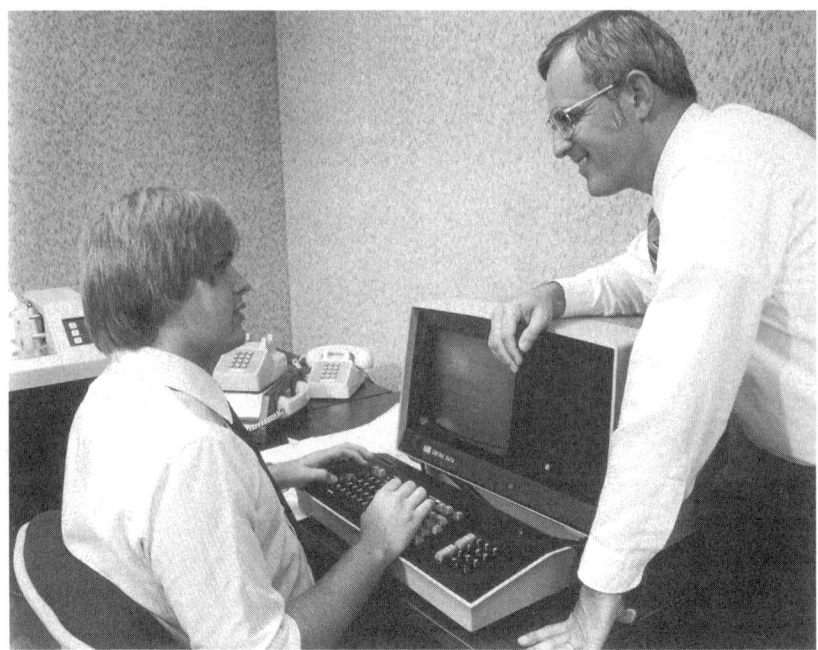

Figure 8.5
Ken Clary (left), an operator at CDC's CYBERNET Services Center in Houston, working with a client (Donald McLain of Tetra Tech, Inc.), circa late 1970s. Courtesy of Charles Babbage Institute, University of Minnesota.

the company to establish the Professional Services Division (PSD). The new division included both the programmers and analysts of the former DSD and those with the sales and marketing organization. This was a major step in the maturation of services at Control Data, and along with unbundling it set the stage for rapid growth of services as a CDC business. Control Data immediately instituted a more aggressive approach to market services to customers of systems and the data centers. Although the "Total Services," strategy and the "Total Marketing" concepts and mantras withered somewhat, some of their attributes—for example, aggressively selling services whenever possible and targeting key industries as well as government—did not. With "Total Marketing" the company sought to analyze customers' needs and put together a particular package of "hardware, software, and service that is best suited to meet those needs," with an "emphasis ... not on pushing our products but on solving their problems."[90] With the start of PSD, CDC also began to break out and publicly report the revenue of its services business. (See table 8.2.)

Table 8.2
Control Data's Services Business.

Fiscal year (ending June 30)	CDC computer services revenue (millions)	CDC's total computer business (millions)	CDC services as percentage of CDC's computer business
1971	$139	$571	24%
1972	$175	$664	26%
1973	$308	$948	32%
1974	$387	$1,101	35%
1975	$445	$1,246	36%
1976	$497	$1,358	37%
1977	$620	$1,513	41%
1978	$747	$1,868	40%
1979	$886	$2,273	39%
1980	$1,032	$2,791	37%
1981	$1,111	$3,101	36%
1982	$1,239	$3,301	38%
1983	$1,251	$3,508	36%
1984	$1,401	$3,756	37%

Source: Control Data Corporation annual reports, 1958–1967 (Charles Babbage Institute).The figures are taken from the reported year with few exceptions—namely when services business revenue was not reported in a given annual report, but a subsequent year's report provided data for the previous year as either a dollar figure or a percentage of the overall computer business (which was always reported). CDC acquired a commercial lending organization (Commercial Credit), which provided loans to businesses. Revenue from this operation fluctuated and at times was quite substantial. Therefore, the computer business (all computer systems, computer services, and peripherals) is listed here. In table 8.1, revenue numbers were for times preceding this acquisition and thus represent only the computer business.

With the 1971 reorganization, Control Data had a renewed and heightened commitment to sell services within its Military Division.[91] This division's first contract in its founding year of 1958, in fact, had been for services—a consultant study of replacing prelaunch fire-control computers for Polaris missiles. The following year, CDC received the prime contract (for $7.4 million) from the US Navy to design and build those computers. Throughout the 1960s and increasingly in the 1970s, CDC sold services along with hardware to government clients, including the Air Force, the Navy, the Army, NASA, NSA, and other departments and agencies.[92]

CDC's computer services thrived in the first half of the 1970s in a difficult overall economic environment characterized by a deep recession resulting from the OPEC oil crisis. Between 1971 and 1975, CDC's services business more than tripled its revenue from $139 million to $445 million, while services went from being one fourth of CDC's overall computer revenue to more than one third. The acquisition of Service Bureau Corporation in 1973 from IBM was important to this rise.[93]

In 1974, as a result of integrating SBC, CDC had data centers in seventy US cities—and multiple data centers in some major cities. CDC's data center services, as well as field consulting, were supplying services to some of the largest industries including banking, insurance, finance, utilities, and retail, as well as government. CDC also rapidly expanded its international services operations in the first half of the 1970s. By the mid 1970s, CDC had substantial data centers and time-sharing infrastructure in Canada, Europe, Brazil, and the Near East.[94] Networking made these operations all the more efficient (more efficient use of computer processing resources) as CDC immediately began integrating its CYBERNET and SBC's computer network.[95]

For a number of years computer services were in the shadows of CDC's supercomputers, but by the mid 1970s the company's leaders began placing it in the spotlight. CDC's 1975 annual report contained an item titled "Services Profits Improve Markedly." "The mature data services," it stated, "produce a higher rate of return than any of our product businesses [and] offer the company our greatest growth potential with less risk than most hardware products now that we have the major part of our world network in place."[96] CDC had thousands of computer services specialists serving customers around the world by the mid 1970s.[97] A 1975 planning document highlighted the diversity of CDC's professional services, which included "data management consulting services, structural engineering, engineering management operations, information services, automated engineering and production consultants, and network consulting services."[98]

While growing markedly in the first half of the 1970s, CDC also suffered some setbacks on delivering on the largest, most complex contracts for programming services. The setbacks highlighted the fact that real-time systems integration (which had long delayed SAGE and SABRE) remained a perpetual challenge in the industry. In the early 1970s, CDC won the contract for the Air Force's Advanced Logistic System (ALS) project, a $250 million multi-year project to build a real-time system (hardware, software, networking, and systems integration) to control logistics (personnel, food, equipment, weaponry, supplies, etc.) for the Air Force.[99] Many of CDC's military projects were successful, but this one was not—and the Air Force had been warned that it was too ambitious by Willis Ware of the RAND Corporation.[100] Ultimately, CDC and the Air Force reprogrammed the ALS just for batch processing, rather than its raison d'être—real-time processing. The company also experienced a setback on a major networked banking system for Union Bank of Switzerland in the first half of the 1970s, further highlighting the great challenges for massive real-time transaction processing systems.[101]

In the second half of the 1970s, CDC's computer services continued to expand quickly. By 1977, CDC had a staff of more than 1,000 who were focused on marketing the services of the company's data center employees and field consultants. That year, in the United States, it provided computer services to more than half of the *Fortune* 500 companies, for eighty leading banks, for seventy-five major finance firms, and for fifty large insurance companies. It also did substantial business servicing the Department of Defense, the Department of Energy, the civilian federal government, and local governments. Its services clients also included 4,500 smaller businesses. Meanwhile, its telecommunications and satellite communications network allowed CDC to serve customers in two dozen countries.[102]

In the early 1980s, Robert Price was named Control Data's president and chief operating officer. William Norris continued as CEO and chairman of the board. During that time period, Norbert Berg, vice chairman of the board, was the third member of the company's leadership. In 1985, Norris retired and Price became CEO and chairman of the board, inheriting the reigns at a difficult time. In 1984, after many consecutive years of growing earnings, CDC lost $45 million. More troubling, the following year it had a decline in revenue and a loss of $563 million, including a $209 million loss from operating activity. (Most of the remainder was a result of restructuring charges—largely the sale of Commercial Credit, a financing subsidiary.) CDC had become too diverse, had too much overhead, and was vulnerable to an economic downturn. When a downturn occurred, events spiraled out of control for CDC and resulted in a liquidity crisis that became all the more challenging when a 1985 public offering of additional stock faltered. Price and CDC focused on trying to shorten receivables while delaying payables, essentially doing whatever they could to conserve cash and stay solvent.[103]

CDC's businesses in mainframe systems and computer peripherals were declining, and the company had invested heavily in unprofitable educational services with PLATO. CDC's computer business (systems, services, and peripherals), now the entire firm with divestment of Commercial

Figure 8.6
Robert Price (foreground) succeeded William Norris (background) as CDC's CEO in 1985. Courtesy of Charles Babbage Institute, University of Minnesota.

Credit, lost another $311 million in 1986 before showing a profit of $57 million in 1987 on revenue that was down by about 10 percent from its 1984 level. CDC achieved earnings of merely $19 million and $2 million in 1987 and 1988 before a disastrous 1989. In that year, revenue was down more $700 million from mid-1980s levels; the company lost $680 million, and that led to changes in top management.[104]

Conclusion

The faltering of CDC's services business by the late 1980s was a part of the company's overall financial struggles. CDC was not alone; the economic environment of the late 1980s was very difficult for the computer business. That environment brought restructuring and multiple years of major losses for IBM (its first losses in nearly eighty years). While growth of personal

Expanding Capabilities

computers played a major part, the overall economic environment did as well. Computer services had long been a standout business for CDC, but it was meaningfully tied to CDC's systems business. CDC's peripherals business was especially hard hit with the rise of the personal computer. In services, much energy and considerable resources had gone into educational services, which never provided a good return on investment for CDC.

In the early 1990s, CDC brought in new management that focused on divestiture and on unlocking the underlying value that remained in the enterprise. The piece with the most promising future— computer and information services—was spun off as Ceridian, a public company that focused increasingly on software and services for human resources departments. In 2007, investors purchased Ceridian for $5.3 billion and took it private.[105]

IBM, as will be discussed in chapter 10, restructured in the late 1980s and the early 1990s. Under CEO Louis Gerstner, it focused heavily on growing its services and software businesses, and avoided CDC's fate of being split apart.

9 Brokering Contractors: Gentry, Inc., Phyllis Murphy and Associates, COMSYS, and the NACCB, 1972–1998

On the morning of February 18, 2010, Andrew Joseph Stack III set his house ablaze. He left it to burn as he drove to Georgetown Airport, two dozen miles north of Austin. Boarding his single-engine fixed-wing Piper PA-28-236, he intentionally flew the small airplane into a seven-story office building in Texas' capital city—a building in which about 200 US Internal Revenue Service employees worked. This horrific act, in the wake of and bearing resemblance to those of 9/11 less than a decade earlier, brought confusion as well as public fears of additional terrorist attacks that morning. Stack's cowardly deed fortunately was a lone effort. Despite considerable destruction to one side of the building, his suicide mission surprisingly was limited to only several fatalities other than himself.[1] Stack had worked for many years as an independent contractor computer programmer, and shortly before his fateful flight he had posted on a website a manifesto that included a strong critique of the IRS and a reference to an obscure part of US tax code: "Section 1706."[2]

Stack's actions that morning prompted Washington-based attorney Harvey Shulman to write an op-ed piece and submit it to the *New York Times*, which published it on February 20.[3] Shulman had served for many years as the general counsel for the National Association of Computer Consultant Businesses, a trade association that represented IT independent contractor brokerages—job shop enterprises that essentially were specialized temporary staffing agencies for independent contracting computer programmers and systems analysts. Shulman, whose brother was an entrepreneur and a past leader of an IT independent contractor brokerage, knew Stack's profession well, but did not know him personally.[4] He began the op-ed piece with a statement that "NOTHING could excuse the murderous plane crash."[5] After fully and unequivocally condemning Stack's actions, Shulman wanted to explain Section 1706, its political origins, and what were, in his opinion, the unnecessary and unfair hardships this "low tech tax code" placed on independent contracting programmers and systems analysts, and on brokerage firms representing them.[6]

Section 1706 of the 1986 Tax Bill—its content and context an important part of the political economy of the IT services industry—targeted IT contractors and removed a "safe haven" provision that previously had protected contractors in the computer services industry, singling out contractors in that industry by removing the safe haven only in the IT field. After Section 1706 was law, the IRS could determine (generally in an audit years later) that an IT services contractor should have been an employee (in view of the role he or she played relative to a client or multiple clients). If such a determination was made, all back taxes and penalties were owed to the government by the brokerage that had placed the independent contractor, or, potentially, by the client organization.[7] This was true even if the independent contractor had dutifully and fully paid the roughly double Social Security assessment (amounting to the employee's and the employer's contributions) of self-employment tax.

The Association of Data Processing Service Organizations, on behalf of mid-size to large computer services companies that used employees rather than contractors, had lobbied for Section 1706. This sizable trade association for computer services and software products companies helped to convince Senator Daniel Patrick Moynihan (D–New York) of the importance of using the tax code to discourage, if not outright restrict, companies from using independent contractors for IT work (rather than hiring employees or contracting for employees from a traditional computer services company with an employee structure). Moynihan succeeded in getting Section 1706 added into the 1986 Tax Bill.[8] As Shulman related in his op-ed piece, Moynihan later recognized his error in advocating inclusion of Section 1706. It restricted independent IT contractors in ways other independent contractors were not restricted. In the years after the 1986 Tax Bill became law, Moynihan and more than sixty other senators spanning the political spectrum—from Ted Kennedy to Jesse Helms—sought the repeal of Section 1706.[9] Once on the books, however, federal tax code is difficult to overturn—especially code favoring large powerful interests (companies like CSC, EDS, ADP, and CDC, represented by their legal and communication departments, and their relatively deep-pocketed, lobbyist-focused trade association ADAPSO) over much smaller, less powerful ones (ranging from brokerages placing several hundred workers to individual independent contractors).[10] Even though a majority of the US Senate recognized the problems with the section, it remained on the books.[11] Section 1706 placed extreme hardships on both independent contractor brokerages and independent contractors themselves. Most companies and organizations purchasing computer services, including some that had previously used contractors, subsequently chose to go the safe route and hire "employees" (either directly or through contracting with a traditional computer services company). Given Section 1706, many former independent contractor brokerages revised their business model to become standard computer services companies with employees in order to protect themselves.

Though a few smaller enterprises and some mid-size companies have been addressed in this book (among them John Diebold and Associates, Canning, Sisson, and Associates, C-E-I-R, Inc., and Computer Usage Corporation), the greatest attention has gone to services enterprises that grew to become giants—IBM, CDC, CSC, ADP, and EDS. More generally, the field of business history has focused far more on large corporations than on small businesses—especially in addressing developments during the twentieth century. There has also been a dearth of historical literature on female entrepreneurs in the twentieth century. This chapter aims to contribute to correcting these biases in business history by concentrating on the brokerage staffing segment of the computer services industry, and by highlighting the major role played by women.[12]

The computer services industry has long included many small companies—companies ranging from enterprises with fewer than 200 or even fewer than 50 workers to one-person and two-person consulting enterprises. Many of the self-employed individuals in IT have far more technical skill as programmers and systems analysts than ability to market themselves to clients (and clients tend to feel more comfortable contracting with an established multi-person small firm rather than with individuals). There are efficiencies in having brokerages handle the business side—the marketing, negotiating, billing, collecting, and other functions—and leaving the contractors to do what they do best: design, program, and fix software code.

Beginning in the 1970s, Grace Gentry pioneered the use of independent contractor IT brokerages to place independent contractor programmers and systems analysts with clients for short, intermediate, or longer terms. After Gentry, Inc., many other small to mid-size brokerages (typically placing 20 to 200 contractors with clients at any given time) formed. This chapter focuses on those brokerages, on the National Association of Computer Consultant Businesses (which represented them to fight Section 1706), and on the particularly important contributions of entrepreneurial women—Grace Gentry, Phyllis Murphy, and Peggy Smith—to this field of IT services.[13]

Grace Gentry and Gentry, Inc.

Grace Marie Hill, born in 1938, grew up in metropolitan Dallas and attended preparatory school in the Dallas area before securing a scholarship to Radcliffe (the women's college of Harvard University).[14] She had been determined to pursue a career as a professor, but she left Radcliffe after a year to marry Richard Gentry, who had recently received a BS degree from Texas A&M University. They soon had a child and moved, first for Richard's Air Force pilot training and then for his doctoral studies in physics at Berkeley. Richard Gentry, finding graduate education less appealing than entering the workforce, soon joined IBM's San Francisco office as a systems engineer. He trained IBM customers in programming and later did services work on behalf of the Alameda County government.[15]

Grace Gentry resumed undergraduate studies at UC Berkeley, taking courses in sociology. She grew disillusioned with the field of sociology as it was taught at Berkeley, however, and switched to statistics. Richard Gentry, knowing that statistics relied heavily on computing, brought home a copy of IBM's Programmer Aptitude Test. Grace Gentry scored high on that test, and on local and federal government service examinations she soon took. She also completed a two-week programming course given by IBM. With high marks on her government service exams, she was offered an attractive management-track position with the Social Security Administration (in metropolitan Washington), but the family was attached to the San Francisco Bay area by that time and she instead took a position as a business analyst for the University of California system, where her job essentially consisted of interviewing administrators, assessing needs, and translating these needs for the programmers and systems analysts who designed and built software systems. On certain efforts, she did computer programming work herself for the university's data processing systems. Having programmed a successful report writer for university admissions, she gained her boss' attention. When she asked the boss how he had heard about her programming, he told her he had heard of it in a conversation in a men's rest room. "That's one of the problems," Grace Gentry promptly responded. "You guys go in there and piss together, and chat and yak. How are we women ever supposed to be successful managers when we can't go in there with you?" Shortly thereafter, a male employee asked Grace Gentry to come into a men's room because he had something to tell her.[16] The joke was enjoyed all around, but was symptomatic of a hurdle for women: Deals, promotion opportunities, and institutional and professional strategies often were discussed by men in spaces that formally or tacitly excluded them—a conference room, a bar, or the back nine of a golf course, if not a men's room.

Deciding in the early 1970s to test the waters in the business world, Grace Gentry took a managerial position with the Bank of America. A senior executive, in the course of hiring her, related that the bank had missed an opportunity in not hiring more minorities and did not want to make a similar mistake with women. Richard Gentry, too, made a career change in the early 1970s. He left IBM for SYSDYN, an entrepreneurial startup, which soon faltered. A talented programmer and analyst, he began work as a programmer/analyst contractor for Alameda County. The county wanted him to incorporate and they also soon had need for additional programmers. Seeing an opportunity, Richard and Grace Gentry formed a brokerage to supply independent IT contractors (programmers and systems analysts) to Alameda County, and soon to other clients.[17] Soon it became Gentry, Inc.

Grace Gentry saw many of the same limitations for advancement for women at the Bank of America as in the University of California system. Possessing strong management and people skills and an entrepreneurial mindset, she decided to leave the bank to devote herself full-time to running and growing Gentry, Inc. She had no direct models, as she knew of no other business in this field of brokering the services of independent contractor programmers and systems analysts.[18] She initially hired five programmers, all women. It was not an intentional effort to promote the opportunities and careers of women; she simply was taking advantage of her network of friends and colleagues in the programming field—people she knew and trusted.

In the early years, some referred to Gentry, Inc. as "Wonder Women, Inc." because all the earliest contractors were women and because of Grace Gentry's management. Some male contractors were eventually brought in, and the firm continued to grow rapidly under Grace Gentry's leadership. Gentry quickly recognized the risk of tying the firm to one client or only a few, and sought a diverse base of customers in government and in a range of industries. On the industry side, she generally targeted small to medium-size businesses that were "under the radar" of IBM, CDC, CSC, EDS, and other leaders in computer services. Another reason for the success of Gentry, Inc. stemmed from Grace Gentry's strong skill for identifying talented individuals and matching them with clients.[19]

One reason the business model of Gentry, Inc. worked so well is that many programmers were independent-minded, creative individuals who prefer the idea of working on projects, and the flexibility it offered, over a traditional long-term employee-employer relationship. Even contractors who ended up working fairly continuously and keeping typical business hours liked the *idea* of flexibility. They could also make substantially more money if they worked the equivalent of full-time hours. At the start, Gentry, Inc. charged a client $12 an hour for the services of a programmer, and in turn paid the programmer $10 an hour. This translated to roughly $20,000 a year gross if the programmer worked about 2,000 hours a year, as was typically specified for salaried employees. In the early to mid 1970s, good programming jobs tended to have annual salaries between $10,000 and $14,000. Even after self-employment tax and other forgone benefits (such as retirement contributions), a programmer could do better financially than a salaried programmer who worked long hours, as many salaried IT workers did. With contracting, the model was set so that all hours were billed. Grace Gentry had the insight to create an employee-like career path. She would help programmers grow in their assignments, and would train them to become systems analysts—the technical role that Richard Gentry played for the company on the most complex projects, while Grace Gentry ran the business.[20]

In the early years, some potential customers wanted to visit Gentry, Inc.'s offices before signing on to use the firm. They wanted to see the operation and, as Grace Gentry put it, "kick the tires." She confidently told them that there was no office. The business side was run out of her kitchen. (See figure 9.1.) She explained that this kept overhead very low and enabled her business to deliver strong talent and results for less than larger firms charged.

Another difference between brokerages and traditional computer services companies (and computer manufacturers selling or bundling services) was that in theory, and for the most part in practice, there was no "bench" time for brokerages. Contractors were paid for the billable hours they worked, not for time between projects. This was in sharp contrast to traditional computer services firms with employees and salaries. Further, Gentry's contractors worked on clients' computer systems—a very different model from running data centers and acquiring expensive computing equipment. Gentry, Inc. had only a small computer workstation, which was primarily to do payroll and accounting for Gentry, Inc.—not programming for clients.

Grace Gentry marketed the business aggressively and successfully and was able to keep good contractors busy on contracts almost continuously, finding a contractor a new assignment as he or she was finishing an earlier one. Keeping contractors satisfied with their desired number of hours of

Brokering Contractors

Figure 9.1
Grace Gentry running Gentry, Inc. from her kitchen table, circa mid 1970s. Courtesy of Grace Gentry.

work was a key to high retention, which Grace Gentry and Gentry, Inc. were quite successful at achieving. The business did require a bit of capital, as there could sometimes be a lag between needing to pay contractors and getting invoiced payments from clients, but the costs of starting and growing the operation were small in comparison with most types of businesses having dozens of employees and expensive office space.[21]

Before the end of its first decade, Gentry, Inc. was regularly placing about a hundred contractors at time. It decided to increase overhead a bit (and open up opportunities to supply contractors who did not want to use a business run out of a kitchen) by acquiring office space in Oakland. Gentry, Inc. sold services to a wide range of local governments and industries, but continued to focus its marketing efforts on small to medium-size enterprises, which often could not afford the services of the largest computer services companies or the computer giants.[22]

By the 1980s there were a number of independent contractor IT brokerages in California, and a smaller number in other states. Leaders of this segment of the industry felt threatened with passage of Section 1706 as part of the larger Tax Bill of 1986. Different firms handled it different ways.

Grace Gentry and some others launched a trade association—the National Association of Computer Consultant Businesses—to fight Section 1706. Harvey Shulman, an attorney and the brother of the leader of an independent IT contractor brokerage business (COMSYS), helped create the NACCB and long served as its general counsel. The NACCB's lone lawyer, who had many other clients, was up against the lobbying of ADAPSO and the push of large-scale services businesses, and Section 1706 remained part of the federal tax code.[23] In view of the risk, successful and established enterprises such as Gentry, Inc. shifted to an employee-based model, and some former contractors acquiesced and became employees. This meant that there was the risk of the bench time that the bigger firms faced, and Grace Gentry focused on minimizing it.

Gentry, Inc. continued to thrive in the 1990s, boosted by a strong economy, by the need for "Y2K readiness" (the need to recode to address pervious use of only two digits for a year), and by the increasingly ubiquitous World Wide Web and the dot-com frenzy. By the late 1990s, Richard Gentry had been doing contract work as a systems analyst for many years and wanted to retire. He and Grace decided to sell the business in 1998. The timing (at least that of the sale itself) was fortunate, insofar as the IT market was about to take a huge hit with the dot-com collapse of 2000. Gentry, Inc. was sold to Personnel Group of America, an employee-based national IT staffing enterprise, for $12.5 million in PGA stock.[24]

Phyllis Murphy and Associates

Over time, especially after the advent of NACCB in 1987, independent IT contractor brokerages would learn from one another. In earlier years, most such enterprises were "invented" by their entrepreneurial founders with little sense of others in their field, or at least not before being in business and experiencing similarly modeled competitors—competition that varied significantly with regard to geography because independent IT contractor brokerages were very much local businesses. This was true of Phyllis Murphy and her firm Phyllis Murphy and Associates, which was launched nearly a decade after Gentry, Inc.

Phyllis Murphy, born in 1939, grew up in Quincy, Illinois, and received an accounting degree from Western Illinois University. She applied to many accounting firms in Chicago for more than a year without a single offer before Draper and Kramer, a small real estate firm on the city's west side, hired her as an accountant. Her boss asked if she wanted to "learn to program," to which she had no immediate reply. He said computers will "do exactly what you tell them to do," which sounded appealing to her. After taking a one-week programming course offered by Sperry-Univac, she was soon "wiring" a Univac 1004. She shifted her career track to become Draper and Kramer's Director of Information Technology.[25]

Wanting to further advance her IT skill set in databases and networked systems, Murphy left the accounting firm and joined Consumer Systems, an employee-based IT services consulting firm with branches in Chicago and Minneapolis. She advanced in the firm and was managing more than 140 staff consultants in Chicago by 1978. Her husband had an attractive career-advancement opportunity in Los Angeles. She agreed to move from the large Chicago Consumer Systems office to head a much smaller one in Los Angeles. She found the IT business culture to be far different in Southern California—more independent contractors, fewer long-term business to client relationships, and sales staff often possessing inadequate skills and training. Disappointed that Consumer Systems would not invest more in growing the Los Angeles operation, she left in 1981 and investigated many possibilities before agreeing to work for an IT placement firm. Working for that firm for several months, she saw firsthand the common problems that plagued many businesses' IT operations, including inadequate training, lack of professionalism, and poor back-office operations. Having made connections, and seeing opportunities to run a business better, she launched her own company to provide programming services. To offer both the flexibility of contracting and the stability of employees, she set it up as a hybrid with roughly 60 percent employees and 40 percent independent contractors. As she later recalled, about 20 percent of the employees and contractors were women. Cognizant of the importance of reputation and name recognition to a small IT services business, she named the enterprise Phyllis Murphy and Associates.[26]

As Grace Gentry had done with Gentry, Inc., Phyllis Murphy initially ran her business out of her home. After about two years, she had more than twenty employees/contractors and began to lease office space and added sales personnel. With major clients, such as Borax, she often was involved in the sales effort. She handled the accounting and the finances entirely by herself. In the early years, Murphy felt that her gender helped her—as a woman IT services business owner in Los Angeles, she stood out. Over the longer term, she believed gender was inconsequential, her business was reputation-based, and what mattered was her coming through for clients. Unlike Gentry, Inc., she targeted larger firms. Although that meant competing more directly with computer services powerhouses, she believed larger companies (generally far exceeding $500,000 in annual sales) were run more professionally and gave her employees and contractors the latitude to succeed rather than trying to micromanage their time.[27]

Murphy, generally, was not concerned with competition from larger computer services consultancies. She related that the larger IT services firms generally entered the area through acquisitions and had little understanding of the business environment in Los Angeles. An example Murphy cited was nationwide Boca Raton–headquartered Compustaff, which moved into the Los Angeles market through the acquisition of smaller IT services companies. Compustaff's efforts, as well as those of other national companies, often were unsuccessful. As she put it, such national IT services providers often had a difficult time securing business and soon shut down their local offices in Los Angeles. She emphasized that larger enterprises would commonly spend considerable sums of money on marketing campaigns, when it was a people and relationship business—sales calls, track record, and networking were the keys to success.[28]

Section 1706 in 1986 led Murphy to quickly join the NACCB and become active with the organization. She started a chain letter opposing the legislation and as a result soon became the "go-to" person for quote-seeking journalists, including in national publications such as the *Wall Street Journal*. With this publicity, and the law in place, she quickly shifted to an all-employee model, believing it to be the more prudent course. It was a wise decision, as four IRS audits, perhaps influenced by her outspoken position, came in rapid succession. While she made substantial contributions to the NACCB, she believed the benefits exceeded what she gave to the organization. The NACCB was launched as a one-issue organization (trying to repeal Section 1706), but soon evolved to become a typical trade association where members shared ideas and gained insights from one another. This was especially fruitful within NACCB, as most were local businesses that operated in different, and hence non-competing, geographies. In Phyllis Murphy's case the NACCB was the key to her establishing group health insurance, as well as adding clauses to protect her company in writing contracts.[29]

Phyllis Murphy and Associates, which has consistently employed more than 100 programmers, analysts, and other IT specialists, continued to grow in the 1990s and into the 2000s. In addition to many mid-size companies and organizations, it has served large companies and institutions including Sony Pictures, Kia, Hyundai, Farmers Insurance, the Walt Disney Corporation, and the University of California at Los Angeles.[30] Though most other early pioneers of independent IT contractor brokerages (including those evolving solely to an employee model in the aftermath of Section 1706) have retired, Phyllis Murphy continues to run this successful company out her Burbank office suite.

COMSYS

Grace Gentry was the earliest known entrepreneur to form an IT services independent contractor brokerage business, and Phyllis Murphy grew a similar enterprise (though with a contractor-employee hybrid model from the start) in the tough business environment of Los Angeles. By the early 1980s, entrepreneurial men and women had launched dozens of independent IT contractor brokerages in the United States. Perhaps none grew larger or more prominent than Fred Shulman's COMSYS.

Fred Shulman was born in Brooklyn, New York. When he was nine years old, in 1965, his father, who worked for the Internal Revenue Service, received a promotion and moved the family—Fred, his mother, and his brothers Alan and Harvey—from Brooklyn to Washington. Shulman was first introduced to computers in a high school mathematics class and was struck by the machines' ability to "respond to commands." At the University of Maryland he took numerous courses in engineering and business before deciding to major in computer science. He learned COBOL, PL/1, and various business applications in finance and accounting, and graduated with a BS in 1978. The summer of his junior year he interned at Sperry-Univac. Upon graduation he was hired by Sperry-Univac. He worked (in Lexington Park, Maryland) on the Department of Defense Worldwide Military Command and Control System, programming drivers for the printers in CMS-2, a precursor to Pascal and C. Wanting to increase his earnings and to live in the District of Columbia, he responded to an employment advertisement placed by the defense contractor Raytheon in the trade publication *Computerworld*. Soon he had a job with Raytheon.[31]

At Raytheon, Shulman was the first employee of a new data center, where he started on a designated "one-year" project doing "specifications, programming, and conversions." Having finished his work in three months, he was commended and told that he had to wait until Raytheon hired a manager of the center for new assignments. Inquiring of a vice president if he could be considered for the manager position, he was told (at age 24) that he was too young. Frustrated at having to wait around for more work, and not being considered for the senior post, he left Raytheon to join computer services company Advanced Technology Systems, which had a contract for switching Amtrak from CDC to IBM mainframe equipment for a seat reservation system. At Advanced Technology Systems he rose through the ranks and ran the project team, which consisted of a group of consultants. He was surprised by the business model of using primarily consultants rather than employees, but at the conclusion of the project he understood the flexibility this offered both to Advanced Technology Systems and the contractors (to jump to another opportunity). During an interview for a job with a Fairchild subsidiary, American Satellite, he asked if that company would hire him as an independent contractor. American Satellite agreed to do so, and Shulman remembered being excited to be in business for himself doing work he "loved to do."[32]

Shortly after college, Shulman had done a short-term project teaming up with a college friend, Howard Stein, to build an inventory control program. Though they discussed the possibility of partnering to start a company, Stein wanted to be a certified public accountant and Shulman had a day job. The idea of partnering to start and run a company, however, remained with Shulman.

The opportunity at American Satellite required a small team. In 1981, Shulman and a former Raytheon colleague, Katherine Griffin, launched COMSYS as an incorporated company (subchapter S) in which the principals would oversee projects and would hire programmers as contractors. Shulman got help in writing contracts from his brother Harvey, by then a successful attorney. The American Satellite project enabled Fred Shulman and Katherine Griffin to rent a small office space in Rockville, Maryland. As the company grew, it relocated to larger and larger quarters in the Washington metropolitan area.

COMSYS initially advertised in *Computerworld* and the *Washington Post* to secure applicants for contractor positions. Most independent contractor programmers had a set rate. COMSYS would add a flat fee per hour. With a $20/hour programmer or a $40/hour senior analyst, COMSYS would add an $8/hour fee and charge the client $28 or $48, respectively. Over time, COMSYS shifted from a flat markup to a percentage, ranging from 25 to 30 percent. Shulman and Griffin would screen candidates and make recommendations, but would let clients meet the candidates and make the final decision on whether or not to bring them onboard.[33]

As Shulman recalled, getting talented and experienced contractors was relatively easy. The independent contractor (self-employed) model often fit the personality type and inclination of successful programmers and systems people. As Shulman put it, "a lot of these workers were independent contractors for a reason. They were independent-minded people. They wanted to move from project to project, not get involved in company politics and just do the work they were tasked to do. … What they didn't like to do or maybe what they lacked doing well was the marketing. We offered them good contacts and they didn't have to market themselves so it was a win-win proposition."[34]

Fred Shulman typically would be the manager on a project in the early stages, but would bring someone else along to oversee it so he could secure the next substantial client. He did this with American Satellite, MCI, GTE, and other clients. This allowed Griffin to focus on the back-office operations of the business. Within several years, Shulman and Griffin grew the business to over $3 million in revenue with about forty contractors at a time, but to expand more they believed they needed a dedicated sales and marketing executive. They looked to Howard Stein, who had left accounting to become a leading marketing executive with the payroll processing computer services specialist ADP.[35]

Stein, who joined the COMSYS management and ownership team in 1984 (receiving a one-third share in the firm to match the shares of Shulman and Griffin), was not well versed in computing and software technology, so Shulman largely stopped working on projects and went out with Stein on sales calls. With a technical leader in Shulman and a marketing leader in Stein, and with Griffin handling back-office functions, COMSYS was highly successful. Shulman would step in and solve problems on projects, either providing technical support or helping to educate contractors on how best to fit in culturally with a corporate or government client, but spent much of his time with Stein marketing COMSYS. COMSYS also hired Fred Shulman's brother Alan, who had long worked at IBM, as chief financial officer, and added several recruiters. Fred Shulman and Katherine Griffin got married.[36]

With these mid-1980s additions, especially Stein's marketing experience and know-how, COMSYS was poised for expansion. Not only did it contract with major firms in telecommunication and many other industries, it successfully was brought on as a sub-contractor by Planning Research Corporation and Computer Sciences Corporation, two of the leading computer services firms specializing in the vast federal government computer services contracting business in the Washington metropolitan area. With Stein taking the lead with marketing, COMSYS also broadened its base. It had to overcome a serious challenge with Section 1706, which Fred Shulman learned of in 1986 from two Washington-area competitors in the independent IT contractor brokerage business, Rich Phillips and Judy Goldberg. Shulman immediately consulted with his brother Harvey.[37]

Soon thereafter, Fred Shulman attended his first ADAPSO meeting and learned firsthand that the organization—representing some of its largest members, including CSC, ADP, and CDC—had pushed Senator Moynihan to add Section 1706 to the 1986 Tax Bill. Shulman saw this as a "serious threat to our business." Advised by his brother Harvey, he realized that COMSYS should immediately change its business model to an employee basis rather than a contractor basis or else it would be at grave risk with the IRS. To Shulman that meant more than just paying a little extra for social security and benefits (expenses that could be passed along) and taking on former contractors as employees; it meant disrupting a culture that worked. With regard to the independent contractors, Fred Shulman later reflected, "We knew their mentality ... . It was like trying to go out to someone who is a 'mercenary' and telling them that they can't be a 'mercenary' anymore. If you want to do business with us you have got to be our employee. We must control you." COMSYS quickly lost about a third of its contractors. Some participants in the industry were less prudent than COMSYS (and Gentry, Inc. and Phyllis Murphy and Associates) and continued with the contractor model, risking a potentially devastating IRS audit. Fred Shulman, working with his brother and with other company leaders in his field, founded the NACCB in January 1987.[38]

Believing there was no rationale for punishing only IT independent contractors and brokerages, but not other industries, Harvey and Fred Shulman expected that Section 1706 could be rapidly overturned. When many senators supported its repeal, they became even more hopeful, but the large companies in ADAPSO and the difficulty of rescinding existing federal tax law carried the day. Within several years, the early optimism of the two brothers at removing Section 1706 waned. Harvey Shulman drew up model contracts that probably would offer some protection in using contractors. COMSYS used these and gradually brought on a small number of contractors. To limit risk in case the model contracts did not offer complete protection, Shulman, Griffin, and Stein mostly hired programmers and analysts as employees.[39]

COMSYS adjusted to the changes and sought to expand to new territories. It did this by seeking to serve existing Washington client corporations and organizations that had major locations and IT needs in other cities. With Honeywell as a major client in the Washington area, COMSYS set up a branch near Honeywell's computer division located in Phoenix. When Honeywell's facility in Phoenix was shut down, the Phoenix office of COMSYS secured business with American Express and Motorola in that area. When MCI (a major customer) relocated its headquarters to Colorado Springs, COMSYS opened an office there. COMSYS also expanded to several cities closer to Washington, including Raleigh and Atlanta. After this expansion, the company had a few hundred employees and dozens of independent contractors—roughly 80 percent employees and 20 percent contractors. In 1994, COMSYS, a company whose revenue and profits had grown steadily, had sales of $55 million a year.[40]

As COMSYS grew, Fred Shulman, Katherine Griffin, and Howard Stein began to think more about cashing out to reduce risk. They had put their life savings and their life's work into the company. They had worked around and succeeded after Section 1706, but what if other unfavorable legislation emerged, not to mention the myriad of other risks all small and medium-size reputation-based companies face, particular ones in the fast-changing field of information technology. Over the years they had entertained a number of offers to buy COMSYS, but all had been stock offers, and they hadn't been interested in trading a risk (lack of diversity with investments) they could potentially control (in running COMSYS) for a risk they could not control (passive investors in a single company). In 1994, they sold COMSYS to GTCR, an equity partner in Chicago, for approximately $20 million. GTCR was consolidating several staffing firms, including a secretarial staffing firm called Talent Tree. The model was to put money into the firms to expand proven enterprises and let the owners cash out while keeping them as executives to run the enterprises. Fred Shulman and Howard Stein signed a two-year contract to stay on, but they left after thirteen months as a result of differences of opinion with management. Shulman and Griffin took their family on a trip around the world while they waited out the remaining nine months of the two-year non-compete clause they had signed. Shulman, Griffin, and Stein then launched a new computer services staffing firm called AETEA.[41]

The Evolution of NACCB and Its Member Companies

Before Section 1706, concerns about statewide legislation restricting independent contractors in information technology led seven IT consultant brokerages in Northern California to form the Software Consultants Broker's Association in 1985. IT independent contractor brokerages in Southern California then launched a Southern California chapter of the SCBA. In late 1986, IT independent contractor brokerages in the Northeast formed the Software Services Association of New England. Some members of the two SCBA chapters and some members of the SSANE then met in New England to discuss efforts to combat Section 1706. A more formal meeting was held in Washington in late January of 1987, when a snowstorm turned a planned two-day meeting into a four-day one. The ten representatives of IT brokerages (Grace Gentry and Fred Shulman among them) worked with the lawyer Harvey Shulman, and a national trade organization—the National Association of Computer Consultant Businesses (NACCB)—was launched. In addition to Grace Gentry, one other woman took part in the founding of the NACCB: Judy Goldberg, from the Washington area.[42]

Tom O'Donohue, the NACCB's first president, worked closely with general counsel Harvey Shulman, who was quite dedicated to the financially struggling organization and who billed only a fraction of the hours he spent on it. In addition to fighting Section 1706, Shulman helped member firms with model contacts, legal advice, wage plans, benefits, insurance, and other matters.[43]

In 1989, Grace Gentry became president of the NACCB. Her innovative leadership helped to advance the scope and usefulness of the association. She presided over the first conference the association held. This marked the formalization and extension of the organization beyond just a lobbying organization to a broader-based trade association that served the function of educating members and spreading best practices throughout the industry. Recognizing that the membership would have to be expanded to relieve the financial burden on the organization and its members (most of which were small businesses), Gentry recruited Peggy Smith to work as a consultant.[44]

In the early 1980s, Peggy Smith worked at Lambda Technologies, a large IT consulting business. Shortly after she joined Lambda, it was taken over by GEIS, General Electric's time-sharing and services company. In 1982, Smith transferred from Philadelphia to set up a GEIS office in Greensboro, North Carolina. The time-sharing firm's top leaders, however, had minimal expertise with the IT consulting business, and their support for the Greensboro operation was inadequate. Disgruntled, Smith joined with two other GEIS employees to launch a new contracting firm. In 1989, after disagreements about the strategic direction of the young firm, Smith was bought out by the remaining partner (the third co-founder had left before Smith).[45]

In 1989, Grace Gentry, aware of Peggy Smith's considerable experience and talent, brought her on as a consultant to expand the NACCB's membership and to improve its financial position. In the mid 1990s, after serving as a consultant since 1989, Smith became the association's first executive director. She worked to help the organization succeed with its growing number of activities and functions, which by then included not only extensive lobbying efforts (which achieved some significant smaller victories despite not being able to get Section 1706 rescinded) but also annual conferences, industry roundtables, the preparation of a Statement of Business Principles, broader participation in IT trade events, marketing, the creation of a benefits committee, and other endeavors.[46] By the mid 1990s, the NACCB had more than 200 members and was more financially stable.[47]

Independent contractors and small businesses brokering their services have been—and independent contractors continue to be—a meaningful part of the IT services industry. Most small computer services companies (including brokerages for independent contractors and independent contractors themselves) have left few documentary records. However, oral history interviews have shed light on some of the entrepreneurial and leadership efforts of a few important figures in small IT services brokerages and in the NACCB, including interviews with Grace Gentry, Phyllis Murphy, Peggy Smith, and Harvey Shulman by the author and an interview with Fred Shulman by Burton Grad.

An NACCB survey conducted in 1996 provides further insight into the makeup of this understudied and underappreciated side of the computer services industry. Responses from all 225 member firms were solicited; responses were received from 119 of them. Eighteen percent of the firms had annual revenue below $3 million, and 31 percent had revenue between $3 million and $7 million. Of the roughly half with more $7 million in annual revenue, 85 percent had revenue under $20 million. A good guess might place the median in the neighborhood of $7 million in annual revenue and the mean around $10 million. The latter suggests that, in aggregate, the firms represented by the NACCB had annual revenue of about $2.5 billion in the mid 1990s—a relatively small but meaningful portion (less than 3 percent) of the revenue of the entire computer services industry.[48]

Most independent IT contractor brokerages (and most employee-based IT consultancies that started as brokerages and remained in the NACCB) were local businesses, especially in their early years. Some extended beyond a particular metropolitan area to include a small region (for instance, Burbank-based Phyllis Murphy and Associates had some business as far south as San Diego, and COMSYS had business in six cities at its peak), but few NACCB companies were national or even broadly regional in scope—most focused on a radius smaller than 50 miles. Three fourths of the responding businesses had only one or two offices, and more than half of them had only one. Years in business was a very meaningful predictor of a higher level of revenue and a larger number of employees (or consultants—though by 1996 few brokerages were challenging the IRS by continuing with a contractor model rather than employee model). These were relationship businesses for which identifying and retaining talent and serving clients effectively and efficiently were important skills. The culture of businesses and regions was important, as Phyllis Murphy pointed out with regard to greater Los Angeles. Larger enterprises seeking to swoop in with acquisitions and grow big quickly in Southern California did not succeed. Small businesses offered greater opportunities for women—both as contractors and as entrepreneurs—than large firms. Some entrepreneurial women had experienced gender bias in working as employees for larger organizations, but had not experienced as much of it in working with clients or in selling their expertise and that of their employees or contractors. In the independent IT contractor brokerage business, the focus was on technical and managerial ability, and what level of service could be provided at what cost. Grace Gentry and Peggy Smith provided strong leadership not only to their companies, but also to the NACCB. Phyllis Murphy also achieved great success in advancing her company and the IT services consulting business.

Conclusion

While the early contributions to the IT brokerage industry by women were exceptional, the contributions of Fred Shulman, Harvey Shulman, Howard Stein, Tom O'Donohue, Dave Cassell, Steve Kenda, and other men were important as well. Like Gentry, O'Donohue, Cassell, and Kenda served as IT brokerage chief executives and as early presidents of the NACCB.

With COMSYS, Shulman, Griffin, and Stein grew one of the largest computer services staffing firms (1994 revenue at $55 million) that had origins in the independent contractor business model. Though smaller, Gentry Inc. was highly successful as well, and Phyllis Murphy and Associates continues to be.

There are many software independent contractors today, though fewer of them find work through brokerages (they individually do not face the risk the brokerages do). That the independent contractor model continues to exist is in part due to the culture and predilections (valuing creativity, flexibility, and independence) of many in the programming field.[49]

III Geographical and Organizational Change

10   Transforming Giants, Offshoring Work, and Creating Clouds, 1982–2016

Historians often are reluctant to write about the recent past, present developments, and future trajectories. The computer services industry underwent massive growth and many dynamic changes in the past several decades—phenomena that cannot be ignored. This chapter surveys the past 35 years through three themes: IT industry giants that underwent major transformations, geographic shifts of IT services enterprises and labor, and the disruptive model of cloud infrastructures and associated services. Future scholars, with more historical distance and with access to a larger store of primary source material, undoubtedly will analyze these issues, and many others, in greater depth.

The market research firm INPUT estimated information technology services industry revenue at $31 million in 1982, with data processing services (table 10.1) and professional/consulting services (table 10.2) the two largest segments.[1] By 2014, the industry had grown by a factor of about 30 to reach annual revenue of $955 billion—roughly the same size as the global pharmaceuticals industry.[2] As was explored in earlier chapters, the trend from the mid 1950s on was for computer services firms to begin by specializing in one segment (programming services, data processing/data centers, time sharing, consulting, or systems integration) and then gradually broaden to build capabilities and businesses in some or all of the other segments. IBM, Hewlett-Packard (HP), Computer Sciences Corporation (CSC), Control Data Corporation (CDC), Electronic Data Systems (EDS), Arthur Andersen/Andersen Consulting/Accenture, and Sperry-Univac and Burroughs (and their combination as Unisys) all became diversified players in the IT services industry—and IBM, HP, and Accenture, along with Japan-based Fujitsu, are the current leaders in the industry. Automatic Data Processing (ADP) stands as an exception, having long remained specialized as an unparalleled leader in payroll processing—it currently serves companies

Table 10.1
The top vendors of data processing services by revenue in 1982.

Rank	Company	Revenue (millions)
1	Automatic Data Processing, Inc. (ADP)	$599
2	Control Data Corporation (CDC)	$590
3	General Electric (GEISCO)	$282
4	Electronic Data Systems Corporation (EDS)	$256
5	Tymshare, Inc.	$178
6	McDonnell Douglas Automation (McAuto)	$177
7	Computer Sciences Corporation (CSC)	$151

Source: INPUT, "US Information Services Markets, 1983–1988: Industry Specific Markets, Volume 1" (1983): 48.

Table 10.2
The top vendors of professional/consulting services in 1982.

Rank	Company	Revenue (millions)
1	Computer Sciences Corporation (CSC)	$420
2	Electronic Data Systems (EDS)	$250
3	Burroughs Corporation	$232
4	International Business Machines (IBM)	$195
5	Arthur Andersen and Company	$187

Source: INPUT, "US Information Services Markets, 1983–1988: Industry Specific Markets, Volume 1" (1983): 58.

and other organizations in more than 100 countries and processes the payrolls of 8 percent of the world's workforce.[3]

In 1982 the top seven firms in data processing services produced $2.2 billion in revenue, while the leading five companies in professional services had revenue (in this area) of about $1.3 billion. Roughly 6,000 other companies, in aggregate, recorded the remaining 83 percent of IT services industry revenue that year—proportions long typical in this highly diffuse industry.[4] While the trade continues to grow rapidly, as do its leading firms, it remains an industry of many small and mid-size companies, and only a handful of very large ones. In 2010 the top five global enterprises accounted for less than 20 percent of the industry's revenue, and many thousands of other companies and independent contractors produced the rest.[5] Over the same period, IBM, the leading IT services firm for more than 25 years, generally recorded 6–8 percent of the trade's annual worldwide

Table 10.3
IBM's global revenue, 1991–2015.

	Revenue from services	Total revenue (billions)	Services as percentage of total revenue
1992	$15.0	$64.5	23.3%
1994	$16.9	$64.1	26.3%
1996	$22.3	$75.9	29.4%
1998	$28.9	$81.7	35.4%
2000	$33.2	$88.4	37.6%
2002	$36.4	$81.2	44.8%
2004	$46.2	$96.3	48.0%
2006	$48.3	$91.4	52.8%
2008	$58.8	$103.6	56.8%
2010	$56.4	$99.9	56.5%
2012	$58.8	$104.5	56.3%
2014	$55.0	$92.8	59.3%

Source: IBM annual reports, 1991–2014, IBM Corporate Archives.

revenue. In 2014, IBM's share of the world market fell to roughly 5.5 percent. (See table 10.3.)

The year 1982 also marked the end of the US Department of Justice's lengthy anti-trust case against IBM—it was dismissed as "without merit."[6] The Department of Justice's lawyers on the case (a group that was perpetually in flux, as heavy workloads and modest government salaries resulted in high turnover) argued that IBM had engaged in subtle anti-competitive practices for many years. The reasons for the abrupt dismissal of the case are not entirely clear, but the Reagan administration, one way or another, wanted to conclude longstanding and costly cases against IBM and AT&T (the AT&T case, which began in 1974, also concluded in 1982, and led to the breakup of the telecommunication conglomerate's local telephone business into the "Baby Bells").[7] A likely factor in the dismissal of the IBM case was that the computer industry looked far different in 1982 than it had in 1969, when the Department of Justice had filed the case—IBM was less dominant, as a result of the growth of mini-computing (a field in which Digital Equipment Corporation, IBM, and to a lesser degree Hewlett-Packard, Honeywell, Data General, Wang Laboratories, Prime Computer, and others had thrived), and of personal computing (in which IBM had many competitors—its early-1980s leadership with the PC was short-lived).

The dismissal gave IBM's leaders an unambiguous green light to focus on computer services as a profit center, which it did when its hardware business struggled severely in the second half of the decade. IBM's transformation propelled it to the forefront of an IT services industry that was undergoing considerable change. Tymshare and GEIS, major specialists in time sharing, would be acquired and then decline as the rapid growth of personal computing sent this segment into a tailspin in the late 1980s and the early 1990s. In 1986, Burroughs and Sperry-Univac, in mutual decline, merged to form Unisys. That combination, continued to provide computer services, as its constituent companies had done, but never thrived. And in 1992, CDC was split apart, its IT services portion, Ceridian, becoming more specialized (in human resources software) and never approaching its earlier glory.

On the other hand, ADP grew steadily as a focused giant, though it was surpassed in overall revenue by a number of diversified IT services companies. And another of the firms that had helped launch the industry, Arthur Andersen (and its Administration Services Division/Andersen Consulting), consistently expanded for many years before its rapid growth in the 1990s, battling the accounting side of the company and becoming a large split-off firm with a new name (in 2001)—Accenture. Another early entrant, CSC, also expanded significantly, especially in its specialty of contracted programming services and systems integration for the federal government. And a major computer hardware industry company, Hewlett-Packard, rapidly developed a sizable IT services business in the 1990s and the 2000s and became (and remains) second to only IBM in revenue in this global industry. Furthermore, in recent years the Japanese computer manufacturer Fujitsu, which first moved aggressively into the services business in the early to mid 1990s (accentuated by the opening of its massive Tatebayashi System Center for outsourcing in 1995), has been among the top four firms in the global IT services industry.[8]

In the second half of the 1990s, the increasing ubiquity of the World Wide Web transformed the IT services industry, which had previously been more closely tied to place. Around 2005, time sharing had a rebirth, reformulation, and renaming as "the Cloud." Cloud services—a fast-growing area—has become a focus of software products leader Microsoft, IBM, e-commerce powerhouse Amazon, search leader Google, and CRM platform specialist Salesforce.com. The World Wide Web and ever-increasing bandwidth were critical to facilitating the expansion of the IT services industry in India (both native firms, such as Tata Consultancy, Wipro, Infosys, and HCL Technologies, and multinationals from the US and Europe—IBM,

Accenture, CSC, Paris-based Capgemini, Cognizant, and others). These native and multinational giants operating in India took advantage of a strong, well-educated, less expensive, largely English-speaking labor pool.[9]

The global IT services industry has grown very rapidly over the past decade and half. In 2003 it was a $570 billion industry. By 2011 it had grown to $848 billion, and in 2014 it reached $955 billion.[10] Behind the strength of cloud services, it is on pace to soon become one of the world's few trillion-dollar industries.

The remainder of this chapter concentrates on how IBM reorganized to become primarily a computer services business, on the transformations of IBM's competitors Hewlett-Packard, Fujitsu, Capgemini, and Andersen/Accenture, on a significant geographical shift of the industry labor to India, and on the rise of "cloud services"—"infrastructure as a service," "platform as a service," and "software as a service."

Transforming Giants: IBM, EDS/HP, and Andersen/Accenture

As I write, in January of 2016, IBM recently reported its fifteenth consecutive quarter of declining revenue. CEO Virginia Rometty took the reins at a challenging time in 2012, and IBM is seeking to concentrate on its "strategic imperatives" of "Cloud, Analytics, Mobile, Social, and Security," double-digit earnings growth areas with major services components, which now represent 35 percent of IBM's revenue.[11] Rometty's predecessor as CEO, Samuel Palmisano, clung to revenue and profits from an earlier model consisting of IBM hardware, customized IT services, and software (with services and software tilted meaningfully, but not exclusively, to IBM hardware systems), which continued even as the cloud-based model (data, software, and services on server farms/data centers rather than localized customer sites) was beginning to take off. This was done because large and reliable near-term revenue streams for IBM otherwise would have been sacrificed to the cannibalizing cloud—a tough proposition for a widely held public company battling to hit or beat quarterly revenue and earnings estimates. IBM hardware at customers' sites remains a focus; as Rometty recently pointed out, these are "critical systems for the world."[12] This current transition is delicate and involves many difficult strategic decisions—IBM's revenue and reputation rest on *both* its existing core and new imperatives. IBM faced similar reorganization hurdles to bolster top and bottom line growth in the second half of the 1980s and in the early 1990s. At that time, unlike CDC, IBM survived its existential challenge, avoided being broken apart, and set a course for two decades of relative prosperity (albeit with a degree of complacency in concentrating on existing businesses).

After the announcement of the IBM 701 (in 1952), IBM's revenue increased every year until 1991, the first year it recorded an annual loss. The $2.86 billion loss in 1991 was followed by a loss of $4.95 billion in 1992 and a loss of $8.10 billion in 1993.[13] These losses were largely attributable to restructuring. IBM's problems, in fact, had begun about ten years earlier. IBM was a huge company built around leasing and selling mainframe computers and midrange systems, and on servicing those machines and the customers' needs with software products and services. Essentially, IBM's services were wedded to its hardware.

Though IBM broke with its usual practice in using a small independent business unit and outside suppliers to quickly design, develop, and release the IBM PC (in 1981), it did not protect its operating system adequately, and Compaq (which in 1982 reversed-engineered BIOS, the one IBM proprietary element), Dell, and other manufacturers of PC clones thrived as the 1980s progressed—as did Microsoft, which had supplied the original operating system for the IBM PC and which went on to supply standard operating systems to the personal computer industry.[14] IBM's organization and cost structure (large staff, relatively high salaries, large facilities, and mainframe-oriented marketing operation) was not competitive in the long term for PCs (including ThinkPad laptops), and in 2005 IBM would sell 81 percent of that business for the relatively small sum of $1.25 billion (plus considerations bringing a truer real price closer to $1.75 billion) to the Chinese computer maker Lenovo.[15] The personal computer was a disruptive technology that hurt IBM, which continued to cling to mainframe and midrange systems and associated software and services revenue in the 1980s. IBM managed to increase its earnings until 1984, then experienced a minor decrease in 1985 and a major one in 1986—the latter year saw a 27 percent decline from the 1984 high.[16] Placing quarterly and annual earnings ahead of disruptive shifts, IBM's leaders delayed making changes, precipitating a more abrupt and difficult transition several years later.

IBM's new CEO in 1993—Lou Gerstner, who had fresh experience as CEO of RJR Nabisco—often has been celebrated for orchestrating IBM's turnaround. Gerstner's memoir, in which he details his leadership principles and practices in making an "elephant" (IBM) dance, became a classic, joining books by Lee Iacocca and others in presenting how a CEO restored a troubled corporation to glory.[17] IBM certainly returned to profitability and growth under Gerstner's watch, and he deserves considerable credit for his leadership in presiding over the financial turnaround. Some of the strategic repositionings that IBM made, however, already were in place in the years immediately preceding Gerstner's arrival, having been initiated by Gerstner's much-maligned predecessor, John Akers, IBM's CEO from 1985 to 1993. Akers appropriately bears the blame for failing to make difficult changes to reposition the company earlier in his tenure and for his fortunately abandoned "baby blues" strategy of breaking IBM into about thirteen independent business units, each which may have had its own stock.[18] Akers' early adherence to IBM's policy of "lifetime employment," whereby attrition and shifts were used but layoffs were generally avoided, made subsequent large-scale layoffs late in his tenure (and under Gerstner) all the more painful. Akers suffered from the burden of history, an entrenched corporate culture, and the legends of Thomas Watson, Thomas Watson Jr., and his other predecessors.

In the late 1980s, with substantial anti-trust concerns behind him, Akers engaged in bold moves with services. First he launched a series of small-scale outsourcing experiments to test strategies and perfect skills. Then, in 1989, IBM contracted to design, build, and manage a "state-of-the-art" data center for Kodak, which was to be located in Rochester. Following a model established by H. Ross Perot (a former IBMer), IBM used excess capacity (computer processing and human resources) to service other clients. Also in 1989, IBM penned a deal to operate all IT services for Hibernia National Bank, and instituted a new business services unit whose purpose was to enable an organization to recover from a disaster (e.g., a fire, a tornado, a hurricane, or a power disruption). These major initiatives in 1989 led to a formalized corporate-wide strategic reorganization in 1991. The cost of the restructuring and the weakening in demand caused by the 1990–1991 recession in the United States (and by longer-lived recessions in the United Kingdom, in Continental Europe, in Japan, and elsewhere) resulted in particularly deep annual losses as IBM made its transition.[19]

In 1991, Akers and IBM's board of directors authorized a new "worldwide services strategy" that was intended to make IBM a "world-class services company" by 1994. The strategy included restructuring the Systems Services Division to become the Integrated Systems Solutions Corporation (ISSC), a wholly owned subsidiary for providing programming and systems integration services to customers. That year, the new subsidiary began serving numerous corporations, including Supermarkets General, Commerce Bancshares, and First American National Bank of Nashville. The following year, IBM launched its Consulting Group, with a staff of 1,500, to provide IT consulting services to customers in the US and in 29 other countries.

With this, IBM was servicing not only customers of IBM hardware, but also clients using a wide range of systems and software from various vendors. In 1992, the ISSC engaged in a major joint venture in networking services with Sears, Roebuck and Company, and signed an $80 million contract to provide IT services to the state government of California for data processing for its welfare system. That year the ISSC also completed a major deal for data centers for United Technologies and a substantial systems integration project for the Czech Republic. Further, it signed on corporate clients for IT for health care systems and for hotel management, two areas that would become increasingly lucrative "verticals" (user industries having specialized needs and requiring specialized expertise). With its growing global services business, IBM established services subsidiaries in Canada, in the United Kingdom, and in Japan in the early 1990s.[20]

Services contracts continued to grow in length and size as IBM became the long-term services provider for data center management for credit bureau Equifax (a $650 million deal) and railway company Southern Pacific Company (a ten-year $415 million contract). While IBM's maintenance services revenue remained roughly constant (between $6 billion and $7 billion) in the first half of the 1990s, its other services businesses (programming services, consulting, data center management, and systems integration) grew rapidly and dwarfed maintenance income by 2000.[21]

Services as an unpriced support to hardware customers had been at the heart of the company's strategy since C-T-R's early tabulation machine days, but the 1989 Kodak data center and others to follow, coupled with the rapid hiring of programmers and consultants (to join the systems engineers, customer engineers, and other staff) soon made IBM principally a computer services company. In 1991, services were one fifth of the business by revenue, with hardware and software products both generating considerably more income. In 1995, IBM officially formed its Global Services Division, which brought in one third of the company's overall revenue in 1998 and more than 42 percent in 2001.[22] In 2002, IBM acquired the Global Management Consulting and Technology Services (GMC&TS) unit of PricewaterhouseCoopers. As IT systems became increasingly critical to businesses, especially in the Web era, technology consulting and management consulting became increasingly intertwined—with finance, logistics and sourcing, simulation and modeling, design, personnel management, knowledge engineering, and many other areas. IBM's existing consultants took a broad view of IT within larger business strategies, much like that of the PwC GMC&TS unit that IBM acquired. The latter, which cost IBM $3.5 billion in stock and cash, brought 30,000 experienced professionals, almost doubling IBM's newly formed Business Consulting Group of its Global Services Division (which also consisted of a similarly large Technology Services Group).[23] Aided in part by this acquisition, IBM's services expanded from $33.2 billion in 2000 to more than $60 billion in 2011, going from 38 percent of IBM's overall revenue to more than 56 percent.[24]

Success often spawns imitation. Hewlett-Packard's management took notice as IBM was transitioning to focus on the fast-growing IT services field at the end of the 1980s. In the 1990s, HP sought to engage in a similar major transformation.

Two Electrical Engineering graduates of Stanford University, William Hewlett and David Packard (counseled by Frederick Terman of Stanford's engineering faculty) formed Hewlett-Packard as an electronics company in Packard's garage in Palo Alto in 1939. Hewlett-Packard, which first produced oscillators, radar-blocking equipment, and other auditory electronics, moved into scientific calculators and minicomputer development in the mid 1960s. Like most manufacturers of mainframes or minicomputers, Hewlett-Packard offered some bundled computer services, but doing so was not a significant business or a core activity for the firm. In the mid 1980s, Hewlett-Packard began building inkjet and laser printers, which, along with producing proprietary ink and toner cartridges, became a major business for the company.[25]

In 1997, Hewlett-Packard created a Software and Services Group for what had been, by the firm's own admission, one of the "least visible" of its business areas. Then, in 2000, it created an IT Services Division to go along with its much larger divisions for Computers and Printers and Imaging. HP's 2000 annual report states that its IT Services Division had $1 billion in revenue that year and began to step up hiring IT services consultants, adding more than 1,700 that year. Its services business at the time was about 3 percent the size (in revenue) of IBM's. HP continued to accelerate its organic growth of services through aggressive hiring in the early 2000s, and in 2002 the company disclosed that it had 65,000 services professionals worldwide—a number that had risen considerably that year as a result of HP's acquisition of Compaq, which had taken over the Digital Equipment Corporation four years earlier.[26] HP was offering consulting services, imaging and printing services, various networking services, and other IT services to small, medium-size, and large companies, nonprofit organizations, and various levels of government in the US and overseas. With the massive and rapid expansion of its services workforce, HP achieved $12.3 billion in IT services revenue in 2003 and $15.6 billion in 2005. The pace of transition to include this new business was quite remarkable—a division that generated 2 percent of the overall revenue for the company in 2000 was producing 18 percent of its revenue only five years later.[27]

In 2002, Carly Fiorina of HP, the first female CEO of an IT giant, orchestrated a controversial $19 billion stock-based acquisition of the personal computer maker Compaq against strong opposition from board member Walter Hewlett, son of the company's co-founder. HP's takeover of Compaq was controversial not only because of Walter Hewlett's opposition, but also because personal computers had become a commodity business with ever-dwindling differentiation, pricing power, and margins. Some insiders opposed to the Compaq acquisition saw services as the more important focus, and although Compaq brought a number of services professionals it primarily was a hardware firm. Timing is everything, and Compaq would not be Hewlett-Packard's last controversial acquisition. In 2008, HP acquired the beleaguered Electronic Data Systems (EDS).[28]

GM divested its EDS computer services division in 1995 and it became an independent public corporation again. It was a far different organization than when it was acquired by GM a dozen years earlier. GM's large and diverse IT needs and guaranteed business to EDS allowed the services business to expand quickly, but there was a clash of cultures between the entrepreneurial, longtime stock-option-incentivized IT services firm and the bureaucratic, slow-moving automobile manufacturer. In 1986 EDS revenues rose to $4.4 billion, with $3.2 billion (more than 72 percent) coming from GM.[29] At the end of that year EDS's charismatic founder Ross Perot was bought out (he was GM's largest stockholder) and forced off the GM board of directors. Two years later he started a competitor in IT services business called Perot Systems. Perot's departure was a further, albeit partly symbolic blow (as he was less active in day-to-day management) to EDS's innovative culture, and left the division to adjust as best it could to the management styles and imperatives of a leading automobile manufacturer. This environment, and overly rapid growth, forced the computer services division to focus on large contracts.

By 1995, when EDS became independent again, it had a data center campus in Plano, Texas (built in 1993) with 37 buildings on 360 acres of land, about 100,000 employees, and ever-increasing pressure to reduce bench time for its large workforce.[30] With the guaranteed GM business gone, the company had to be more and more aggressive in bidding on large-scale, riskier contracts. On some of these major contracts it lost substantial sums of money—accurately costing and effectively bidding on major programming and systems integration business remains highly difficult to this day. It also led EDS to focus on Y2K readiness/compliance contracts to keep its large workforce busy. What was a blessing for emerging Indian IT giants (discussed later in this chapter) was a curse for EDS. It had carved out a place on the lower-margin, lower-skill side of the computer services business (for a sizable portion of its employees) at a time when IBM was specializing in higher-margin specialized programming for verticals in finance, healthcare, and other industries. The GM years, followed by US-based Y2K contracts, led to a predominantly US-based EDS workforce at a time when IBM and Andersen Consulting were downsizing in the US and expanding their operations in India and in other countries with lower labor costs. EDS was struggling mightily in 2008, when HP acquired it for $13.9 billion—a price that was a mere 63 percent of trailing year's sales (0.63 price to sales ratio). Valuations under 100 percent of annual sales (or a less than one price-to-sales ratio) are generally considered quite low valuations (especially in the IT field), and an acquiring firm almost always has to pay a substantial premium to the target firm's pre-acquisition stock price.[31]

In 2007, IBM had IT services revenue of $54 billion; EDS was second in the global industry at approximately $22 billion.[32] By acquiring EDS, HP (which had $16.1 billion in IT services revenue in 2007) immediately became the clear second in the industry, with an eye on dethroning IBM from the top spot. The EDS workforce had grown to about 140,000 employees at the time of the acquisition. HP had 176,000 employees overall, but it had fewer than EDS on the IT services side. IBM and HP were generating roughly similar operating margins of slightly over 10 percent, which was strong performance at the time for an industry that had long been growing fast but had been characterized by modest margins. EDS's operating margin was considerably lower—about 6 percent. HP was purchasing a relatively expensive workforce that was heavily US based and had many thousands of employees in a single location—Plano, Texas.[33] The major challenge ahead for HP was integrating the two businesses. At this time, IBM and the new HP were facing steadily increasing competition from emerging Indian IT services firms, which were operating with lower costs because a large percentage of their employees were in India.

HP also was having reputational problems resulting from a high-profile boardroom scandal and "bad press" concerning underperforming acquisitions. With the former, in an effort to contain ongoing corporate leaks, chairwoman Patricia Dunn authorized the hiring of outside investigators, who proceeded to spy on other board members, employees, and journalists. She and four others were indicted on felony charges in California for "using false pretenses to obtain confidential information from a public utility, unauthorized access of computer data, identity theft, and conspiracy."[34] Dunn had a recurrence of cancer, and the state dropped the charges against her "in the interest of justice."[35] The news story, in the reputation-based business of IT services (particular for a case involving computer crime), hurt HP. Troubling news remained in the headlines for HP as CEO Mark Hurd, who had replaced a fired Carly Fiorina in 2005, resigned after a scandal concerning Hurd's alleged sexual harassment of an HP contractor/greeter, former adult film actress and reality television contestant Jodie Fisher. An HP inquiry found that Hurd was not guilty of sexual harassment but had violated "HP's Standards of Business Conduct"; through mutual agreement with the board of directors, Hurd resigned in August 2010.[36] Hewlett-Packard then looked to Margaret (Meg) Whitman, who had successfully led the rapid growth of the auction and electronic payments firm EBay, to get HP back on track. She signed on as HP's CEO in 2011. However, the company's financial struggles, and the woes that ensued from its acquisitions of Palm, 3com, Autonomy, and other companies between 2009 and 2012, persisted.

In 2015, Whitman and the board of directors decided to spin off the enterprise business (IT services, software, and networking) from the computer and imaging/printer business. Whitman provides leadership for both—as chair of the board of Hewlett-Packard (computer/printers), where Dion Weisler is now the Chief Executive, and as CEO of Hewlett-Packard Enterprise.[37] HP Enterprise has its work cut out for it as it, like IBM, seeks to grow higher-margin businesses in cloud services and analytics. In late May 2016, HP Enterprise announced that it would merge with Computer Sciences Corporation, creating a larger entity to better compete with IBM and Accenture. Less than a year earlier, CSC spun off its lucrative government services businesses to merge with SRA in an independent public company called CSRA.[38]

This (first) separation/change for HP, to create HP Enterprise and have a fully independent IT services consulting business, was also an issue that had played out with Arthur Andersen's Administrative Services Division (ASD)—though it was a quite different situation as the separation was from Andersen's accounting side, not a computer hardware business. Arthur Andersen's ASD had the advantage of never having been part of a hardware company in which computer services were partially or fully tied to the parent firm's equipment—as it had been for many years at IBM, Sperry-Univac, Control Data, Burroughs, Honeywell, Hewlett-Packard, and other companies. This was more of a burden for IBM's competitors than for IBM, thanks to IBM's share of the market for mainframes (though it became a challenge for IBM in the PC era when it had less dominance in the hardware market and it was trying to grow services as a core profit center).

The structure of Arthur Andersen's ASD made for less bureaucracy and rigidity with hiring, promotion, and launching and relocating new offices—relative to giant mainframe firms with large facilities, such as IBM. By the 1960s, the ASD's two largest offices were in Chicago and New York City, but it set up many smaller offices as business developed in various US cities and overseas.

Although twelve years had passed before Andersen had hired its first female consulting employee, Andersen/Accenture became committed to gender diversity. In 1965, the Chicago office hired Susan Butler, a graduate of Purdue University. Butler worked on programming a new IBM System/360 at American National Bank in Chicago and went on to be a manager in the Andersen ASD's Detroit office and then the first female partner in 1979, and shortly thereafter the first managing partner.[39] She paved the way for other female executives at Andersen/Accenture, including Gill Rider (who became Accenture's Chief Leadership Officer in 2002) and Jane Hemstritch (who was named Accenture's managing director for the Asia Pacific region in 2004).[40] Today about 130,000 of Accenture's 360,000 employees are women, and there is a corporate goal of going from 36 percent to 40 percent by 2017.[41]

By the early 2000s, Andersen/Accenture was very much an international operation. Andersen's ASD expanded aggressively in Europe in the 1970s, setting up its international umbrella organization called Arthur Andersen and Company Société Coopérative in Geneva in the mid 1970s. Later, Geneva would become Andersen/Accenture's headquarters. During the 1970s and the 1980s it expanded its offices in Paris, London, and other major European cities. With growth overseas and in the US, Andersen had about 10,000 IT consultants by 1987.[42]

Arthur Andersen's ASD grew faster than the accounting business in the 1980s and in subsequent years, in parallel with and contributing to the rapid growth of the IT services industry. In 1989 the ASD changed its name to Andersen Consulting to better reflect its activities and to help with branding. By the mid 1990s it made up the larger share of the firm's earnings, and on a per-person basis it contributed far more to revenue and earnings. Many of those in Andersen Consulting resented the level of transfer payments to the parent company. The ongoing conflict played out frequently in the press. Andersen Consulting's dynamic and successful CEO George Shaheen led the effort for independence. In 2000 an arbitrator ruling led to formally splitting the two enterprises. Andersen Consulting, in 2001, changed its name to Accenture to further distance itself—with a name combining "accent" or "accentuate" and "future." This change proved fortunate, as the Arthur Andersen name soon was tarnished as a result of the revelation that it had shredded documents (and, allegedly, had earlier been complicit and/or negligent) in the Enron scandal, in which accounting fraud had led to the energy services company's bankruptcy and dissolution.

The Enron debacle precipitated the end of Arthur Andersen's accounting business; it relinquished its license to practice public accounting and auditing in 2002.[43] That same year, Accenture moved its headquarters from Switzerland to Bermuda. In 2009, it moved its headquarters to Dublin. Its longstanding international focus is serving it well in the global IT services industry. Though all major IT services companies now have a substantial presence in India, Accenture moved quickly to expand its business there. It also has a substantial workforce in the Philippines. Having a substantially lower cost of labor due to its substantial work force in lower cost countries—India and the Philippines—has helped it consistently grow revenue, while achieving roughly 10 percent growth. This is a feat exceeded only by some of the leading India-based IT services giants that have benefited from the massive increase in the outsourcing or offshoring of IT services to India by the US, European countries, and developed Asian countries, and from the lower cost of labor in India relative to the US and Europe.[44]

Before considering the phenomenal growth of IT services in India, we should take note of two other IT services giants outside the US that are important to understanding the makeup and the dynamics of the global computer services industry over the past several decades: Capgemini and Fujitsu. Capgemini—founded by French entrepreneur Serge Kampf—has similarities to EDS and CSC as a startup services specialist that steadily increased the scale and the scope of its services businesses for years before rapidly expanding in the past two decades. Fujitsu has more in common with IBM and HP than Capgemini has—it is a highly successful vertically integrated computer manufacturer that long provided services, but increasingly it made this one its core businesses (moved from bundling services to hardware, in favor of separately pricing services).

Serge Kampf, born in Grenoble in 1934, received degrees in economics and law before joining the General Direction of Telecommunications in Paris in 1960. Soon thereafter he moved into the computer industry, taking a position with Compagnie des Machines Bull, a firm that had been founded in 1931 to manufacturer punch card tabulation machines and had then (like Remington Rand and IBM) made a transition into computers in the early post–World War II era.[45] In 1967, Kampf left Bull and founded a computer services firm called Sogeti SA (convincing three of his Bull colleagues to join him).

Sogeti's originally location was a two-bedroom apartment in Grenoble that Kampf converted to an office. By the end of its first year, with Kampf as president, the firm had 27 employees, primarily engaged in advisory and programming services, and had made 1.5 million francs. In 1969 it had 49 employees and earned 4.2 million francs (worth about $815,000 in 1969). By that time it had set up a Swiss subsidiary and an office in Lyon. By 1972 it had offices in a dozen cities in France and elsewhere in Europe.[46]

In 1973, Kampf successfully led Sogeti's hostile takeover of the French computer services company CAP. In 1974, it acquired the New York–based computer services firm Gemini Computer Systems. The company was renamed Capgemini Sogeti in 1975, and shortened its name to Capgemini in 1996. With the merging of these three firms, Capgemini Sogeti became the leading French IT services enterprise, with roughly 2,000 employees and with sales exceeding 180 million francs in 1975. It provided consulting services, programming services, facilities management, and systems integration to a wide range of industrial and government organizations in France and throughout Europe.[47]

In the early to mid 1980s, Capgemini Sogeti made its first major expansion in the largest IT services market, the United States, by acquiring firms that it grouped into Capgemini Inc. in 1986. In the previous year it had boosted its resources for both internal expansion and further acquisitions by going public—with its stock being listed on the Paris Exchange. Back in 1982 it had crossed annual revenue of 1 billion francs and less than a decade later, in 1990, it grew 9-fold to exceed annual revenue of 9 billion francs. By the latter year it had 16,500 employees, aided by its 1990 acquisition of the Hoskyns Group (a prominent IT services firm in the UK) for approximately 2 billion francs.[48]

The many European acquisitions Capgemini Sogeti made in the 1970s and the 1980s gave it a large presence on the Continent (where collectively it held 7 percent of the market), but its market share in the US was only about 1 percent at the start of the 1990s. To provide funds to enable more rapid expansion in North America and elsewhere, Kampf and his board agreed to sell 34 percent of Capgemini Sogeti to the automobile company Daimler-Benz in 1991 for 5 billion francs, with an option for Daimler-Benz to acquire majority ownership by 1995. At the start of the 1990s, however, the global recession (1990–1991 in the United States, 1990–1993 in parts of Europe, and the start of Japan's "Lost Decade") led to the company posting its first annual loss in 1992, which was followed by losses the next two years before reorganization of the firm and improving economic conditions returned the company to profitability in 1995. This included the rapid growth of its facilities management business as well as greater integration internationally to more effectively meet global demands of clients. While funds from Daimler-Benz (and serving this automobile manufacturer's international IT needs) proved important to reorganization and new initiatives, it was also challenging for both sides. Daimler-Benz's new leaders in the mid 1990s decided against boosting its stake and instead divested its Capgemini holdings in 1997. Though shorter-lived and less complete than GM's purchase of EDS, it proved to be a similarly problematic ownership experiment between an automobile manufacturer and IT services provider, where different corporate cultures created hurdles to successful integration.[49]

In 2000, Capgemini engaged in by far its largest deal to date, acquiring Ernst & Young Technology. The cash-and-stock deal netted the accounting firm $11 billion. For Capgemini, it added 18,000 (mostly consultants) to its labor force, and strongly boosted its resources and capabilities at the intersection of IT consulting and management consulting. After the merger, the newly named Capgemini Ernst & Young had 40,000 employees. For the Big Five accounting firm Ernst & Young, the split (like that of Arthur Andersen with Andersen Consulting/Accenture) reduced real and perceived conflict of interest between its accounting and IT services operations.[50]

In the early 2000s, Capgemini Ernst & Young accelerated its global delivery of services from various locations and labor forces to enhance efficiency while maintaining attention to the management culture, setting, and needs of clients. In 2003 it trademarked its global delivery system of services as the "Rightshore" concept. This involved major expansion into and continually growing the company's workforce in India.

By 2008, having shed its awkward three-word name to become Capgemini, the company was operating in thirty countries with four primary business units—Consulting Services, Technology Services (systems integration), Outsourcing Services (including facilities management), and Local Professional Services (tailored to local needs for infrastructure and operations). Its Outsourcing Services grew quickly in the early 2000s to produce 35 percent of the company's revenue. Like other large IT services companies, Capgemini increasingly focused on sector verticals to meet the needs of major industries and industry clusters. It concentrated on Public Sector; Energy, Utilities, and Chemicals; Financial Services; Manufacturing, Retail, and Distribution; and Telecommunications, Media, and Entertainment. By 2008, Gartner, Inc. listed Capgemini as the sixth largest global IT services company, the five ahead of it being its primary competitors (IBM, HP, Accenture, Fujitsu, and CSC). Other meaningful competitors included the leading Indian firms, Tata Consultancy, Wipro, and Infosys. France and Morocco accounted for 24 percent of Capgemini's revenue, the United Kingdom and North America adding 22 percent and 19 percent respectively.[51] By 2015, like its major competitors, Capgemini was particularly focused on growing its cloud services business. It currently operates in more than forty countries throughout the world and has a global workforce of 180,000, about 85,000 of them in India.[52]

While Capgemini is the leading European computer services company with reach throughout the world, Fujitsu is the leading Asian computer services company and has a large global presence—fourth in the world in the international industry. Fujitsu began in 1935 as Fuji Tsushinki Manufacturing Company, a 700-person, 3-million-yen spinoff of Fuji Electric (a partnership between Furukawa Electric and Siemens AG of Germany formed in 1923). Fuji Tsushinki Manufacturing consisted of the fast-growing telecommunication, radio, and switching equipment business of its parent company, Fuji Electric. In 1954, enterprising engineers in Fuji Tsushinki Manufacturing's R&D Department, led by Toshio Ikeda, designed and built the FACOM 100, Japan's first relay-based electronic computer—an operable, experimental machine that led to the commercial relay-based FACOM 128A Scientific Computer, delivered to the Ministry of Education's Institute for Statistical Mathematics in September 1956.[53] With this, Fuji Tsushinki Manufacturing established itself in the emerging Japanese computer industry, which by the following year included electronic digital computers from two major electronics companies, Hitachi and NEC.

Much like IBM, Fuji Tsushinki Manufacturing bundled engineering, systems, and programming services with early mainframes. In 1963, three years after IBM but earlier than any of IBM's other competitors, Fuji Tsushinki Manufacturing established a formal program for "Systems Engineers" to aid organizational customers with computer installations. Four years later the company changed its name to Fujitsu Limited (Fujitsu Kabushiki Kisha). Without the anti-trust legal pressures faced by IBM, it continued to bundle the work of Systems Engineers and other computer programming and system support throughout the 1970s and the 1980s.[54] This changed in June 1992, when Fujitsu introduced its PROPOSE ("Professional Total Support Service") framework for delivery of computer services.[55] Fujitsu's PROPOSE, similar to IBM in creatively offering and marketing "solutions" to customers in government and industry, received an award from the Japanese Ministry of International Trade and Industry. Fujitsu's history in the 1990s and in subsequent years also parallels IBM in that it offered mainframes (including supercomputers), midrange systems, personal computers, and laptop computers and also focused on rapidly growing services as an increasingly important profit center.[56]

Fujitsu boosted its computer services (and hardware) infrastructure and geographic presence through acquisition and internal expansion. Back in 1972 it formed an equity and strategic partnership with US computer company Amdahl Corporation, founded two years earlier. Gene Amdahl, who had been one of the lead architects for IBM's System/360 series in the mid 1960s, launched Amdahl Corporation (in Sunnyvale, California in 1970) to compete with IBM by making IBM-plug-compatible machines. With the early 1970s recession, and short on funds, Amdahl Corporation partnered with Fujitsu. Fujitsu took a 24 percent ownership stake in Amdahl Corporation and provided manufacturing for Amdahl computers. Amdahl Corporation—like many computer manufacturers—bundled some services into hardware contracts. While Amdahl is nearly always mentioned in the computer industry literature as a hardware competitor to IBM, it in fact evolved to become primarily a computer services enterprise by the mid 1990s. Fujitsu had increased its minority ownership stake in Amdahl Corporation to 42 percent by this time period and in 1997 paid $850 million to acquire the remaining 58 percent ownership of the Silicon Valley firm. Fujitsu's leaders made this move believing their firm needed more than "an arms-length relationship" with the US company and to gain a meaningful foothold in computer services in the US market.[57]

Fujitsu's leaders also looked to European markets for expansion through acquisition in the 1990s. In 1990 Fujitsu acquired an 80 percent ownership stake in British national champion International Computers Limited (ICL—the product of a series of government-encouraged mergers of early British computer companies British Tabulating Machine Company, Powers-Samas, Ferranti, English Electric, Elliott Automation). ICL—like other global mainframe firms IBM, Sperry Rand, and Control Data—had significant computer services operations and infrastructure. This partial acquisition provided a major boost to Fujitsu's services business in Great Britain and continental Europe, which was furthered by taking over the remaining 20 percent of ICL in 1998.[58]

In 1995, Fujitsu opened a major data center, Tatebayashi System Center, in Gunma Prefecture, 120 kilometers northwest of Tokyo, to greatly expand its data processing and facilities management/outsourcing services.

Figure 10.1
Gene Amdahl, circa 1980s. Courtesy of Charles Babbage Institute, University of Minnesota.

Two years later, the company opened another large data center, the Akashi System Center, in Hyogo Prefecture, less than an hour by train from Kyoto. These two data centers and Fujitsu's early anticipation and strategic push toward cloud computing and the "internet of things" have helped the company achieve its current position as number four in the global computer services industry.

In 1999, Fujitsu announced its "Everything on the Internet" business strategy, and the following year its president, Naoyuki Akikusa, and executive vice president, Yuji Hirose, published on the topic at the IEEE's Ninth International Symposium of Semiconductor Manufacturing. One component of this strategy was to become the top "provider of internet solutions that merges platforms, information, electronic devices, and services."[59] A major part of these efforts has been a dedicated commitment to reliability. From the start of the "Everything on the Internet" strategy in 1999, its Tatebayashi System Center and Akashi System Center have been central and have undergone substantial expansion and refinement. After the 2011 earthquake, tsunami, and power plant disaster, Fujitsu added two new large buildings to its Akashi System Center, an Earthquake Resistant Datacenter and a Seismic Isolation Datacenter.[60]

Fujitsu also established more than a hundred smaller centers within Japan (roughly two thirds) and throughout much of the world (about one third) in the succeeding two decades. In India, Fujitsu Consulting India Private Limited has IT consulting services centers in Pune, Bangalore, Noida, and Hyderabad that serve customers in 36 countries. In addition to the sizable workforces of IBM, Accenture, Fujitsu, HP, and other multinationals in India, India's own large-scale computer services companies serve customers all over the world.[61]

Offshoring, Outsourcing, and India's Big Five

In early 2016, while visiting IBM's facilities in various Indian cities, CEO Virginia Rometty spoke in Bangalore. "We are today mostly a software and services company," she proclaimed. "But we have to transform—in this transformation, we will emerge as a cognitive solutions and cloud platform company ... . Everything we do is part of that strategy. ... This century—the 21st century—will be the Indian century ... . I truly believe that India will be the center of this fourth technology shift [meaning a shift to IBM's analytics/AI, or, as it is currently marketed, "Watson"]."[62]

IBM's accelerating investment in its Indian operations (new facilities and rapidly growing workforce) over the past couple of decades—since it returned to the country (after India's "Economic Liberalization") with a Tata-IBM joint venture in 1992—is striking. IBM hit an all-time peak employment of 434,245 workers in 2012 (through growth outside the US, especially in India), which declined more than 12 percent to 379,592 by the end of 2014.[63] The trade journal *Computerworld* reported that in 2012 IBM India's headcount probably had exceeded its US headcount—IBM stopped releasing country employment numbers in 2010. By that time, there was increasing media, public, and government scrutiny of major US corporations' offshoring of information technology jobs. A union figure published in 2012 placed IBM US employees at less than 93,000.[64] *Computerworld* obtained an internal IBM document (that it stated the company would not verify for the trade journal) that placed IBM India employment at 6,000 in 2002 and 112,000 by 2012—and reported the average salary for IBM India employees in 2012 to be about $17,000 a year (a highly competitive IT industry salary in India at the time).[65] Most IBM India employees are on the services side, and Rometty's remarks probably were made to reassure traditional IT services employees about their place (and India's) in the company's future—an enterprise undergoing a major transformation in which services remain central but the types of services and the skill sets that are sought are changing with the increased concentration on cloud computing and on analytics.

In the 1990s and the early 2000s India's IT services industry (including its workforce at multinationals such as IBM, Accenture, CSC, and Capgemini, as well as emerging Indian giants Tata Consultancy, Wipro, and Infosys) was engaged primarily on the lower-end, commodity side of the computer services industry—debugging for Y2K, and later the outsourcing of fairly basic business processes. The lower end of the industry was derisively referred to in the global industry and trade press as commoditized "body shops"—implying "jobs that just about any IT worker could do." At that time, that India might be at or near the forefront of innovation for a legendary US-based global IT company (IBM) might have seemed hard to believe. Since then, however, a significant portion of Indian IT workers at multinationals and native giants have moved beyond "body shop" services to more complex and lucrative "verticals" of specialized software systems catered to specific industries and company needs. In view of this and the comments Rometty made during her 2016 visit to India, IBM's strategy for Watson in India seems not just plausible but probably advisable, owing to the high level of talent and the lower salaries. How did India make the transition so quickly?

The historian Ross Bassett, in his 2016 book *The Technological Indian*, helps answer this important question by looking at deeper, long-standing foundations. Bassett insightfully examines Indian engineering graduates from MIT—a modest number of late-nineteenth-century and early-twentieth-century pioneers before India's 1947 independence and many hundreds since—and the outsized influence of these often visionary individuals on ideas about technology and on Indian educational institutions (the Indian Institutes of Technology, particularly IIT-Kanpur), and Indian companies (especially those in the Tata Group, which in the late 1960s launched the first Indian international IT services company, Tata Consultancy). Bassett demonstrates how the bonds of technological elites in the US and India were deep, even when relations between these two democratic nations were strained during the early years of the Cold War.[66]

Thanks to MIT's unrivaled early presence in computing in the 1950s and the 1960s (with Whirlwind, SAGE, CTSS, Project MAC, Multics, and pioneering work in artificial intelligence research), a number of Indians studying at MIT were exposed to (and contributed to) important advances in computer and software technologies.

Tata Consultancy owed its origins to three MIT Project MAC graduates. In 1963, Lalit Kanodia, who had a degree in mechanical engineering from IIT-Bombay, left India for graduate study in industrial engineering at MIT. Exposed to computing through graduate assistantships at the well-funded and rapidly growing Project MAC, he became deeply interested in computing and its potential applications, doing consulting work at both Ford Motor Company and Arthur D. Little. On his return to India in 1965, a neighbor introduced him to Rustom Choksi, an executive with the Tata Group (a giant multinational industrial conglomerate). The ambitious young graduate proposed to Choksi and others at Tata ways the Tata Group could improve efficiency in its dozens of businesses by managing electrical loads and financial accounting by means of computer applications. Forward-thinking Tata executives gave Kanodia an unusual opportunity to return to MIT for a doctorate and then to launch (with two other Indian MIT graduates, Ashok Malhottra and Nitin Patel) a venture in computing to help Tata better apply IT internally. In September 1967, a Tata Computer Centre was launched with the installation of an IBM 1401. In the center's first year, the three young engineers gave many talks to managers at various Tata companies on possible computer applications. All too often, long-serving managers accustomed to existing systems and methods were unreceptive. Top Tata executives, however, saw a potential for outside business, and in 1968 the Tata Consultancy was born. Kanodia, with an entrepreneurial mindset and wanting to own and run a firm rather than to head what was at the time a tiny business group within a giant conglomerate, left the following year and founded Datamatics Staffing, a recruitment company that offered computer education courses. After obtaining an IBM 1401 in the late 1970s, Kanodia founded Datamatics Consultants to concentrate on financial and accounting business data processing.[67]

With Kanodia's departure, F. C. Kohli of Tata Electric, who held an MS degree from MIT, became general manager and managing director of Tata Consultancy Services (TCS). In 1969, TCS secured a contract to handle computer-generated billing for the Bombay telephone system with 140,000 customers. With international aspirations from the start, TCS partnered with Burroughs Corporation on small contracts, and obtained some independent business, but overall the 1970s were characterized by slow growth, and crippling government bureaucracy and conditions—in particular, tough exchange rates and related restrictions on importing computing equipment. And beyond government hurdles, other hurdles were presented by the policies of IBM, the world's leading vendor of mainframe computers to India and rest of the world.[68]

The government of India had ongoing but strained relations with IBM, which had begun doing business in India in 1951. The government (and Indian industry) grew to resent IBM for selling only its older computers in India—systems often possessing little value in developed markets. By the late 1960s this meant importing used IBM 1401s. Soon IBM set up factory facilities in India to repair, refurbish, and rebuild 1401s. These machines were then leased to Indian banks, insurance companies, steel companies, airlines, public utilities, and other firms and organizations. The largest customer was India Railways, which ran fourteen IBM 1401-based computer centers to enhance operational efficiency. IBM's strategy of selling only older systems in India continually antagonized that country's government, and government leaders proceeded to investigate amending rules and restrictions on international companies operating in India. In the mid 1970s, IBM executives and the government of India engaged in increasingly tense negotiations. The IBM negotiating group made some concessions (regarding newer machines and a proposed new software center), but the concessions were too little and too late. In 1977 the government of India demanded that IBM relinquish majority ownership of the Indian operation to it. IBM's CEO, Frank Cary, was unwilling to substantially dilute IBM's equity in India and to relinquish control, and IBM officially left India in June 1978, selling already installed (but at the time leased) machines at a great discount to Indian customers.[69] IBM did not return until 1992 (a year after the start of India's economic liberalization), when it entered into a partnership with the Tata Group. Though the small joint venture grew, a half-decade passed before IBM moved into the market more substantially with a second initiative. The latter joint venture was 80 percent IBM Global Services and 20 percent Tata Group. In 1998, IBM established an IBM Research Laboratory at IIT-Delhi. Tata divested itself of both joint ventures with IBM in 1999, and the new IBM India has grown rapidly ever since.[70]

In the late 1970s, another MIT graduate, Narendra Patni, who had previously worked with Jay Forrester and at Arthur D. Little, launched a small computer enterprise, which then partnered with Data General to market that company's minicomputers in India. To meet the foreign exchange requirement, Patni's company hired a dozen programmers to add value by writing software for the Data General systems. The lead programmer, Narayana Murthy, had gone to IIT-Kanpur to get an MS in electrical engineering in preparation for a career in the electrical power industry. At IIT-Kanpur, Murthy was captivated by computers and developed strong programming skills. Murthy left Patni's enterprise in 1981 and (along with a half dozen others) formed an IT services company called Infosys, which with Tata Consultancy Services and Wipro would make up the triumvirate of emerging IT services giants in India. Wipro, in contrast to the IT startup Infosys, evolved into a computer services firm as its IT services division greatly outpaced its other businesses.[71]

Mohamed Hasham Premji launched Western India Products Limited as a vegetable oil firm in 1945. After several decades of focusing on producing vegetable oils and related chemical products, he expanded and diversified the business, first into industrial equipment (hydraulic cylinders) in the mid 1970s, and then into computing. In 1980 he founded Wipro Information Technology Limited (Wipro ITL) in Bangalore. Wipro, which was similar to the Tata Group but not nearly as large, was a conglomerate of businesses, one being Wipro ITL. This computing firm provided hardware, peripherals, and services (programming, maintenance, and training).[72] Whereas the Tata Consultancy, though very large, is just one among other giant companies in the Tata Group, Wipro Information Technology Limited (later shortened to Wipro Technologies) came to be Wipro's dominant business, and its rapid growth helped to make Bangalore the IT center it is today.[73]

Two later entrants, Satyam Computer Services Limited/Cognizant and HCL Technologies, rounded out what would become India's Big Five. Satyam Computer Services was founded in Hyderabad in 1992. Dun & Bradstreet (D&B), a credit reporting and business information specialist based in New Jersey, formed a major partnership with Satyam in 1996. That year, D&B spun off its internal IT unit (which had been created to deal with Y2K readiness) to engage in contract external IT services. The Indian entrepreneur Chintalapati Srinivasa Raju (widely known as Srini Raju) and his D&B IT unit partnered with Satyam to become Dun & Bradstreet Satyam Software (DBSS), an IT services firm three fourths owned by D&B and one fourth by Satyam (D&B later bought out Satyam's share). For Dun & Bradstreet, DBSS was a means of quickly acquiring an Indian IT services workforce. (Eventually, roughly two thirds of the workforce of DBSS was based in India, which led many analysts and journalists to consider it an India firm.) In the late 1990s, Dun & Bradstreet spun off DBSS, which changed its name to Cognizant Technology Solutions Corporation.[74]

HCL Technologies (originally HCL Overseas Limited), the smallest of the five IT services giants, used a joint venture to advantage, as Cognizant did. Formed in 1991, HCL Technologies grew slowly before partnering with Perot Systems in 1996. Until the early 2000s, HCL Technologies also focused on lower-value basic business process outsourcing (BPO) and related services. Since then, it has built a highly skilled workforce. It leveraged Perot Systems customer connections to provide higher-value advanced application system services to major multinationals such as the KLA-Tencor Corporation, NCR, and Toshiba.[75] More generally, Indian IT firms reinvested profits to expand their capabilities to begin to transition from exclusively lower-level work (such as BPO) to a mix that included assignments further up the value chain.

Throughout the 1980s, Tata Consultancy Services, despite international aspirations and a partnership with Burroughs, primarily served Indian customers, as did Wipro and Infosys. Travel outside India was expensive, and the telecommunications infrastructure for international computer networking was limited. Furthermore, international telecommunication-based data transmissions were expensive. Texas Instruments, struggling in the early 1980s in the US, established operations in India in 1982, and in 1985 set up a satellite-based data communication infrastructure to connect its Bangalore Software Development operation to London, and from London to the US. For IT enterprises, this established a model for, and demonstrated the feasibility of, international data networks connecting a developing country with developed countries.[76] In the early 1990s, with this model in place, India's economic liberalization quickly brought new IT opportunities to India's hardware, software, and IT services businesses.

The government of India and Indian IT services firms saw providing software and programming services to overseas clients (particularly in the United States) as a major opportunity, forecasting that the export business of roughly $164 million in 1991 could grow to $1 billion a year by 1996.[77] However, programmers' skills varied widely, and project management experience and expertise often were in short supply. Quality-related problems associated with programming projects and with debugging efforts were not unique to India; they also existed in the United States and throughout the world. The US Air Force, having fallen victim to a number of failed internal and contracted programming projects (such as the Advanced Logistic System project in the 1970s), sought to develop a framework for certifying software/system quality standards.[78] It contracted with Carnegie-Mellon University's Software Engineering Institute (SEI) and with the MITRE Corporation to develop the SEI Capability Maturity Model (CMM), with Watts

Humphrey (the founding director of the SEI Software Process Program, and a former director of the IBM Systems Research Institute) leading the effort.[79] The first version of the CMM was circulated for feedback in 1991 and was then published in a 1992 technical report as Version 1.1. At the CMM's lowest levels (levels 1 and 2), processes are unpredictable and reactive; at levels, 3, 4, and 5, respectively, processes are defined, quantitatively managed, and fully controlled and optimized.[80]

Motorola, which had suffered problems with software quality, utilized a new Motorola India Electronics Limited development group as a test bed for instituting CMM Version 1.0 in 1991. Using CMM, that group achieved level 5 in its first year. The CMM is focused on complete documentation and process controls—something far easier to start anew with a dedicated, talented group than institute at an established facility that has set practices and workplace cultures, and is prone to cut corners with documentation. The publicity of Motorola's success led Wipro ITL and other IT services firms to invest heavily in CMM. Wipro was the second company to achieve level 5 (in 1998). In February 2000, India had ten level 5 firms and fourteen level 4 firms.[81] CMM was used in the US and around the world, but nowhere was it embraced as quickly, fully, and as successfully as in India. It was a mechanism for assuring quality and encouraging US clients to sign contracts with Indian computer services firms; it also was a mechanism for improving processes, which could help with efficiency, meeting delivery schedules for working systems, and long-term cost containment (through minimizing expensive and laborious debugging). Indian IT services corporations also enthusiastically embraced other quality metrics, methods, and credentialing, such as ISO 9000 and Total Quality Management (TQM).

The "Millennium Bug" or "Y2K problem" resulted from the use of only two digits for a year in COBOL and other programming. (The practice dated from the late 1950s, when computer memory was limited and expensive.) Before the 1990s, use of only two digits for a year was extremely common. The legacy of two-digit code for years led to widespread fears that computer systems would go down or would operate erratically at the turn of the millennium. Hundreds of billions of dollars were spent worldwide in the second half of the 1990s (mostly between 1996 and late 1999) on "Y2K compliance." Leading market research firms' estimates of the costs of Y2K readiness varied widely. The Gartner Group's estimate of the worldwide cost, which was roughly in the middle, was $600 billion. Merrill Lynch alone spent more than half a billion dollars on Y2K readiness. Few US corporations ever revealed their expenditures on Y2K readiness. The US government, local governments, corporations, and other organizations probably spent well in excess of a hundred billion dollars.[82]

Efforts to become Y2K compliant caused a labor shortage of people skilled at debugging COBOL code. Some Indian firms well up the ladder of certification on the CMM, Wipro among them, stepped up. They could not have done so without the strong, educated workforce the IITs had provided or without the networking infrastructure Texas Instruments and other companies had pioneered.

Tata Consultancy set up a "Y2K factory" in the Indian city of Chennai (called Madras before 1996). There about a thousand programmers went through approximately 2 million lines of code *per day* and made fixes. As the deadline grew closer, Tata Consultancy Services contracted out some of the work to seven other computer services firms in India. Infosys and Wipro also made substantial commitments and benefited from large-scale Y2K compliance work, most of it for US organizations. For Infosys, 23 percent of 1998 revenue and 20 percent of 1999 revenue was from Y2K compliance code checking and fixes.[83] These efforts not only produced revenue that enabled Indian companies to expand; they also provided opportunities to demonstrate reliable performance on technical work (albeit on a rather low level) and to establish relationships with American and multinational corporations. This helped set up India's largest computer services providers to be leaders in business process outsourcing in the early 2000s and in subsequent years.

Business process outsourcing is a broad category referring to accounting systems, payroll systems, human resources systems, supply chain management systems, and many other types of IT or computer services. It also can refer to the business activity of taking calls from customers and providing customer service (on billing or other questions)—not primarily an IT service, with the exception of help-desk work that assists customers with IT equipment, networking, or software. The outsourcing of business processes by US corporations and organizations to India involved both traditional computer services (back-office processing) and customer-service call centers.

In a 2005 book titled *The World Is Flat: A Brief History of the Twenty-First Century*, Thomas Friedman discussed the phenomenon of the United States and other developed nations outsourcing IT services and call centers to developing countries, particularly India. According to Friedman, it was during a trip to India shortly before he began writing the aforementioned book that he came up with the metaphor "the world is flat" to characterize computer networking precipitating a large-scale round of globalization applying to information and computer services—essentially, technology leading to a level playing field for work and commerce.[84] Although the metaphor is clever, it is somewhat distorting. The world may be leveling to a degree, but because of different geographies, resources, populations, histories, cultures, customs, governments, laws, and other factors it will never be truly "flat"—that is, equal in opportunities for enterprise and wages. And frequent in-person interaction with clients or team members can make a difference.[85] Direct personal interaction can be quite helpful in applications programming and systems integration projects, whether a particular project involves an independent contractor providing networking support services to a small nonprofit organization, or a sizable company installing a new system for management of its supply chain, or a massive real-time system for the US Department of Defense. The business of computer services has always been a reputation-based business, and face-to-face interaction often can be helpful. The movement toward cloud services (discussed more fully later in this chapter) certainly increases opportunity for work at a distance, but also has produced substantial physical infrastructure. Thus far, many of the largest data centers and server farms are in the United States, Great Britain, Germany, France, Japan, and Australia—some of the most expensive countries in terms of land and labor. And those facilities require considerable work to build and to maintain.

Journalistic reports of the outsourcing of IT (including Friedman's op-ed columns in the *New York Times* and his book *The World Is Flat*) have led to a backlash in the popular press, among politicians, and among the public concerning the loss of well-paying jobs in an industry to which the US has contributed more innovation than any other country. Although some US-based jobs have shifted overseas, there has been considerable growth in US computer systems and IT services jobs in the past two decades, including in the half-dozen years after Friedman's book. The bursting of the dot-com bubble caused the only multi-year downturn (2001–2003) in computer system and IT services employment in the past 25 years. The recent Great Recession brought only a one-year decline of 1 percent (in 2009). As the Bureau of Labor Statistics noted in a 2013 report on employment in IT services and in computer system design, from 2003 to 2011 US labor force in those fields grew by 37 percent (despite the Great Recession), exceeding 1.5 million employees by 2011.[86] Thus, even though IBM and some other of the largest IT services companies have reduced their US labor forces in recent years and increased their labor forces in India, many IT jobs remain in the US, and many more jobs have been created at smaller, mid-size, and large firms (including Google, Facebook, Salesforce.com, Amazon, and Rackspace) in the broad IT services industry.

In 2012, Tata Consultancy Services was the sixteenth-largest global IT services company in revenue, at $10.9 billion. Within three years, its $15.5 billion in annual revenue placed it in the top ten.[87] In 2014 the Indian IT services industry had revenues of more than $70 billion.[88] Tata Consultancy's workforce of more than 319,000 is larger than that of HP Enterprise and roughly comparable to that of Accenture and (probably) to that of IBM's Global Services Division.[89] Like Wipro, Infosys, HCL, and Cognizant, TCS has made significant shifts to cloud services, analytics, and other higher-margin areas. But TCS's revenue is only about half of Accenture's and less than a third of IBM Global Services. It has higher margins, but to secure business it still charges far less (and pays significantly less in total compensation) than the global average for IBM and Accenture. We are still far from a truly "flat" world. And, as the convergence lessens the labor cost divide (to become a bit flatter), the pace of outsourcing to India is likely to slow. Already there has been a shifting geography of IT labor to some other low-cost countries, including China, Argentina, the Philippines, Egypt, and Bulgaria.[90]

Today in India, the salaries of programmers with little experience (less than five years) tend to be quite low by US standards—under $12,000 a year. However, often such a programmer can double his or her salary (or increase it even more) after a decade of experience and a willingness to leave for another job.[91] And based on purchasing power parity, the differences in compensation between the United States and India in IT services are less stark.[92] Turnover is high in the Indian IT industry, a constant struggle for Indian firms (where the annualized percentage of those voluntarily leaving Tata Consultancy is now around 15 percent and numbers have been higher at Infosys).[93] To keep senior highly talented software developers, salaries in India at times have begun to approach pay levels in the US or European countries. Clearly higher-end software development and systems integration work is being outsourced, not only the low-level debugging and basic business process work. At the same time, this comes at a substantially higher price (to clients) than "body shop" labor and puts increasingly critical systems outside the ready reach of companies in the US, Europe, and developed countries in Asia. The overall IT services wage gap in the early years of outsourcing (late 1990s and the early 2000s) could result in savings that amounted to being five times cheaper in view of the disparity between labor costs in India and those in the United States. Now the differences are far less than twice for experienced software developers and analysts, and that disparity in compensation is continuing to diminish.[94] India has had higher single-digit inflation for years and talented programmers and systems analysts can command salary increases far in excess of cost of living growth—to keep top IT talent sometimes requires annual increases of 10–30 percent.[95] Job hopping is a frequent tool for experienced and talented IT professionals (and for less experienced people) to increase their salary.

The easier jobs to outsource were outsourced in the early 2000s. Now managers, in deciding about the more complex ones, are faced with diminished savings that do not always seem to justify real or perceived higher risks. Some large US corporations have been keeping more work in house in recent years, often working locally with IT services providers in order to have better control, or at least a better sense of control. Others are focused on shifting IT to cloud providers. All the Indian majors, like the multinationals, are concentrating on security as never before, but high-profile breaches at corporations can potentially make a services supplier halfway around the world less attractive. And while all the Indian giants have become cloud services players, Amazon, Microsoft, IBM, Accenture, HP, and others have been aggressive in shifting to this new paradigm and at present appear to be extremely well positioned for this field. The geographical and demographic labor impact with the switch to cloud services remains to be seen.

Creating Clouds

In late 1965, as the Rolling Stones released "Get Off of My Cloud" (a follow-up single to "Satisfaction"), time sharing—a precursor of "cloud computing" with a lineage broken apart by a two-decade preference for using local computer resources (PCs and local servers)—was just getting started commercially (by GE and Tymshare) and certainly was not on the minds of Mick Jagger and Keith Richards,[96] the song was about wanting solitude and not being hassled or held back—being left alone and free in a private, peaceful place.

Time sharing of the mid 1960s through the 1980s—as a technology and a business—has important commonalities with the cloud computing of the past decade. These include information technology services companies selling remote computer resources and associated services—storage, processing, and software—over a network either to supplement or in place of directly possessing such resources. The time-sharing segment of the computer services industry had a deeper history dating back to origins of computer networks—to Whirlwind, SAGE, the first proprietary networks, and the ARPANET. And computer networks had a deep connection and built upon much earlier infrastructural networks (to lay cable, create network exchanges, and install other equipment)—railroads, telegraphy, sewers, highways, and so on. After all, the Internet provider Sprint was a 1978 spinoff of a railroad company—its name stands for Southern Pacific Railroad Internal Network.

Time sharing also had stark differences from what we now call cloud computing. It was of more limited scope (used heavily for education, science, design, and engineering applications); it was less software infrastructure/ platform oriented; and security, while not wholly ignored, was less of a concern. Perhaps the greatest difference with cloud is scale (big data) and the ability to make use of large data stores through real-time global access and analytics. Storage of data is quickly becoming a commodity, and in the future it may not command any higher profit margin (less than 5 percent) than outsourcing storage of paper records. Alternatively, customizing cloud application software, a heavy concentration on computer security, and developing software systems that adapt and learn from ever-increasing data stores present exciting new opportunities and rewards—intelligent, smart, or "knowing" clouds.

Four years after "Get Off of My Cloud," Joni Mitchell sang of clouds as "ice cream castles in the air" and "feather canyons," and also of the dark side of clouds: sometimes they "block the sun" and "rain and snow on everyone."[97] For Mitchell, the lyrics were a meditation on myth and reality in love and life. "Cloud computing," a term of uncertain origin, gained currency in the early 2000s for offsite (generally outsourced) computer servers for memory, platforms, infrastructure, and software applications. Clouds, of course, feature not only in the lyrics of popular songs but also in oral traditions, mythology, religion, poetry, literary fiction, paintings, drawings, photographs, plays, and films.[98] For computing, "cloud" is an appropriation of an idyllic, peaceful, serene, heavenly, and euphemistic metaphor generally bringing to mind a clean, bright blue open sky peppered with soft, puffy white clouds—not huge server farms requiring many acres of land, new roadways, mammoth amounts of electricity for operating the computer systems and air conditioners, technical and managerial workforces, security infrastructure and personnel, and the underlying materiality of the Internet (massive Internet exchanges, satellites, hundreds of thousands of miles of twisted pair, coaxial, and fiber-optic cables, countless switches and routers, and so on). The electricity for server farms and data centers, in much of the world, is still generated by burning coal and thereby emitting carbon dioxide, the primary greenhouse gas we as humans can control. And the cloud metaphor of space and serenity is far from networking realities of congestion, vulnerabilities, and crime. Meanwhile, and closely correlated, mobile computing (access to data on the go) has escalated at a breakneck pace with the near ubiquity of wireless networked computing devices—smartphones—and with the availability of highly portable tablets and thin, lightweight laptops. With such devices, both business data (accounting, engineering designs, logistics, etc.) and personal data (images, text, and video on Facebook or Google) are sent and retrieved from remote servers.

In some respects, cloud computing with massive server farms/data centers is a classic case of taking advantage of economies of scale, with server infrastructure, maintenance, and expertise—rather than inefficiently having servers and maintenance at countless organizations worldwide (and the people to manage them), they become more centralized and larger. While recognizing the challenge of improving "an imperfect and evolving portfolio of metrics," several leading energy scientists and computer engineers conducted a study at Lawrence Berkeley National Laboratory and concluded that data center services hold promise for reducing societal carbon emissions. To best realize these energy optimizing opportunities (with ample "free" cooling and/or near low-carbon electricity grids), proper locational and equipment decisions are necessary. To achieve this on a wide scale, appropriate incentives from policy makers and legislators probably will be required.[99] These scientists at Lawrence Berkeley National Laboratory are focused on the important issues of energy usage at data centers and the carbon imprint of that energy based on its source (such as burning coal or a nuclear power plant)—the supply side. On the demand side is our culture of consumption. What is not environmentally friendly are potential changes in consumption that the cloud might bring—such as "bitcoin mining," where the e-currency is generated with complex, computer processing-intensive equations.

If companies build massive energy-intensive infrastructure for the mobile streaming of countless hours of ultra-high-definition video—be it movies, television, or peer-to-peer video platforms—the public probably will consume it. The Carbon Trust, an international consultancy with headquarters in London, examined the carbon footprint for watching a soccer match. It found that a phone or a tablet can be much more energy efficient if watched at home (by means of Wi-Fi or cable) on a screen smaller than that of a typical present-day television, but less efficient than if watched on a TV with a group at a pub. And if mobile data is used, this is the most carbon intensive of all (more than 10 times the carbon imprint of a broadband connection).[100] Choices by individuals and on behalf of organizations regarding access and use of data, along with the energy source, are at the core of how cloud computing will affect atmospheric carbon and climate change.

For a firm or an organization, cloud computing can offer a number of economic and operational advantages. Organizations are not burdened with the up-front costs of extensive servers, processing, and storage resources. With cloud computing, much as was the case with time sharing, companies can pay as they go for only services that are used. Scaling up or scaling back is easy and is not required in larger discrete increments, as is often the case with in-house technology and personnel resources. It also allows organizations to benefit from the analytics technologies of providers as well as the expertise of their personnel who work with and learn from serving many different customers, often including some of the same industry verticals. Cloud is a more complete outsourcing than traditional IT services (outsourcing programming, maintenance, and systems integration) since it also includes hardware and middleware. In some respects, it is part of a greater outsourcing and pay-as-you-use trend in our economy and culture—where music streaming, movies, Zipcar, Uber, Lyft, and renting (rather than buying) residencies have all become increasingly prevalent.[101]

At the same time, cloud computing may also have some drawbacks. In an era of unprecedented computer crime and violations of privacy, we continually increase our dependence and information infrastructure on transmission of data over an Internet designed for interoperability rather than security. The most secure data (such as top secret US government classified data) either needs to be maintained in house on secure offline systems or outsourced to isolated off-the-Internet servers with similar physical security (a service offered by some large providers of cloud services). A survey conducted in September 2011 by Cisco Cloudwatch reported that 76 percent of corporate and organizational decision makers in the UK cited security and privacy as the "top barrier" to adoption of cloud computing.[102]

Like the "wild, wild West" of the Internet and dot-com frenzy of the late 1990s, cloud computing has attracted established players in IT services to the business—IBM, Accenture, HP, Dell/EMC, CSC, Fujitsu, Capgemini, and India's Big Five—as well as many startups, with some of the latter emerging as new, path-breaking powerhouses in certain spaces (Salesforce.com for CRM and Rackspace for infrastructure).[103] It has also spawned quick-to-go-public companies with flighty valuations reminiscent of the IT bubble of the late 1990s.

At the end of 2013, Gartner Inc. predicted that one fourth of the top 100 cloud companies would either be acquired or go bankrupt within a few years.[104] The rate of bankruptcies in this field, of course, will be higher among newer, smaller players with fewer resources to protect the data and interests of clients. Thus, bankrupt cloud companies present clients with considerably greater risk than other types of services providers, as they have their clients' proprietary, valuable, and private data. These risks are likely to push organizational customers to the largest, most stable IT firms rather than small providers of cloud services. Shifting to cloud services can create path dependencies, making decisions more difficult (one reason why in 2016 cloud services make up considerably less than 15 percent of the IT services industry). Transitioning to a cloud provider can result in loss of IT organizational capabilities—which positively allows organizations to focus more on their core, and negatively makes them more reliant and presents challenges if IT services are ever brought back in house.

Cloud computing services typically are divided into three broad categories—from the most basic and flexible to most advanced and structured— "infrastructure as a service" (IaaS), "platform as a service" (PaaS), and "software as a service" (SaaS). IaaS provides storage and a virtualized interface; the customer brings the operating system and applications. PaaS includes the operating system and certain applications such as databases. SaaS refers to standard and customizable advanced applications such as systems for CRM, ERP, logistics, etc. Some cloud services providers specialize in just one of these three areas, while others in two or all three. The US Department of Commerce's Cloud Computing report of June 2015 cited market research firm Forrester placing the cloud computing global market at spending of $72 billion in 2014, and cited market research firm IDC's projection of it becoming a $191 billion market by 2020. Other projections are significantly higher—in early 2015, Gartner, Inc. projected cloud revenue to reach $240 billion by 2017.[105] In roughly five years, it may constitute one fifth of the IT services industry, and substantially more than that a decade from now. Cloud is a dominant trend, but will take time, as migrating from existing IT environments is difficult and disruptive.

In addition to established IT services companies, powerhouses from e-commerce (particularly Amazon) and from software products (especially Microsoft) made early moves to capitalize on the fast-growing field of cloud services to complement their lower-margin "cash cow" businesses. Cloud computing includes the roughly decade old B2B cloud segment of the computer services industry as well as the business-to-consumer (B2C) cloud outside of it, primarily in the social networking field. The B2C cloud—which operates on an advertising revenue model, typically provides "free" platforms, and has different industry dynamics, strategies, and metrics—is becoming well-established.[106] While the B2C social networking cloud (outside the scope of this book) arguably is changing social practices, communication patterns, and interaction (and its advertising certainly can influence consumers), the enterprise cloud is likely to be where many of the greatest economic effects will occur, with large corporate users and government departments and agencies as drivers.[107]

In view of its precursor in 1960s time sharing and its dependence on much longstanding computer networking, memory, and processing infrastructure, there is no readily identifiable start date for cloud computing. But in terms of its becoming a recognizable industry segment, Jeff Bezos' public announcement of Amazon Web Services (AWS) in March 2006 is a useful demarcation. Roughly a half-dozen years earlier, Salesforce.com was launched and initially provided cloud CRM services to primarily small enterprises (additionally, in the latter 1990s, some Internet providers offered infrastructure with virtual private network, or VPN, services).[108] Before 2006, however, larger IT firms were not reporting on cloud initiatives, and the largest dedicated cloud firm, Salesforce.com, had revenue under $300 million. In 2008, Gartner, Inc. placed global cloud revenue at less than $47 billion, and it had grown considerably between 2006 and 2008—it probably was less than a $20 billion industry segment in 2006—well short of 4 percent of the overall IT services industry.[109]

In publicly removing the veil over AWS in early 2006 (for an initiative probably in the works many months if not a few years earlier), the world's leading e-commerce firm clearly had become committed to leveraging its already tremendous infrastructure and know-how in storage, databases, and analytics as a new higher-growth, higher-margin business. Amazon was not only leveraging know-how, but also balancing loads from its e-commerce business. The setting for Amazon's announcement was MIT's annual Emerging Technologies event—specifically, a keynote address by Amazon's founder and leader, Jeff Bezos. Amazon Web Services—with large margins (unlike Amazon's ecommerce retail business, which has focused on revenue and market share far more than on profits)—has been the primary reason for the company's recent multiple consecutive quarters of positive net earnings.[110] And although Amazon announced AWS in 2006, the company—notorious for keeping financials for its businesses such as Kindle e-readers under wraps—did not break out its revenue and earnings for AWS until 2014.

Amazon, like its fellow IT giant Apple, stands out for its ability to keep its product and services strategies secret, as well as for its well-coordinated promotional efforts. Along with Amazon's 2006 launch of AWS in the enterprise world and its consumer-focused Kindle e-reader the following year, the company even managed to generate a huge buzz for services it has not yet offered. The announcement of Amazon's research on drone delivery on the CBS television program *60 Minutes* in late 2013 amounted to 20 minutes of free advertising for Amazon as an innovation leader for research on a delivery mechanism never likely to be feasible and practical on a wide scale. It was a brilliant public relations maneuver right before "Cyber Monday."[111] Although Jeff Bezos is less charismatic than Steve Jobs was, he shares with the late Apple Computer co-founder an ability to skillfully envision products and services appealing to consumers (to see beforehand what they will want or to create a perceived need), and use spectacle to effectively market a leading IT brand. It was Bezos' bold vision that resulted in the seemingly disconnected cloud business of AWS, one that on the surface seemed far afield from (primarily B2C) e-retailing. AWS, in reality, was deeply connected by infrastructure and know-how to Amazon's core business, which helped it to rapidly emerge as the world's largest provider of enterprise cloud services.

In early 2015, Bezos disclosed that Amazon's AWS was a $6 billion-per-year business that had grown about 50 percent from the previous year and contributed quarterly earnings of $265 million.[112] As each quarter brings major growth, AWS is now generating more than $10 billion a year.[113] This makes it the leading cloud enterprise company by revenue. If AWS were an independent company, it would probably be valued at more than $100 billion, about a fourth of Amazon's overall valuation.[114] To this point, Amazon has focused on providing IaaS and AWS has aggressively plowed resources into adding and expanding data centers and offering additional services—including encryption and "machine learning."[115] With its long-time focus on IaaS, Amazon has unrivaled computing power—according to *Fortune*, Gartner, Inc. recently reported that Amazon's AWS had ten times the computing capacity of the next fourteen cloud companies combined.[116] Amazon has an estimated 57 percent of enterprise IaaS customers.[117] With revenue, the dominance for AWS is there but not quite as pronounced—it probably controls between a third and a half of global IaaS.[118]

AWS has recently targeted the largest banks—such as Citigroup, J.P. Morgan Chase, and Capital One—many of which have long had their own data centers and contracted with IBM and other IT services powerhouses for support. In 2015, Capital One signed on with AWS, seeking to reduce its number of data centers from eight to three; others in the financial industry may follow suit. AWS has made major investments in security, and in 2013 the Central Intelligence Agency selected AWS as a cloud services provider.[119] Among AWS' many large corporate customers with large IT infrastructure needs are Netflix, Expedia, Comcast, Pfizer, and Samsung Electronics.[120] While Amazon strongly leads in IaaS, Google is also a major competitor, and there are other key participants in the high-margin SaaS field, in which the leaders are Microsoft, IBM, Salesforce.com, and Oracle.

Google, formed in 1998, the longtime leader in Internet search (the driver for its primary revenue source—advertising or sponsored search), in the 2000s successfully moved into providing operating systems for mobile devices (Android phones and tablets) and netbooks (Chrome), and mobile devices (Nexus). Google operates in a number of cloud spaces, including many successful B2C-focused services provided "free" to consumers such as search, Gmail, and Google + (paid for by advertising). With these it primarily is competing with Facebook, Yahoo, and other B2C-focused businesses. On the enterprise side, Google Cloud Platform provides IaaS and PaaS to corporations, government, and other organizations—directly competing with Amazon and Microsoft. In IaaS and PaaS for enterprise, Google recently signed on to provide service to Northrop Grumman to prototype software used for genomics research, and serves Coca-Cola, SnapChat, and many other corporations. In the growing music streaming delivery space, Google recently signed a deal to provide cloud services for Spotify, a customer leaving AWS.[121] Whether this is a trend and Google is beginning to gain ground from its third place position in cloud revenue behind Amazon and Microsoft remains to be seen.

Microsoft's Intelligent Cloud Group (with services branded as Microsoft Azure), is second in the cloud services segment and was generating about $6 billion in revenue in 2015 (and likely is around $10 billion annually today).[122] This group's success has been a significant factor in the stock valuation of Microsoft more than doubling over the past four years. Microsoft has invested heavily in security in recent years, including a new Redmond, Washington, Cyber Defense Operations Center—cloud is one reason the firm is now making security a high-profile priority. The software and now cloud giant has not disclosed its cost of establishing the center, but stated it spent $1 billion on security in 2015 and has roughly 3,500 computer security specialist employees worldwide.[123]

Microsoft's cloud business is representative of increasing "co-opetition" (cooperating competitors) in cloud space. It has partnered with longstanding rival Oracle Corporation. And Oracle, which is transitioning to and cannibalizing existing CRM software revenue streams (that had been aided by enterprise software acquisitions of Siebel Systems in 2005 and BEA Systems in 2008—both firms' capitalizations at the time they were acquired were at a small fraction of their lofty peaks during the dot-com bubble) for new cloud-based ones, has even partnered to integrate certain enterprise software with Customer Relationship Management (CRM) cloud competitor Salesforce.com.[124] Oracle founder and Executive Chairman Larry Ellison and Salesforce.com co-founder and CEO Marc Benioff's relationship is complex and characterized by rivalry and fierce competition, as well as with respect and co-opetition.[125]

Salesforce.com was launched in 1999 by Marc Benioff (along with Parker Harris),who left Oracle to deliver CRM software over the World Wide Web. This startup's rapid growth, particularly after its 2004 IPO allowed for rapid expansion, demonstrated the viability of cloud computing and influenced IT giants to transition into this new business. In 2005, Salesforce.com recorded $176 million in revenue.[126] A cloud SaaS business, Salesforce.com grew its revenue to more than $5.3 billion in 2015—a figure placing it in the top four of cloud firms and a notch higher than IBM in this field.[127] In CRM, cloud clearly is the way of the future and Salesforce.com has a meaningful head start and currently has a lead over Oracle and others. By 2018 more than 60 percent of all CRM software is estimated to be cloud-based.[128]

IBM recorded $4.4 billion in annual cloud revenue in 2015 (but likely will double that for 2016 and be near Microsoft for the second spot in the global cloud computing industry segment). IBM is particularly focused on data analytics and artificial intelligence applications to allow customers to make better strategic decisions. In 2015 it opened a new major cloud data center in Chennai, India—one major piece of its recent $1.2 billion cloud expansion. As further evidence of the prevalence of co-opetition in cloud, IBM has formed partnerships with longtime IT services competitors Accenture and CSC.[129] IBM is betting heavily on analytics, which it markets as Smarter Cloud, an extension of its Smarter Planet campaign to improve efficiency with energy and other infrastructures. Expanding its AI/analytics research and development—which was famously demonstrated with supercomputer "Watson" beating top human contestants on the TV game show *Jeopardy!* in 2011—IBM invested more than a billion dollars and established a Watson Group in 2014. This group, with team members worldwide, is headquartered at IBM's facility in New York's Silicon Alley and has a staff of 2,000.[130]

It is possible that US firms are becoming even more dominant in cloud computing than their sizable majority share of the overall IT services industry. This is because some of the leading firms in other IT areas—electronic commerce (Amazon), personal computers and peripherals (Dell), and software products (Microsoft and Oracle) have moved fast to invest in developing cloud services businesses. Moreover, some of the leading newer firms formed to compete in this space are based in the US, such as the CRM-leader Salesforce.com and the IaaS specialist Rackspace. Nonetheless, Fujitsu, Capgemini, and the Indian Big Five are all investing aggressively in their cloud services infrastructure and services. In view of its early investment in cloud services, Fujitsu probably is in the top five in this area.

On October 12, 2015 Michael Dell and partners, who took Dell, Inc. private in 2013, announced the agreement they made to acquire EMC, the leading data storage company. This $60 billion acquisition, completed late in 2016, was the largest IT acquisition in history and is propelling Dell into one of leaders in storage and cloud computing.[131] EMC, founded in the late 1970s, initially was a memory component supply firm to Prime and other computer companies. In the mid 1980s, it moved into data storage and grew that business rapidly to become the industry leader in the 1990s. In the early 2000s it began to concentrate on network data services, IaaS, and became a significant cloud industry provider by decade's end. Dell, in seeking to add a higher-margin business as a complement to its commodity PC business, launched Dell Business Services, partnering with Salesforce.com, to enter aggressively into the cloud enterprise space in 2011. Dell's deal with acquiring EMC is to rapidly become a large-scale provider of cloud services. The combination seeks to create a new major integrated computer, software, and services company in both the business to individual consumer (PCs) and enterprise spaces. Because Dell is a private company, it will not be subject to the quarterly financial reporting scrutiny that often makes long-term strategic transformations for public companies more difficult.[132] Interestingly, Dell's integrative approach is a trend reversal for the IT industry—IBM sold its personal computing business in 2005, and HP divided into two companies in 2015—one for end market computers and printers (HP Inc.) and the other, the new HP Enterprise, focused on services.

IBM and HP (with HP Enterprise now merging with the commercial IT services side that remains at CSC, after CSC spun off its government business) are the two leaders in the global IT services industry. At one time, the fast growth of India's Big Five seemed to indicate that Tata Consultancy or one of the others might grow to rival or exceed the size of IBM in services revenue. Now that seems less clear. The Indian giants still enjoy larger profit margins and have lower costs, but the converging of labor costs may result in far less US IT work being sent offshore, or it might be outsourced to countries with costs lower than those in India. IBM, HP, Accenture, Tata Consultancy, Wipro, Cognizant, and Infosys all have hundreds of thousands of services workers. For many years, the Big Five in India were seen as "body shops" for low-end, low-value chain work. While they have grown beyond just lower-level services for basic BPO, India has a very large IT services workforce—by some estimates, more than half of those in the world in this field—for a country that brings in less than 10 percent of global IT Services revenue. The Indian IT services giants still have a substantially higher portion of their work at the lower end of the value chain compared to IBM and Accenture. All of India's Big Five are engaged in growing cloud infrastructure and services, but they will face great challenges (perhaps greater than IBM, HP, and Accenture) with this transition as a result of their existing workforces and a greater portion of their personnel focused on lower-end BPO business that probably will decline with the cannibalizing cloud where new platforms, software products, and software services are offered. Virginia Rometty is right to see opportunity for IBM India to help lead the way with analytics, but this will be a disruptive change. The coming decade is the first one in the history of the IT services industry when there is a possibility that (because of cloud and analytics being more platform, system, and software focused) the number of global IT services industry workers might decline even as revenue continues to climb. Smarter systems that adapt and learn might lead to a situation where to make IT work for organizational clients might mean less IT work for IT workers. Seemingly countering this trend, Accenture hired a staggering 100,000 employees in 2015 (90 percent of them college graduates of the "millennial generation") for positions that will bolster analytics, mobile, and other newer focal points. This, however, is not a net figure (it does not take into account lost employees through attrition), does not specify geography of new hires, and probably is a one-time major hiring blitz.[133]

Trying to understand an industry, industry segment, or new industry paradigm in its first decade or so is at once a useful and important endeavor, and a bit of a fool's errand. In the first decade of the automobile industry (defined by the introduction of the Benz Velo, the first production model, in 1894), some technological elements of automobiles, their production, the commercial trade, and the technological system of automobiles/"automobility" were already evident—Benz's company, the Ford Motor Company, the Olds Motor Works, reliance on component supply networks, trade associations, trade publications, print advertising, auto racing as promotion, and the awarding of the Selden patent. Much infrastructure was missing or still in quite infant form that first decade, such as quality roads, filling stations, service stations, state issued licenses (only required in several states), dealer networks, and much more. There were many startups, and most of them failed.

The elements needed to understand the longer-term trajectory of the automobile industry—the implementation and success of Ford's moving assembly line, the highly standard and thus more affordable Model T, Ford's $5-a-day wage (turning workers into potential consumers), the combination that resulted in General Motors, the fate of electric vehicles, the outcome of Selden patent litigation, the formation of the Society of Automobile Engineers, Alfred P. Sloan and GM's marketing a vehicle for "every purse and purpose," the rising power of the United Auto Workers, the relative decline of public intra-city and intercity transportation in the US relative to automobiles, interstate highways, suburbanization, anti-lock brake systems, the success of Japanese imports, the increasing dependence on foreign oil, and more recently, the diminishing influence of the UAW, computerized diagnostic systems, the coming of hybrids, the return of electrics, R&D in self-driving cars, fracking and a global oil supply glut, and peer-to-peer ride sharing cloud platform companies such as Uber—were in the unknowable future. The cloud computing momentum of Amazon Web Services, Microsoft, IBM, HP, Salesforce.com, Accenture, Oracle, Fujitsu, Capgemini, and India's Big Five will probably ensure that these are all cloud services providers a decade or two from now, and data analytics will probably become more and more important. Given the size, talent, and financial resources of these companies, it is less likely that a startup will enjoy the success that relative newcomer Salesforce.com has had in pioneering the CRM cloud space, but some important domain specialists will undoubtedly arise and thrive. Security will become more and more critical, both as a source of revenue and (like analytics) as a field in which providers can achieve differentiation. AI, a field that for more than fifty years was long on promises for the future but far shorter on real-world deliveries, is becoming a critical tool to achieve new things, to generating revenue and net earnings, and to potentially precipitating for the first time a modest reduction in the number of IT services professionals.

The cloud segment (including analytics) clearly is gaining momentum and the industry has its first vestiges of an emerging form and cast of important players, but it is also rapidly evolving and much is in flux. There undoubtedly will be many surprises in cloud computing and the IT services industry. Two decades from now—with the benefit of hindsight—we might reflect that, in the words of Joni Mitchell, we really didn't know clouds at all.

Conclusion

The computer services industry is by far the largest of the three major IT industries. It generates more revenue than the other two (computer hardware and software products) combined. At $955 billion in 2014, it is fast becoming one of the world's few trillion-dollar industries—similar in size to the global pharmaceuticals industry, the alcohol industry, and the telecommunications industry.[1] Despite this, it is far less conspicuous and is vastly underappreciated relatively to both the computer hardware and software products industries. Historians, economists, management theorists, sociologists, and other scholars (as well as journalists) largely have ignored the computer services industry and its history.[2]

Why has this industry—which is fundamental to making IT work for corporations, governments, and other organizations—remained "under the radar"? Several factors stand out. First, computer services (commonly referred to as IT services in recent years) are primarily for organizational customers. These services, and the companies providing them, are advertised to the broader public (on TV, on radio, in the popular press, or on the World Wide Web) less frequently than products or services designed for individual consumers. To a degree this is blurred by companies that provide cloud business-to-business and business-to-consumer platforms and services, such as Google and Microsoft, but historically the IT services industry has been—and for the most part it remains—an enterprise-focused business-to-business industry. Second, customer organizations—large corporations, small businesses, government departments and agencies, and nonprofits—tend not to highlight the fact that they outsource some or all of their IT. There is no incentive to highlight that, and (particularly for large corporations or the federal government) doing so might convey a lack of expertise, or an inability or unwillingness to maintain hands-on control and protection of customer/client/constituent information. Third, with Ross Perot from the late 1960s into the 1990s the one possible exception (primarily because he ran for president in 1992 and 1996), it hasn't had leaders as well-known as those in the computer, software products, and semiconductor industries—for example, Thomas Watson, Thomas Watson Jr., Steve Jobs, Bill Gates, Robert Noyce, Gordon Moore, William Norris, Seymour Cray, Andy Grove, and Larry Ellison.[3] Fourth, the industry and its companies tend to gain the interest of journalists and the public only when major government-funded IT services projects (usually large-scale systems integration for the defense or intelligence sectors) are partial or complete failures and the press runs stories focused on taxpayers' being fleeced. Fifth, it is an industry of many players, with no company ever possessing a market share of 10 percent or more. Much of the work is done by smaller enterprises and even by individual independent contractors. Women made major contributions to launching the industry's independent contractor brokerage side, and women and smaller businesses have relatively seldom been the focuses of historians of business and technology. Sixth, one component of services is maintenance, which is often overlooked—whether maintenance of machines (repairs and troubleshooting) or code through debugging. Ignoring maintenance and maintenance laborers is endemic in all areas of the history of business and technology. Seventh, IT services is a fluid, rapidly evolving industry, and its boundaries are more porous than those of the hardware products industry or the software products industry (though looking at which companies compete against one another offers important clues and makes defining its segments possible). Eighth, though many of its principal players have substantial presences in Silicon Valley, and though HP was founded and remains headquartered in Silicon Valley, IT is a global, geographically dispersed industry with no one dominant innovative region—leading companies' headquarters are located in many different places, including Armonk (IBM), Palo Alto (HP), Seattle (Microsoft), Dublin (Accenture), Washington (CSC), Paris (Capgemini), Tokyo (Fujitsu), Mumbai (Tata Consultancy), and Bangalore (Wipro). Large data centers are being built rapidly, across the United States and around the world, and access to land and energy grids are fundamental to locational decisions. And finally, much of what IT services creates—custom software code—is ethereal; it lacks materiality. Software products were boxed and packaged and sold on disks for decades; services are more elusive. Developers of software products (Seymour Rubenstein, Bill Gates and Paul Allen, Mitch Kapor, Bob Frankston, Dan Bricklin) have become famous to varying degrees; leaders of custom applications programming efforts, however large, never have become famous.

One custom programmer, Ellen Ullman, achieved considerable name recognition not from her programming acumen, but by writing a book, *Close to the Machine*, in which she articulated how the labor done by people who work in programming services (especially for smaller organizations) is invisible to outsiders, oscillates between exhilaration and frustration, and reflects both the creativity and fallibility of its human designers and developers. "The project," Ullman notes, "begins in the programmer's mind with the beauty of a crystal … . Code seems lovely in its structureness [sic] … . The world is a calm, mathematical place … . Human and machine seemed attuned to a cut-diamond-like state of grace … ." Next she conveys how this is fleeting: "[S]omething happens … . The irregularities of human thinking start to emerge … . There are dark, unspecified areas." And finally, "a process of frustration": "[The] programmer goes back to analysts with questions, the analysts go to the users, the users to their managers, the managers back to the analysts, the analysts to the programmers … . Things are not understood … . The whole graceful structure loses coherence … . A state of grace reveals itself to be a jumble. … [The] human mind is messy."[4]

Before writing *Close to the Machine*, Ullman worked—sometimes alone, sometimes on a small team—mainly on programming efforts for nonprofit organizations. For larger projects, the challenges can be even greater. In general, larger programming services and systems integration projects are extremely difficult; that is why a substantial portion of them come in late, over budget, and requiring many hours of debugging, and some are abandoned. This is particularly true for new types of applications in challenging areas such as simulation, health care, advanced real-time logistics, and other fields. From SAGE and SABRE in the late 1950s and the 1960s to a six-year $1 billion plus abandoned (in 2012) Air Force Logistics real-time system modernization effort in the 2000s and the 2010s, software (especially on a grand scale) has proved time and again to be quite hard.[5]

Computer services differ from hardware and software products in blurring the line between producer and users. In consulting and programming services, there is a partnership seeking IT solutions. This is true to a more limited extent with facilities management. With data processing and service bureaus, the services firm often is the end user of the technology even though it is not the end user of the resulting information or data. With many types of computer services, interaction between providers of services and user organizations is frequent and ongoing. Thus, this book contributes to our understanding of computer users and uses, but in a far different way (a different vantage point) than earlier surveys (Cortada's *Digital Hand* trilogy) and monographs (JoAnne Yates' *Structuring the Information Age*) that are focused on user industries, firms, and other organizations. This book also has some similarities to, but more differences from, *FastLane: Managing Science in the Internet World*, a book that Thomas Misa and I wrote on the history of the National Science Foundation's FastLane system. In that book, we closely examined a client (NSF) of a programming services provider (Compuware), but we concentrated heavily on external end users of FastLane (the scientific research community proposing to or reviewing for NSF) and the legacy internal users (NSF staff) of the overall FastLane platform. Compuware is not a primary focus of the book after our brief discussion of its role in programming the earliest iterations of FastLane.[6]

Tabulation rooms at corporations and outsourced pre-computer data processing were the norm when machines for scientific computation—ENIAC, UNIVAC, and IBM's Defense calculator—entered the scene in the mid 1940s and the early 1950s. The UNIVAC, and the even more business-friendly (in lease price, size, and connection to existing equipment in tabulation rooms) IBM 650, had the potential to serve as tools for data processing, but organizations had to learn to use digital computers for data processing, or learn to effectively make outsourcing decisions (choosing what to outsource and what companies to outsource to). Here, computer consultants were critical and represent the true start of the computer services industry.

The continuities of pre-computer and computer era consulting are clear in the cases of IBM, the accounting firm Andersen and Company's Administrative Services Division, and the mid-1950s startup consultancies Diebold and Associates and Canning, Sisson, and Associates. The aforementioned consultancies were already well acquainted with the pre-computing tabulation machine data processing technology that many organizations had in place, and were capable of advanced analysis of the economics of computer installations. Andersen and Company, Diebold and Associates, and Canning, Sisson, and Associates often recommended that client organizations continue with pre-computing technology for a while. By providing detailed and trusted technical and economic analysis for clients, they survived and thrived. The computer consulting field of the 1950s certainly included charlatans, with minimal knowledge, who promoted computing in an unrestrained way, but their enterprises did not survive. The reputable consultancies constructed and projected expertise, staying at least one step ahead of the client and putting in the hard work and the study that enabled them to achieve true expertise in digital computing and its applications. Much of their knowledge was gleaned from astute examination of many current and potential client organizations and of ideas and practices that were circulating among clients' industry peers. Andersen's Administrative Services Division quickly added a programming services business; Diebold expanded more slowly and diversified at a measured pace. Richard Canning's business (Roger Sisson left after several years) continued to consult into the early 1960s, but Canning didn't want to grow the firm and become a manager. Instead, he became the publisher of a monthly newsletter, *EDP Analyzer*.

The early computer services industry saw the emergence of new segments, new types of services, and new delivery mechanisms. Specialist Automatic Payrolls/ADP stands out for remaining highly specialized in data processing and particularly in payroll processing. Eventually it added transaction processing for stock brokerages as a new and important business area, but payroll processing has always been its main business.

Meanwhile, emerging programming services firms such as Computer Usage Corporation (CUC), C-E-I-R, Inc., and Computer Sciences Corporation (CSC) diversified into service bureaus, data processing outsourcing, and systems integration. Overly rapid growth, cash-flow challenges, and lack of focus (entering hardware and other businesses) outside of IT services contributed to CUC's downfall. Similarly, a lack of focus and liquidity issues became problematic for C-E-I-R, which was acquired at a time of relative weakness by the computer manufacturer Control Data Corporation. Control Data treated computer services as an important profit center (in contrast to IBM) from shortly after its formation in 1957. That same year, IBM had to split off its service bureau as a subsidiary corporation as a result of a 1956 consent decree, and through the 1960s it bundled its services (offered as "free," embedded within hardware contracts) as it focused on maximizing revenue and earnings from its hardware business. Owing to concerns about pending Department of Justice antitrust litigation, and recognizing the existence and rapid growth of the computer services industry and the emerging software products industry, IBM executives decided to unbundle many of the company's software products and some of its services in late 1968. (The unbundling was announced in 1969 and was partially implemented in the early 1970s.)

Selling Service Bureau Corporation to Control Data in a portion of the settlement of CDC's lawsuit against IBM certainly did not mean that IBM was exiting from services, but it was both a real and symbolic move de-emphasizing services as a recognized business. Yet IBM continued to provide customer engineers, systems engineers, and other services professionals in ever-greater numbers to help customers with IBM equipment and to develop new applications for IBM systems (in order to sell more hardware). Computer services was, without question, a growing activity for IBM, even when it was not a core business for the firm; this has been overlooked in the voluminous secondary historical literature on IBM. And IBM's Federal Systems Division also provided programming and systems integration services, but much of these services were classified and less visible. IBM's leaders undoubtedly had less concern that Department of Justice regulators would be scrutinizing services it provided to NASA, the Department of Defense, the Department of Energy, the National Security Agency, and the Central Intelligence Agency.

On the hardware side, at the start of the 1960s, IBM launched its Components Group, soon to be renamed the Components Division. That division manufactured memory and logic semiconductor components for IBM's systems.[7] With its computer business and its business and support activities in software, services, and components, IBM was (and remains) a prototypical "Chandlerian" firm—a company with a continuing propensity toward greater vertical integration and toward controlling key (upstream) inputs (materials and components) and (downstream) marketing of products.[8] In essence, Thomas J. Watson Jr. and his immediate successors strongly favored managerial hierarchies over markets when allocating resources. In the 1960s and the 1970s—like IBM, though to a lesser degree—the entire mainframe industry (including Control Data, Burroughs, GE, Honeywell, and Sperry-Univac) tended toward vertical integration. Though Sperry-Univac outsourced semiconductors, it had a services business, as did CDC, GE, Honeywell, and Burroughs.[9]

In showing contrast to IBM and the mainframe industry's focus on managerial hierarchies, Dell Computer (later Dell Inc.), as some business historians have highlighted, contributed strongly to and exemplified the use of advanced networking technology for information and logistics to achieve a market-oriented "virtual integration" in the 1980s.[10] The vertical integration of IBM in the 1960s and 1970s and the virtual integration of Dell and lack of integration at its clone manufacturing peers with personal computers in the 1980s and 1990s has become the default time frame for when the computer industry shifts from favoring managerial hierarchies (an earlier dominant period of mainframe manufactures) to a new, post-Chandlerian, market-oriented or network model for personal computers. For years Dell was more innovative with mass customization, supply chain efficiency, and direct marketing than its peers, but it is often overlooked that the whole PC-clone industry used markets or networks for inputs for a product that quickly became a commodity. In this environment, branded semiconductor and software components—"Intel Inside" and "Microsoft Windows"—defined personal computers as much as the assembler's own branding (Dell, Compaq, Packard Bell, Gateway, Toshiba, Hitachi).[11]

The timing of this perceived shift from hierarchies to markets in computer "manufacturing" coincides with shifts to market models by the American auto industry and other major industries in the 1980s. The 1980s brought a heightened interest in "lean" production and in the Toyota Production System, exemplified by the joint venture by General Motor and Toyota to partner in building certain GM-branded and Toyota-branded cars in California (which continued until the government-funded post-bankruptcy reformation of GM in 2010).[12] These changes often have been highlighted as the exemplification of an accelerating shift from a manufacturing economy to a service-based economy in the United States and (to a lesser degree) in other developed countries. The IT industry—from hierarchies of mainframes (in the 1960 and the 1970s) to markets for PCs (in the 1980s and beyond)—conveniently fits into such meta-narratives, but only if the focus is on IT's most conspicuous segment: computers. In computer services—what became IT's largest industry—changes were underway much earlier.

The computer services industry has always been a mechanism and a force for using markets rather than hierarchies for IT resources. In fact, pre-computer service bureaus facilitated the outsourcing of data processing between the 1930s and the 1950s, before the advent of computer-based business data processing. In the computer era, ADP provided outsourcing of payroll. IBM's Service Bureau Corporation and other service bureaus offered outsourcing of this type as well as many other business data processing applications. Programming services and systems integration services supplied corporations, government, and other organizations with outsourced custom applications in the 1960s and beyond as IT became more and more important to engineering, design, manufacturing, human resources, logistics, simulation, and many other fields.

Emerging time-sharing firms of the mid 1960s, among them GEIS and Tymshare, allowed users to either supplement or fully outsource their computer processing. Meanwhile, Ross Perot and Electronic Data Systems (EDS) pioneered facilities management as a partial corrective to earlier misallocations of resources for expensive computer systems (internal data processing departments) at customer sites. In essence, EDS and later providers of facilities management represented market mechanisms for more efficient use of computer time among a cluster of computer services' customer organizations by balancing loads. While IBM was heavily integrated, and its mainframe competitors meaningfully integrated, these companies, and some independent firms, were supplying the computer services or outsourcing options to corporations, the federal government, and other organizations. To be sure, many large companies had substantial computing facilities, some of which did not turn things over to a facilities management services specialist. But even here it is important to recognize the substantial role of external (to the firm) IT labor and expertise. IBM's share of the computer industry was well over 60 percent. Although many large corporations took possession of IBM systems (generally through leasing), this model was actually a hybrid of hierarchies and markets: IBM's onsite customer engineers and system engineers provided maintenance, systems integration, analysis, programming services, and debugging.

By the mid 1980s, IBM's staff of highly trained Systems Engineers numbered more than 20,000 and was focused on making systems useful for IBM customers. In short, while many large companies were buying or leasing hardware and maintaining IT staffs, these firms certainly were outsourcing IT expertise and work, at least in part, to IBM—and, to a lesser extent, to other computer manufacturers, including Control Data, Burroughs, and Honeywell, that provided either bundled or contracted services.

The IT services industry was a facilitator and an exemplar of the great diversity of coordination mechanisms in the US economy (in general, providing strong market options) throughout the entire second half of the twentieth century and beyond—even in the early postwar decades, an era generally associated with the growing managerial hierarchies of major corporations. Ironically, Dell Computer—the company that historians, management scholars, and economists have highlighted as an assembler relying on the market for all the major components of its computers—now is moving partially toward a more integrated model (at least regarding a newfound focus on cloud services) with its $60 billion purchase of the leading data storage company EMC. This purchase (completed in 2016) was the largest IT acquisition in history.[13] Reminding us of the IT industry's diversity of coordination, rather than a trend toward greater integration, HP's management recently headed in the opposite direction, dividing the longtime IT giant into two separate corporations—HP Enterprise (services) and HP Inc. (computers and printers). HP Enterprise has sought to increase its scale and customer base in merging with the remaining commercial computer services business of CSC. And IBM, while still heavily integrated, is partnering with Apple and other former competitors and is concentrating more and more on cloud services and data analytics. Twenty years from now, it probably will be a very a different company—it may have retired completely from manufacturing hardware, and its organizational structure may look more like the split-off HP Enterprise or the consulting-focused Accenture than like today's IBM, with its divisions for services, hardware, and software.

Thanks to low barriers to entry, IT services has long been a low-margin, fast-growth business. Parts of the industry's contributions that long ago were considered "cutting edge" (for example, business data processing for accounting, payroll, and human resources) have become commonplace and less profitable. Areas ripe for innovation are industry-specific verticals where new software and customized systems are created for the financial, petroleum, health care, aerospace/defense, biotechnology, and many other industries.

Much of the quite small management-and-economic theory-focused literature on innovation in IT services is presented more broadly under the rubric of "services innovation." In an influential article published in 1986, Richard Barras posited that services might have a "reverse product cycle" characterized by efficiency gains, quality gains, and new services.[14] That idea could be useful in regard to the business data processing previously done in tabulation rooms and later accomplished by computers, but it seems far less useful for understanding major systems integration efforts such as SAGE and SABRE. IT not only provides opportunities for efficiency gains and quality control; it also creates new possibilities. Increasingly, many of those possibilities appear to be at the intersection of mobile computing, cloud services, and the application of artificial intelligence in the dynamic field of data analytics. Here innovations are not merely for process improvements but for engineering new knowledge to help firms and organizations.

Further, cloud platforms are blurring the gap between what is a software product and what is a service. Is a base of standard code always a product if interaction with the supplying IT company's staff inevitably is critical to its implementation with a particular organization and its unique setting and circumstances? In one sense it is assistance with a product; in another sense it is creating something unique from a base where all is intertwined in a system of interaction and modification by a services provider working closely with a client. Applications that are available on the go—such as Salesforce.com's cloud CRM platform and services—bridge mobile, cloud, and software platforms and customizable applications to offer new possibilities. This has made Salesforce.com one of the fastest-growing IT start-ups in recent decades and by far the largest firm specifically founded to provide cloud computing enterprise services. Cloud services, in general, bring together data as never before, continually creating new possibilities for big data analytics, a differentiation area already yielding high margins and likely to continue to do so.

The number of IT services employees in the US may soon reach a plateau, and some of the largest players have decreased their domestic headcounts as they have been expanding their workforces based in India, the Philippines, and other lower-labor-cost countries. IT services, however, is a diverse industry, and many smaller players in IT services have grown domestically. Except during the Great Recession (2008–2009), domestic employment in this industry has gradually increased, and a major portion of higher-margin work in cloud computing and in analytics for US firms is done by IT services companies in the US—the host nation of many data centers and server farms. This bodes well for talented and location-flexible US-based IT professionals. To what extent US multinationals will look to lower-cost labor in India and in other countries for IT services for cloud businesses and infrastructure remains to be seen. (Amazon, Microsoft, and other once overwhelmingly domestic employers are now increasing their operations in India.) It also remains to be seen whether disturbing trends with the declining levels of women's participation in IT over the past three decades (both in education and industry) can be reversed.

Some of the major players in the IT services industry—including IBM, HP, and Accenture—have formal efforts/programs/goals to boost the workforce percentage of women workers/professionals/managers. To facilitate positive outcomes, there likely will need to be changes within corporations, government, colleges, and universities (where there is a far lesser percentage of women majors in computer sciences after a late 1980s peak), K–12 education, professional organizations, trade associations, and other institutions, with regard to gender and underlying cultures of computing (and more broadly, technology). This book has shown that women had some opportunities and made major contributions to computer services but also faced some considerable hurdles in reaching upper levels of technical and managerial leadership in the industry (especially at midsize and large firms). Computer Usages' first four (hired) programmers (in 1955) were all women, yet six years later, as the programming force expanded to dozens and five hierarchical job classes had been formed in programming, the programmer labor force at the firm was predominantly male and exclusively male within the top job classification of "principal analyst." And five years after that, the one woman who had been on the board, Louise Greene was no longer a board member and the company had an exclusively male board. Before seeking to control their careers as entrepreneurs,

Grace Gentry, and others were discriminated against at large organizations. Sometimes this was explicit, other times it took the form of conversations and informal meetings in male gendered spaces. Hopefully in the future there will be deeply researched studies focused on gender and the various IT industries (computer, software products, and computer services) in the US and worldwide, as well as internal IT operations of corporations, government, and other organizations. Marie Hicks' recently published book *Programming Inequality: How Britain Discarded Women Technologists and Lost Its Edge in Computing*, provides a rich example in analyzing the history of women IT workers in Great Britain and the consequences of discrimination.

In 2003, Nicholas Carr published in the *Harvard Business Review* a controversial article which he expanded into a short book the following year.[15] The provocative and somewhat deceptive titles of the article and the book were, respectively, "IT Doesn't Matter" and *Does IT Matter?* In the book, Carr looked at the approximately $2 trillion spent annually on IT (including both the major industries of IT—computers, software products, computer services, networking/communications—and in-house IT personnel) and argued that IT's strategic importance for corporations had dissipated. Carr's argument—a meaningful departure from the general question "does IT matter?"—really focuses on whether investing in IT for competitive advantage makes sense for most companies in view of the competitive landscapes of their industries.[16] In *Does IT Matter?* and in a subsequent book, *The Big Switch: Rewiring the World, From Edison to Google*, Carr characterizes cloud services as a massive commodification of IT to the point where it essentially is a utility—like electricity.[17] There is certainly a continuing trend toward cloud services; however, as is true of computer services overall, the cloud space is extremely complex and varied, and it has quite strong elements of *both* commodification and differentiation. Ever since Automatic Payrolls/ADP's largely commodity payroll processing and SDC's highly differentiated programming and systems integration work on SAGE in the second half of the 1950s, in the early 1960, and in subsequent years, the computer services industry has always been a mixture, and it probably will remain so for a long time.

Carr cites a 1997 survey of CEOs and directors of large American and European corporations in which respondents indicated that they believed that, by the year 2000, 60 percent of their IT initiatives would be for "achieving competitive edge," not just "catching up or staying afloat."[18] Realistically, how many leaders of very large corporations that are spending hundreds of millions or billions of dollars annually on IT are going to admit to themselves or others that their firm is just trying to tread water, or remain in the middle of the pack, with IT (whether internal, outsourced, or a hybrid) or any other high-expenditure function? When the survey Carr cites was conducted, the dot-com bubble was expanding rapidly. Attitudes were temporarily at a fever pitch; they probably would have been more subdued after the dot-com bust of the early 2000s. In 2004, when Carr's book was published, the (information) technology-laden NASDAQ market was still down 60 percent from its March 2000 high. In areas of considerable investment in technology and labor, company leaders want their firm to achieve excellence and to have a competitive edge (if only a small one), even when not engaging in outsized IT investment relative to competitors. Keeping up with IT in industries such as finance, petroleum, automobiles, banking, insurance, and retail will always be very expensive, even when a corporation is not trying to one-up its rivals with IT. Productivity gains from overall IT spending by corporations and other organizations have long been debated (and isolating one factor's role with productivity is notoriously difficult), but questions about the productivity paradox that were widely cited in the 1990s (questions as to whether IT investments led to comparable gains in productivity) have seemingly been put to rest by studies showing substantial returns on IT investment.[19]

Carr's analysis also pays relatively little attention to the financial and reputational costs that major breaches of computer security can bring.[20] Over the past decade, high-profile security breaches or hacks have hurt Sony Pictures, Home Depot, Target, AshleyMadison.com, eBay, JPMorgan Chase, and many other corporations. Significant breaches also have been suffered by the Department of Defense, the Internal Revenue Service, the Office of Personnel Management, many state and local governments, and nonprofit organizations.

Appropriate information technology infrastructure and investment (for a particular organization in its particular industry or environment), the right technological tools coupled with proper procedures and practices to best ensure security, and taking part in an ecosystem characterized by knowledge-circulating conduits for information technology all are extremely important. The expertise and knowledge-circulating functions of providers of IT services—giants such as IBM, HP, Accenture, Fujitsu, and Tata Consultancy, smaller players, and specialists in security products and services such as Palo Alto Networks—can be critical to organizations and may require major IT investments. In short, IT, understood within its broader systems of organizational structures and users, most certainly does matter.

Conclusion

Carr cites Max Hopper—American Airlines' longtime top IT manager, who directed the evolution of SABRE in the 1970s—as "bravely" and insightfully seeing the writing on the wall in 1990 by predicting that IT would come to be seen as "more like electricity or the telephone network than as a decisive source of organizational advantage .... In this world, a company trumpeting the appointment of a new chief information officer will seem as anachronistic as a company today naming a new vice president for water and gas ...." Max Hopper (who I had the privilege of meeting and chatting with in 2004)—unlike Nicholas Carr or myself—was old enough to remember, and probably was influenced by, the extensive writings and rhetoric about a future computing utility in the late 1960s, when time sharing was still a young technology and was sometimes spoken of in idyllic, far-reaching, and utopian terms. Several decades later, Hopper felt these utopian dreams were coming true.

Chief Information Officers, however, are more important than ever, regardless of where the boundaries of the firm are drawn with IT—outsourcing wisely, like internal IT operations, takes great knowledge and skill.[21] And in addition to CIOs, most major corporations now have Chief Security Officers. Both CIOs and CSOs will become increasingly important to corporations and other organizations in the years and decades ahead.

The trend toward outsourcing of IT continues to gain momentum, and for decades the IT services industry has grown far faster than the overall economy. Organizations will purchase a range of standardized information technology platforms and applications, as well as custom solutions. Cloud computing unquestionably includes some characteristics of the old computer utility concept, particularly infrastructure as a service, but many customized cloud services will be purchased by organizations, many to their advantage, but undoubtedly some purchases will be misallocations of resources. Cloud analytics, which probably will continue to grow rapidly, offers new possibilities for making the most of ever-larger stores of data. Computer security will remain a challenge, and IT services companies will continue to invest heavily in tools and techniques to protect themselves and their clients. Computer services, as has been true since the start of the industry more than six decades ago, will be characterized by some homogeneity (commodification), much heterogeneity (differentiation), a range of successes and failures, many uncertainties, and very rapid change.

Notes

Introduction

1. https://www.gartner.com/doc/3032029/market-share-analysis-it-services (accessed December 26, 2015).

2. Arthur Koestler, *The Ghost in the Machine* (Macmillan, 1967); Frederick P. Brooks Jr., *The Mythical Man-Month: Essays on Software Engineering* (Addison-Wesley, 1975): 7. The Hungarian-British literary author and critic Arthur Koestler borrowed the phrase from the British philosopher Gilbert Ryle (who had used it in reference to Cartesian mind–body dualism) for his 1967 book on philosophical psychology. Koestler's book inspired the rock group the Police to give a 1981 album the same title, further popularizing the phrase. The final two quotes come from IBM OS/360 project manager and computer scientist Frederick P. Brooks Jr.

3. Kenneth Lipartito, "Review Essay—Reassembling the Economic: New Departure in Historical Materialism," *American Historical Review* 121, no. 1 (2016): 101–139.

4. "Healthcare.gov" was deeply politicized—often the early critique of the IT system and the Affordable Care Act were deeply intertwined. See https://www.bloomberg.com/news/articles/2014-09-24/obamacare-website-costs-exceed-2-billion-study-finds (accessed December 20, 2015).

5. Problems often arise from continually changing IT system functionality desired or demanded by clients.

6. Nathan Ensmenger richly documents the evolving understandings of the art of programming and efforts of programmers to gain professional status in his book *The Computer Boys Take Over: Computers, Programmers, and the Politics of Technical Expertise* (MIT Press, 2010).

7. Martin Campbell-Kelly, *From Airline Reservations to Sonic the Hedgehog: A History of the Software Industry* (MIT Press, 2003).

8. Most of these companies—such as the software product pioneers Applied Data Research and Informatics, Inc.—continued to maintain their computer services business/division for many years.

9. Martin Campbell-Kelly and Daniel D. Garcia-Swartz, "Economic Perspectives on the History of the Computer Time-Sharing Industry, 1965–1985," *IEEE Annals of the History of Computing* 30, no. 1 (2008): 16–36; Jeffrey R. Yost, "Maximization and Marginalization: A Brief Examination of the History and Historiography of the US Computer Services Industry," *Entreprises et Histoire* 40 (November 2005): 87–101. Recently, Martin Campbell-Kelly and Daniel D. Garcia-Swartz published *From Mainframes to Smartphones: A History of the International Computer Industry* (Harvard University Press, 2015), a short synthetic overview of all IT industries (computers, software products, and computer services) in a global context; I discuss it further in a note to the conclusion of this book. Three internally produced, self-published books on computer services companies have come out: Claude Baum. *The System Builders: The Story of SDC* (System Development Corporation, 1981); Richard L. Foreman. *Fulfilling the Computer's Promise: The History of Informatics, 1962–1968* (Informatics, Inc., 1985); Scott McMurray, *Values. Driven. Leadership: A History of Accenture* (History Factory, 2005). Though these three books are informative, they are highly celebratory in nature, often lack deep analysis, and do little to contextualize these enterprises in the broader industry.

10. In this book, for the purposes of continuity and connections with the longstanding trajectory of the computer services industry (as well as business models of suppliers of services and the relationship between suppliers and consumers), I focus on the B2B, or the enterprise, side. In the cloud era (and to a lesser extent, its precursor with time sharing), distinctions between software products and software services are increasingly blurred. Microsoft's supply of Software as a Service (SaaS) has led to it becoming one of the leaders in cloud services in recent years. This is both on the enterprise and business to consumer sides. With this, there potentially is communication and direct interaction between supplier and consumer—a longstanding attribute of computer services. It is discussed in chapter 10. Alternatively, social networking businesses—Facebook and Twitter—focus heavily on end consumers (individual consumers), and the revenue model is far different (primarily advertising). Quite simply, Facebook is not a competitor of computer services leaders such as IBM, HP, Accenture, Fujitsu, and Tata Consultancy, and thus Facebook and all B2C cloud enterprises are outside the scope of this book.

11. http://www.forbes.com/billionaires/list (accessed July 1, 2016).

12. Ken Follett, *On Wings of Eagles* (Morrow, 1983); Doron P. Levin, *Irreconcilable Differences: Ross Perot Versus General Motors* (Plume, 1990); Gerald Posner, *Citizen Perot: His Life and Times* (Random House, 1996). Perot stands out as the only company founder and/or chief executive in the computer services industry who became a widely known public figure—in contrast with the computer and software products

industries, in which Thomas Watson, Thomas Watson Jr., Steve Jobs, Steve Wozniak, Ken Olsen, William Norris, Seymour Cray, Carly Fiorina, Bill Gates, and others became famous.

13. The Charles Babbage Institute at the University of Minnesota has long had some important archival materials on computer services (System Development Corporate Records and the Control Data Corporation Records) and recently has added substantial new collections (ADAPSO Records, Diebold Group, Inc. Client Reports, Milton R. Wessel Papers, Gartner Group Records, etc.) important to the study of this historical topic. The trade press, especially the trade journal *Datamation*, is a rich but underutilized source of valuable content documenting aspects of the computer services industry. Oral histories I recently have conducted, and a handful of oral histories by others over the past decade, are also a major resource for this book, especially in analyzing smaller firms and industry segments and trade associations.

14. Accenture, which originated with the Chicago-based Arthur Andersen spinoff Andersen Consulting, continues to have a major facility in Chicago (though it has been incorporated in Bermuda, and is now incorporated and headquartered in Dublin).

15. Campbell-Kelly (2003). Unlike consulting services firms like Diebold, Campbell-Kelly does briefly write about service bureaus.

16. Some did this through internal expansion, others by acquisition and integration of existing firms.

17. One-person consulting enterprises that left no document trail may have preceded Diebold and Canning, Sisson, and Associates by months or a year, but it is less likely that any consulting firms with multiple consultants existed before these two firms.

18. Louis Galambos, *Competition and Cooperation: The Emergence of a National Trade Association* (Johns Hopkins University Press, 1966).

19. Thomas David Haigh, Technology, Information, and Power: Managerial Technicians in Corporate America, 1917–2000, dissertation, University of Pennsylvania, 2003; Ensmenger (2010); Atsushi Akera, "Voluntarism and the Fruits of Collaboration: The IBM User's Group, SHARE," *Technology and Culture* 42, no. 4 (2001): 710–736.

20. Larry D. Browning and Judy C. Shetler, *Sematech: Saving the US Semiconductor Industry* (Texas A&M University Press, 2000).

21. In *IEEE Annals of the History of Computing*, historian Thomas Haigh published three concise and insightful pieces on ADAPSO in three different eras—its service bureau origins (beginning in 1961), time sharing of the late 1960s to the mid 1970s, and its increased political and lobbying role from the mid 1970s to the mid 1980s. These are cited in chapter 5.

22. Although ADAPSO was formally designated as an organization for service bureau firms at its start, pioneering programming services companies generally had service bureau divisions by 1961, and the trade association quickly became a broad-based organization to include computer services firms of other segments of the industry.

23. MIT scientists invented time sharing and published papers on it in the open community. In the closed, classified community of the National Security Agency, several experimental predecessors to NSA's early-to-mid-1960s RYE time-sharing system (ROGUE and ROB ROY) were developed (in the mid to late 1950s) well before CTSS. See Samuel S. Snyder, *History of NSA General-Purpose Electronic Digital Computers* (Department of Defense, 1964) (declassified in 2004).

24. Beginning in the late 1960s, some larger minicomputers were also used as the basic resource-sharing computers for time-sharing systems.

Chapter 1

1. Konrad Zuse's Z4 was an electronic digital computer, built in Germany, that preceded the ENIAC. It was a smaller, less powerful machine that was completed in the first half of 1945. Germany's faltering war effort—with frequent Allied bombing raids and the advancing front—resulted in multiple efforts to transport and protect the Z4, and it had no significant applications before the completion of the ENIAC in November 1945. Howard Aiken partnered with IBM in the design and development of the electromechanical computer that has been referred to as the Mark I, the Harvard Mark I, and the Automatic Sequence Controlled Calculator. "ASCC" was IBM's name for the machine. At the dedication of the system, Thomas J. Watson expressed displeasure with how Aiken failed to appropriately credit IBM's important design and engineering contributions, and with Aiken's having named the machine "Harvard Mark I."

2. Thomas Haigh, Mark Priestley, and Crispin Rope, *ENIAC in Action: Making and Remaking the Modern Computer* (MIT Press, 2016). In this book, the authors analyze this pivotal computer system and its technical, organizational, social, and scientific contexts.

3. T. R. Kennedy Jr., "Electronic Computer Flashes Answers, May Speed Engineering," *New York Times*, February 15, 1946.

4. The ENIAC, unlike many dedicated analog and electromechanical machines, was a general-purpose computer that could be programmed (through laborious manipulation of patch cords to a desired configuration) for many different calculating tasks. For a discussion of the December 1945 "Los Alamos Calculations" in support of the hydrogen bomb development effort, see Haigh, Priestley, and Rope (2016): 78–83.

5. In 1949 Edmund Berkeley wrote the first popular book on digital computers, *Giant Brains, or Machines That Think* (Wiley, 1949). "Electronic brain" was the metaphor that took hold with the popular press and was used in countless news articles during the second half of the 1940s and throughout the 1950s. See Gladwin Hill, "'Electronic Brain' Able to Translate Languages Is Being Built," *New York Times*, May 31, 1949; Hill, "Scientists Confer on 'Electronic Brain,'" *New York Times*, June 23, 1949; "Electronic Brain Does Research," *New York Times*, July 3, 1949.

6. "Electronics Seen Directing World," *New York Times*, February 2, 1949.

7. Ibid.

8. Ibid.

9. Thomas Haigh, "The Chromium-Plated Tabulator: Institutionalizing an Electronic Revolution, 1954–1958," *IEEE Annals of the History of Computing* 23, no. 4 (2001): 76.

10. The most important examination of IBM's pre-computer office machine competitors is James W. Cortada's *Before the Computer: IBM, NCR, Burroughs and Remington Rand, and the Industry They Created, 1865–1956* (Princeton University Press, 1993).

11. The best study of Eckert-Mauchly, Engineering Research Associates, Remington Rand, and the early history of Sperry-Univac is Arthur L. Norberg's *Computers and Commerce: A Study of Technology and Management at Eckert-Mauchly Computer Company, Engineering Research Associates, and Remington Rand, 1946–1957* (MIT Press, 2005).

12. Cuthbert C. Hurd, "Early Computers at IBM: Edited Testimony," *Annals of the History of Computers* 3, no. 2 (1981): 172.

13. Franklin M. Fisher, James W. McKie, and Richard B. Mancke, *IBM and the US Data Processing Industry: An Economic History* (Praeger, 1983).

14. Haigh (2001).

15. Ibid.

16. JoAnne Yates. *Structuring the Information Age: Life Insurance and Technology in the Twentieth Century* (Johns Hopkins University Press, 2005).

17. James W. Cortada, *The Digital Hand: How Computers Changed the Work of American Manufacturing, Transportation, and Retail Industries* (Oxford University Press, 2004); Cortada, *The Digital Hand*, volume 2: *How Computers Changed the Work of American Financial Telecommunications, Media, and Entertainment Industries* (Oxford University Press, 2006); Cortada, *The Digital Hand*, volume 3: *How Computers Changed the Work of American Public Sector Industries* (Oxford University Press, 2008).

18. Haigh (2001). Haigh's aim appears to be to highlight the "revolutionary" fervor for computing and electronics in the mid to late 1950s (to which consultants most assuredly contributed), not to broadly characterize early computer consulting.

19. "Arthur Andersen and Company" is often shortened to "Arthur Andersen" or just "Andersen." In the rare cases in which I am referring to the man (in mentioning the company's founding and his death), this is made clear. All other references to "Arthur Andersen" and "Andersen" are to the company. The same is true of the use of "McKinsey" to refer to McKinsey and Company.

20. Christopher D. McKenna, *The World's Newest Profession: Management Consulting in the Twentieth Century* (Cambridge University Press, 2006). Booz, Allen, and Hamilton and McKinsey and Company were founded in 1914 and 1926 respectively, but consisted of few people beyond the founders until they grew quickly after the passage of the Glass-Steagall Act in 1933. McKinsey and Company had 25 employees by 1936 and had opened a second office in New York. McKenna astutely argues that, over time, the management consulting profession that originated in the early twentieth century owed far more to accountants and engineers at accounting firms than to Frederick Winslow Taylor's scientific management. He identifies four distinct research strands of scientific management or Taylorism—worker psychology, workplace and tool design, wage systems, and cost accounting. Only the last and least recognized strand of the Taylorist system—cost accounting—has connections to management consulting as practiced in the 1930s and later.

21. Bundling was among the central issues of the Justice Department's ongoing investigations of IBM in the 1950s and the 1960s. Partially as a preemptive move, IBM announced "unbundling" of software and services in 1969, months before the Justice Department filed an anti-trust case against IBM that dragged on for more than ten years. Probably the largest private-sector efforts in programming services in the 1960s was IBM's work for American Airlines in creating the real-time airline reservations system SABRE. IBM also had a Federal Systems Division (FSD), which did services for the government. IBM did not enter into the consulting field in the early 1990s (as McKenna suggests); it merely redirected long-established organizational capabilities in consulting services that had contributed to the business fundamentally through being bundled with hardware sales for decades. It did create new data centers, but even here there was distant precedent from Service Bureau Division/Service Bureau Corporation. (The latter was sold in 1973.)

22. Frank Grippo, "Leonard Spacek: Ahead of His Time, Relevant Today," *CPA Journal*, March 2004: 16–17.

23. McMurray (2005): 6–8.

24. Ibid.

25. Ibid.

26. Ibid.: 10–22.

27. John A. Higgins and Joseph S. Glickauf, "Electronics Down to Earth," *Harvard Business Review*, March-April 1954: 97–104.

28. Roddy F. Osborn, "GE and UNIVAC: Harnessing the High-Speed Computer," *Harvard Business Review* (July-August 1954): 99.

29. The verb "outsource" has long been used to describe utilizing an outside supplier for goods or services. In recent years the term has seen heavy use specifically referring to outsourcing overseas, such as outsourcing manufacturing to Chinese firms or IT services to Indian firms (including Business Process Outsourcing). My use of the term throughout the book refers to the general case of outside the firm, not necessarily outside the country. For the latter I either specify to make it clear or use the verb "offshore."

30. Osborn (1954): 106.

31. McMurray (2005): 20–23.

32. Ibid.: 29.

33. David Leonhardt, "Andersen Split into Two Firms by Arbitrator," *New York Times*, August 8, 2000. The number of consulting personnel at IBM at that time is not known, but the total workforce of IBM (which each year became more focused on consulting services) was substantially larger in 2000.

34. Wilbur Cross, *John Diebold: Breaking the confines of the Possible* (Heineman, 1966): 21–49. ARDC would go on to achieve fame and great financial success in providing seed funding for the early growth of 1957 startup Digital Equipment Corporation.

35. Cross (1966): 30–33.

36. Ibid.

37. John Diebold, *Automation: The Advent of the Automatic Factory* (Van Nostrand, 1952).

38. Like a number of management consulting firms (such as James O. McKinsey and Co.), John Diebold and Associates soon dropped the first name of its founder.

39. Delmar S. Harder, the Ford Motor Company's vice president for manufacturing, is generally credited with coining the term "automation" in the second half of the 1940s to refer the art of "applying mechanical devices to manipulate work pieces in timed sequence with production equipment" at Ford. The first published use of the term in this context was an article by Rupert Le Grand on Ford's automated production processes: Rupert Le Grand, "Ford Handles by Automation" *American Machinist* 92, no. 22 (1948): 107–122.

40. Diebold Group, Inc. Reports, 1957–1990 (Charles Babbage Institute, University of Minnesota—hereafter CBI). This collection provides rich breadth and depth to understand the work of Diebold's firm and interactions with clients. However, it does not contain any records other than the consulting reports, so specific data on the number of associates, organizational routines, managerial practices, and finances of Diebold and Associates/Diebold Group, Inc. remain largely unknown. Rough approximations and understandings of some of these topics can be gleaned from examinations of the client reports.

41. Diebold Group, Inc. Reports, 1957–1990 (CBI). There are some clients that Diebold's firm was serving by 1960 that it did not produce reports for until 1961—and thus, they do not appear on this list. Client reports generally have completion dates listed, but not necessarily start dates and generally no information on contractual compensation to Diebold and Associates.

42. Diebold Group, Inc. Reports, 1957–1990.

43. ALWAC (all capitals) was an abbreviation of the name of the firm's first computer, the Axel Lennart Wenner-Gren Automatic Computer. Alwac was the name of the corporation.

44. "Alwac Corporation, Operations Management Volume 1: Progress Reports," n.d.; "Alwac Corporation, Operations Management Volumes 1–2: Progress Reports" n.d.; "Alwac Corporation" Product Planning Study Volume 1–2: Recommendations—Economic Data" December 1957 (Diebold Group, Inc. Reports, 1957–1990, CBI).

45. "Alwac Corporation" reports (listed in preceding note).

46. "Mead Johnson & Company, Data Processing Study" (Diebold Group, Inc. Reports, 1957–1990, CBI).

47. Ibid.

48. Ibid

49. "Mead Johnson and Company, Advanced Computer System Opportunities and Current Activity Survey," 1966; "Mead Johnson and Company, Evaluation of Office Machine Usage" 1967; "Mead Johnson and Company, A Role in the Medical Information Field," 1968 (Diebold Group, Inc. Reports, 1957–1990, CBI).

50. "American Hospital Association, Hospital Data Processing: A Case Study" (1959) (Diebold Group, Inc. Reports, 1957–1990, CBI).

51. "American Hospital Association, Hospital Data Processing: A Case Study" (1959).

52. Joseph November. *Biomedical Computing: Digitizing Life in the United States* (Johns Hopkins University Press, 2012).

Notes to Chapter 1

53. "Communications Workers of America. Automation: Impacts and Implications—Focus on Developments in the Communications Industry," 1964 (Diebold Group, Inc. Reports).

54. Diebold Group, Inc. Reports, 1957–1990.

55. Jennifer Bayot, "John Diebold, 79, a Visionary of Computer Age Dies," *New York Times*, December 27, 2005.

56. Ibid.

57. Despite "and Associates" in the company's name, it originally consisted just of Richard Canning and Roger Sisson.

58. Richard Canning oral history (interview conducted by Jeffrey R. Yost, Vista, California, August 22, 2002) (CBI).

59. Ibid.

60. Richard G. Canning, *Electronic Data Processing for Business and Industry* (Wiley, 1956).

61. Ibid. Canning's book came out a year earlier than most other high-quality examinations and was rivaled only by Ned Chapin's *Introduction to Automatic Computers* (Van Nostrand, 1955), which was equally reasoned and insightful and which came out slightly before Canning's book. Also see William D. Bell, *A Management Guide to Electronic Computers* (McGraw-Hill, 1957); Esther R. Becker and Eugene F. Murphy, *The Office in Transition: Meeting the Problems of Automation* (Harper, 1957).

62. Canning (1956).

63. Richard G. Canning, *Installing Electronic Data Processing Systems* (Wiley, 1957).

64. Canning oral history (2002).

65. Launched in 1949, the NMAA became the Data Processing Management Association (DPMA) a decade later, the name change recognizing the central role of computers for accounting applications.

66. Margaret Milligan, "Data Processing Digest: Thirty Years Before the Masthead," *Annals of the History of Computing* 7, no. 3 (1985): 245–250.

67. Canning oral history (2002).

68. Milligan (1985).

69. Canning oral history (2002). All information in this and the immediate prior paragraph is from this interview. Canning long suspected that photocopying had occurred, because he would get calls from secretaries about a missing issue—telling him they needed it to make and distribute the copies. Though he might have

preferred that this never occur, he generally was content. He did not want to upset his loyal reader base, and *EDP Analyzer* (which stood for Electronic Data Processing Analyzer) became a lucrative small enterprise for him and his family.

Chapter 2

1. To refer to securing services from a foreign nation, I either use the term "offshoring" or explicitly state outsourcing overseas or outsourcing abroad.

2. Tabulation Machines continued to be used at some bureaus throughout the 1950s and in the early 1960s. Generally bureaus in larger cities were early adopters (in the second half of the 1950s) of digital computers.

3. This includes a number of impressive works, including the following: Emerson W. Pugh, *Building IBM: Shaping an Industry and Its Technology* (MIT Press, 1995); Robert Sobel, *IBM: Colossus in Transition* (Times Books, 1981); James W. Cortada, *Before the Computer: IBM, NCR, Burroughs and Remington Rand and the Industry They Created, 1965–1956* (Princeton University Press, 1993).

4. Meaningful attention to and analysis of service bureaus is absence from the historical literature on pre-computer punch card tabulation and computers. No earlier studies significantly address service bureaus in either of these (for a time overlapping) technological eras (tabulation machine/computers). One book that does devote roughly a half dozen pages to the history of computer service bureaus concentrates primarily on time-sharing companies of the 1960s (one type of service bureau and a later type), and ignores the second half of the 1950s and also ignores the Service Bureau Corporation (the IBM subsidiary that was by far the largest and most important service bureau. See Franklin M. Fisher, James W. McKie, and Richard B. Mancke, *IBM and the US Data Processing Industry: An Economic History* (Praeger, 1983). In the literature on the history of software, service bureaus get minimal attention. In his excellent book on computer programmers, Nathan Ensmenger discusses the adoption of computer installations by scientific bureaus (such as IBM's at Columbia University) in the 1950s and earlier, but doesn't discuss service bureaus focused on business data processing. Ensmenger does use a 1964 SBC ad to highlight the difficulty of finding good programmers in the 1960s. Nathan Ensmenger, *The Computer Boys Take Over: Computers, Programmers, and the Politics of Technical Expertise* (MIT Press, 2010): 46, 119. Martin Campbell-Kelly's *path-breaking* book on the software industry (which, as I discussed in my introduction to this book, focuses most on software products) devotes just a paragraph to service bureaus (before digital computers), and presents their activities as merely processing courier delivered punch cards of customers—see Campbell-Kelly, *From Airline Reservations to Sonic the Hedgehog*: 59, 61. As I note in this chapter, their range of services were often more varied. And for IBM, Service Bureau Division/Corporation played an important synergistic role in the data processing giant's overall business.

Notes to Chapter 2

5. Martin Campbell-Kelly, William Aspray, Nathan Ensmenger, and Jeffrey R. Yost, *Computer: A History of the Information Machine, Third Edition* (Westview, 2014): 13–14.

6. Geoffrey D. Austrian, *Herman Hollerith: Forgotten Giant of Information Processing* (Columbia University Press, 1982): 1.

7. Campbell-Kelly et al. (2014): 13–18.

8. Ibid.: 43–44.

9. Austrian (1982): 1; Campbell-Kelly et al. (2014): 16–17.

10. Austrian (1982): 70.

11. Ibid.: 69. Campbell-Kelly et al. (2014): 18.

12. Austrian (1982). Foreign census work using Hollerith machines is detailed in Austrian's chapter 7. The different organizational customer applications (uses of Hollerith equipment by different clients) are detailed throughout his book.

13. For several years Hollerith was an advisor to C-T-R's management, but his role was quite limited by 1914, when Thomas Watson was hired.

14. Cortada (1993): 66–74.

15. Campbell-Kelly et al. (2014): 33.

16. Thomas J. Watson, "IBM Is Not Merely an Organization of Men, It Is an Institution That Will Go On Forever," *Business Machines*, February 16, 1926: 2 (IBM Corporate Archives).

17. "Service," *TM Business Record*, July 21, 1917: 1 (IBM Corporate Archives).

18. Hollerith's 1906 Type I Tabulating Machine was the first to use a wiring panel to configure tabulating jobs.

19. Clarke Hayes, "What We Mean by Service," *International Business Machines*, March-April 1921: 1 (IBM Corporate Archives).

20. Arthur Andersen was providing advisory services on tabulation machine operations out of its Administrative Accounting Division in the 1930s, but not as early as C-T-R/IBM.

21. C-T-R/IBM's first efforts to hire female sales/service professionals began in the mid 1930s—a topic discussed in chapter 8.

22. "Frederick Nichol Appointed Assistant to the President" and F. W. Nichol "The Part that Business Machines Have Played in the Development of American Industry," *Business Machines* (October 26, 1926) (IBM Corporate Archives).

23. IBM's longest running internal trade newspaper/magazine, *Business Machines*, contained countless articles on the graduating classes of the various IBM sales and

services schools, often quoting from talks at these events by Nichol. I carefully surveyed all issues of *Business Machines* from the 1920s through the 1950s at the IBM Corporate Archives.

24. "Organization and Operating Plans for a Tabulation Machine Division" (April 25, 1922), IBM Corporate Archives [individual document provided to me by IBM's Corporate Archivist Paul Lasewicz].

25. Pugh (1995): 322.

26. Sources: author's analysis of data presented in Pugh (1995): 323; annual reports of C-T-R and IBM for years mentioned.

27. Chapter 5 of Pugh (1995) provides a good overview of Watson's daring investment during the Great Depression but does not highlight or meaningfully discuss the Service Bureau Division in that context.

28. Cortada (1993): 153.

29. Pugh (1995): 62–65.

30. Ibid.: 323.

31. "Users As Well As Non-Users Employing TM Service," *Business Machines*, October 27, 1932: 1, 5.

32. "IBM Offices Have Increased in Size, Number, and Beauty Since the Early Days," *Business Machines* (February 12, 1954): 4–9 (IBM Corporate Archives).

33. "Users as Well as Non-Users ..." (1932).

34. Pugh (1995): 323–324.

35. On the history of Eckert-Mauchly, ERA, and Remington Rand (in computing), see Arthur L. Norberg, *Computers and Commerce: A Study of Technology and Management at Eckert-Mauchly Computer Company, Engineering Research Associates, and Remington Rand, 1946–1957* (MIT Press, 2005).

36. Jeffrey R. Yost, *The IBM Century: Creating the IT Revolution* (IEEE Computer Society Press, 2011): 10. Memoirs by both Birkenstock and Hurd in this volume discuss efforts to sway Watson, and the significance of the letters of intent.

37. IBM Corporate Archives online resources—https://www-03.ibm.com/ibm/history/exhibits/650/650_ch1.html (accessed January 17, 2015). Watson Jr. became "President" of IBM in 1952, but his father effectively remained in charge and had the role of chief executive and officially chairman of the board until 1956.

38. Thomas Haigh, "The Chromium-Plated Tabulator," *IEEE Annals of the History of Computing* (October-December 2001): 84; IBM Corporate Archives, https://www-03.ibm.com/ibm/history/exhibits/650/650_ch1.html (accessed August 20, 2014). The "Model T" analogy was used both with the IBM 650 and (later) the IBM 1401.

39. "Electronic Data Processing Service Is Organized at WHQ," *Field WHQ News* (supplement to *Business Machines*), March 1, 1955: 1, 3 (IBM Corporate Archives).

40. Thomas Watson Jr., undated memo on Service Bureau Division (from context mid 1950s), IBM Corporate Archives.

41. IBM annual report, 1957.

42. IBM annual reports, 1960–1965.

43. IBM annual reports, 1960–1965.

44. E. Drake Lindell Jr., "IBM/CDC Pact: No User Disruption, Service Bureaus Sold by IBM," *Computerworld* 7, no. 4 (1973): 1–2.

45. "Henry Taub, 1927–2011," http://www.hamiltonpartnership.org/es/node/611 (accessed August 25, 2014); ADP, Inc. *ADP 50th Anniversary, 1949–1999* (Automatic Data Processing, Inc., 1999): 2 (made available by ADP, Inc. Investor Relations in early 2000s).

46. ADP, Inc. (1999): 2.

47. Duff Wilson, "Henry Taub, a Founder of a Payroll Firm That Became a Global Giant, Dies at 83," *New York Times*, April 4, 2011.

48. Emma Brown, "Senator Frank Lautenberg, Five-Term New Jersey Democrat Dies at 89," *Washington Post*, June 3, 2013.

49. Frank Lautenberg oral history (interview conducted by Paul Ceruzzi, Washington, May 3, 2002) (CBI).

50. Ibid.: 2.

51. Ibid.: 4.

52. Ibid.

53. Ibid.

54. ADP, Inc. (1999): 7.

55. Lautenberg oral history. In recent decades the firm's name has been simply ADP; it is considered an acronym only in the historic sense.

56. ADP, Inc. (1999): 7.

57. Lautenberg oral history.

58. Ibid.: 5.

59. Ibid.

60. ADP, Inc. (1999): 10.

61. Ibid.: 12.

62. Around the time IBM first brought the IBM 701 to its New York City World Headquarters (scientific) Service Bureau facility, and its first IBM 650s to a handful of existing bureaus, Remington Rand initiated Univac I–based bureaus in both New York and Los Angeles. Little exists to document the history of Sperry Rand's service bureau operations (Sperry Corporation merged with Remington Rand in 1955). Other IBM mainframe competitors probably had even less service bureau infrastructure than Sperry Rand into the late 1950s and were relatively small divisions or businesses in comparison with IBM Service Bureau Corporation. In the 1960s and the 1970s, Control Data, through internal expansion and acquisitions of C-E-I-R in 1967 and SBC in 1973, had a substantial service bureau infrastructure in the US and overseas. Many small independent bureaus used IBM computers and operated one or several facilities in one large US city.

63. ADP, Inc. (1999).

64. The services (both those bundled and those sold separately) offered by IBM from 1960s through the mid 1980s, and the services offered by IBM's competitor Control Data, will be explored in chapter 8.

65. India's substantial and fast-growing IT services industry and labor force in recent decades will be examined in chapter 10.

Chapter 3

1. Arthur Andersen's Administrative Services Division first engaged in programming services during 1954 (having done some advisory consulting in 1953) for GE, though that division was not predominantly a programming services supplier in the mid to late 1950s.

2. CSC broadened its repertoire in computer services.

3. Most organizational customers were end users, however, some important programming services work was done by independent computer services firms in tandem with if not directly for computer manufacturers—for system software, compilers, and other programming projects. Such efforts are described in this chapter. Unlike CUC and C-E-I-R, CSC continued and thrived for many succeeding decades.

4. Elmer C. Kubie, "Recollections of the First Software Company," *IEEE Annals of the History of Computing* 16, no. 2 (1994): 65–71.

5. Cuthbert C. Hurd, "Early IBM Computers: Edited Testimony"; James W. Birkenstock, "Pioneering: On the Frontier of Electronic Data Processing, a Personal Memoir," in Jeffrey R. Yost, ed., *The IBM Century: Creating the IT Revolution* (IEEE Computer Society Press, 2011).

6. Kubie (1994). TCB was connected to both the Applied Science Department, and as a scientific/technical service bureau, part of its wider Service Bureau Division operations. It is discussed more extensive in chapter 2, which is in large part on IBM's pre-computing, and computing Service Bureau Division/Corporation business.

7. Kubie (1994).

8. Ibid.

9. Nathan Ensmenger, *The Computer Boys Take Over: Computers, Programmers, and the Politics of Technical Expertise* (MIT Press, 2011): 66–67.

10. Ensmenger (2011): 67–70. Dallis Perry and William Cannon were SDC psychologists.

11. Dallis Perry and William Cannon, "Vocational Interests of Computer Programmers," *Journal of Applied Psychology* 51, no. 1 (1967): 28–34; "Software Gap—A Growing Crisis for Computers," *Businessweek*, November 5, 1966: 126–133. Nathan Ensmenger (2011): 70–71.

12. Ibid.

13. Jennifer S. Light, "When Computers Were Women," *Technology and Culture* 40, no. 3 (1999): 455–483. The most important study of human computers is David Alan Grier's well-researched and elegantly written book *When Computers Were Human* (Princeton University Press, 2007).

14. Though undoubtedly some of the four women (and probably all four) worked on the 1955 California Research Corporation project (as they were the only four programmers on staff that year). This project discussed in text of this chapter.

15. "CUC Roster of Personnel" (July 1961). Computer History Museum (henceforth CHM).

16. "CUC Roster," July 1961.

17. Kubie (1994).

18. Ibid.

19. Ibid: 67.

20. Kubie (1994). The vast majority of the more than 2,000 IBM 650 machines installed in the five years after its introduction in 1954 were in business settings. However, a considerable number went to universities and laboratories. Some scientists did their first programming on an IBM 650, and a significant percentage of early computer services industry contracts secured by pioneering firms such as CUC were for scientific computing.

21. George R. Trimble Jr., "A Brief History of Computing: Memoirs of Living on the Edge," *IEEE Annals of the History of Computing* 23, no. 3 (2001): 44–59. In this memoir, Trimble refers to himself as the seventh employee, though Kubie's memoir names seven (including himself) before Trimble. One of the programmers may have left by that time (two of the four were not on the 1961 roster), so Trimble may have been one of seven when he signed on to join CUC, but he was not officially "the seventh" to join the firm.

22. Trimble (2001).

23. Cuthbert C. Hurd oral history (interview conducted by Charles R. Fillerup, Portolo Valley, California, August 28, 1995) (CBI).

24. Later renamed Federal Aviation Administration.

25. Hurd oral history.

26. The Staff of the Computer Usage Company, *Programming the IBM System/360* (Wiley, 1966).

27. Fernando J. Corbató oral history (interview conducted by Arthur L. Norberg, Cambridge, Massachusetts, April 18, 1989 and November 14, 1990) (CBI).

28. Jeffrey R. Yost. *The Computer Industry* (Greenwood, 2005): 66–67.

29. Hurd oral history.

30. Jeffrey R. Yost, "Manufacturing Mainframes: Component Fabrication and Component Procurement at IBM and Sperry-Univac, 1960–1975," *History and Technology* 25, no. 3 (2009): 219–235. Drawing from organizational and transaction cost economics, business historians, economists, and management theorists have long focused on the strategic question of markets versus hierarchies. This, however, has been grossly understudied in the history of computing. This is particularly true on the software and services side, which this book, in part, seeks to address. Regarding hardware (semiconductor) components, my study as well as the following two studies are the principal works addressing the topic. Richard N. Langlois, "Computers and Semiconductors," in Benn Steil, David Victor and Richard R. Nelson, eds., *Technological Innovation and Economic Performance* (Princeton University Press, 2002): 265–284; Guido Krickx, "Vertical Integration in the Mainframe Computer Industry: A Transaction Cost Interpretation," *Journal of Economic Behavior and Organization* 26 (1995): 75–91.

31. Kubie (1994).

32. Computer Usage Company annual report, 1961 (Computer History Museum, hereafter CHM).

33. Kubie (1994).

34. Computer Usage Company annual report, 1961 (CHM).

35. Trimble (2001).

36. Hurd oral history.

37. "Environment Description," *Computer Usage Communiqué* 1, no. 5 (1961): 2 (CHM).

38. *CUC Capabilities Report* (1965) (CHM).

39. Ibid.

40. Computer Usage Company annual report, 1967 (CHM).

41. Computer Usage Company annual report, 1965; Computer Usage Company annual report, 1967 (CHM).

42. Hurd oral history.

43. Kubie (1994).

44. C-E-I-R, Inc. annual report, 1959 (CBI). This was the company's first annual report, and it provides some brief information on the history of the Council of Economic and Industry Research, and first several years of C-E-I-R, Inc. On C-E-I-R's early years and transformation into a for-profit enterprise, see Herbert Robinson oral history (interview conducted by Bruce W. Bruemmer, Bethesda, Maryland, July 13, 1988) (CBI).

45. Robinson oral history.

46. C-E-I-R, Inc. annual report, 1959 (CBI).

47. Both the agency and the act use the same acronym: DPA.

48. Robinson oral history.

49. Ibid.

50. Ibid.

51. C-E-I-R, Inc. Records, Burroughs Corporation Records (CBI).

52. Robinson oral history.

53. Ibid.

54. Ibid.

55. Ibid.

56. C-E-I-R, Inc. annual report, 1959 (CBI).

57. Robinson oral history.

58. Harry M. Markowitz oral history (interview conducted by Jeffrey R. Yost, San Diego, March 18, 2002) (CBI).

59. Orchard-Hays went on to publish a major text on linear programming, *Advanced Linear-Programming Computing Techniques* (McGraw-Hill, 1968).

60. Robinson oral history.

61. Herbert R. J. Grosch, *Computer: Bit Slices from a Life* (1992). http://www.columbia.edu/cu/computinghistory/computer.html (accessed October 17, 2014)

62. Grosch (1992); C-E-I-R, Inc. annual report, 1960 (CBI).

63. C-E-I-R, Inc. annual report, 1960 (CBI).

64. Ibid.

65. "Motor Vehicle Registration System," C-E-I-R, Inc. Records, Control Data Corporate Records (CBI); Peter Harris oral history (interview conducted by Thomas Haigh, Mountain View, California, November 19, 2004) (CHM).

66. C-E-I-R, Inc. annual report, 1960 (CBI).

67. Ibid.

68. Robinson oral history.

69. C-E-I-R, Inc. annual report, 1963 (CBI).

70. Robinson oral history.

71. George Dick oral history (interview conducted by William Aspray, Arlington, Virginia, February 19, 1988); C-E-I-R, Inc. annual report, 1962; C-E-I-R, Inc. annual report, 1963 (CBI).

72. C-E-I-R, Inc. annual report, 1961 (CBI).

73. Emerson W. Pugh. *Building IBM: Shaping an Industry and Its Technology* (MIT Press, 1995): 236–237.

74. C-E-I-R, Inc. annual report, 1963 (CBI). Dick oral history.

75. Ibid.; C-E-I-R, Inc. annual report, 1963; C-E-I-R, Inc. annual report, 1964 (CBI).

76. C-E-I-R, Inc. annual report, 1963 (CBI).

77. C-E-I-R, Inc. annual report, 1964 (CBI).

78. Robinson oral history; Dick oral history.

79. C-E-I-R, Inc. annual report, 1965 (CBI).

80. C-E-I-R, Inc. annual report, 1966 (CBI).

81. Ibid.; Oliver Goodman, "Control Data Plans to Acquire C-E-I-R in $28 Million Deal," *Washington Post*, August 16, 1967; "Control Data to Buy C-E-I-R for Less Than

Announced," *Wall Street Journal,* October 2, 1967. "Control Data Joins C-E-I-R in Talks," *New York Times,* August 16, 1967.

With fluctuating stock prices of the two companies and modest adjustments in shares before the deal occurred in late 1967, the end price was reported to be closer to $36 million worth of CDC shares than the originally reported $28 million as merger talks began. The number rose to $38 Million at one point; hence the title of the October 2 article that cites $36 Million in its text. For smaller enterprises in the IT field, like C-E-I-R, capital investment (facilities and equipment) in a given year had a great effect on earnings, so price/earnings at a single point in time often is a less revealing valuation measure than price/sales (as long as a company is solvent and making money and has sufficient adequate cash flow, all of which apply to C-E-I-R at the time of the acquisition).

82. Ibid.

83. John Brooks, *The Go-Go Years* (Weybright and Talley, 1973).

84. Nutt tried to persuade his colleague at United Aircraft Peter Harris to join him, but Harris stayed with United several more months before going out on his own to form DATA-TECH (discussed above). United Aircraft had been the ninth firm to get an IBM 701 (delivered in October 1953) and along with Nutt and Harris had considerable computing and programming talent. Harris would regret not leaving for CSC, as the wildly inflated C-E-I-R, Inc. stock he received for DATA-TECH plummeted in value before he sold it. Source: Harris oral history, IBM Corporate Archives, https://www-03.ibm.com/ibm/history/exhibits/701/701_customers.html (accessed December 21, 2014).

85. "One Compiler, Coming Up: Jones, Nutt, and Patrick Form Computer Sciences Corporation," *Datamation* 5, no. 3 (1959): 15.

86. Martin Campbell-Kelly, *From Airline Reservations to Sonic the Hedgehog: A History of the Software Industry* (MIT Press, 2003): 53.

87. UNIVAC LARC advertisement, *New York Times,* October 28, 1956. The ad mentions "opportunities to associate with the prominent pioneers of computer science."

88. Louis Fein, "The Role of the University in Computers, Data Processing, and Related Fields," *Communication of the ACM* 2, no. 9 (1959): 7–14.

89. Campbell-Kelly (2003): 52–54.

90. Harold Bergstein, "Computerized Reflections at CDC: An Interview with Fletcher Jones, President Computer Sciences Corp," *Datamation* 9, no. 3 (1963): 39–42.

91. Bergstein (1963): 39.

92. Sam Wyly oral history (interview conducted by David Allison, Washington, December 6, 2002) (CBI); Dennis Hamilton, *Beyond Tallulah: How Sam Wyly Became America's Boldest Big-Time Entrepreneur* (Melcher Media, 2011).

Chapter 4

1. Although IBM's exact numbers of programmers are not available for specific years, an advertisement run in late 1960 ("'Ghost Writers' for Computers," *Datamation* 6, no. 6: 11) claims that IBM's programmer force had grown to more than "300 men and women." System Development had about 800 employees at that time.

2. Though System Development Corporation was spun off as a separate entity in early 1957, it retained some connections to RAND—including some RAND senior officers sitting on its board of directors. While C-E-I-R was initially nonprofit, this was before it became a computer services company.

3. This differed markedly from the computer services industry in general. The only computer services company somewhat close to SDD/SDC's dependence on one project at the start was Computer Sciences Corporation with its Honeywell FACT compiler project—an eighteen month contract. SDD/SDC had only one major project (with two facets) during its first four years—*programming/integrating* an IT system for air defense, and the human-factors side of *training* US Air Force personnel to use the system.

4. Martin Campbell-Kelly, William Aspray, Nathan Ensmenger, and Jeffrey R. Yost. *Computer: A History of the Information Machine* (Westview, 2014): 176. The increase to 700 was over several years, so the estimate of 1,200 programmers nationwide would probably be considerably higher by the time SDC actually reached 700 programmers. Nevertheless, no other enterprise grew even close to having 700 programmers in the late 1950s or the early 1960s.

5. Campbell-Kelly et al. (2014): 143–144. Flight simulators were developed to advance the speed and efficiency, and to reduce risk (to loss of life and equipment) of training pilots. A universal flight simulator could be "programmed" to simulate many different types and model of aircraft, thus offering far greater efficiency than dedicated flight simulators based on a single model of aircraft.

6. Campbell-Kelly et al. (2014): 143–144.

7. Ibid.: 143–152.

8. Kent C. Redmond and Thomas M. Smith, *From Whirlwind to MITRE: The R&D Story of the SAGE Air Defense Computer* (MIT Press, 2000). The resulting general-purpose, real-time computing system of Forrester and Everett's project became known simply as "Whirlwind."

9. Kent C. Redmond and Thomas M. Smith, *Project Whirlwind: The History of a Pioneer Computer* (Digital Press, 1980).

10. Steven J. Zaloga, *The Kremlin's Nuclear Sword: The Rise and Fall of Russia's Strategic Nuclear Forces, 1945–2000* (Smithsonian Institution Press, 2002).

11. Redmond and Smith (2000).

12. This lack of integration and resulting problems for Remington Rand, and later its successor, Sperry Rand, is detailed in Arthur L. Norberg's book *Computers and Commerce: A Study of Technology and Management at Eckert-Mauchly Computer Company, Engineering Research Associates, and Remington Rand, 1946–1957* (MIT Press, 2005).

13. Campbell-Kelly et al. (2014).

14. David A. Rennels, "Fault-Tolerant Computing—Concepts and Examples," *IEEE Transactions on Computers* 33, no. 12 (1984): 1116–1129. There were only a few efforts at fault-tolerant computing in the 1950s. Most notably Antonin Svoboda's work on the SAPO project in Prague. Work in fault-tolerance gained momentum in the United States at the start of the 1960s, and NASA was a principal early sponsor of research.

15. Emerson Pugh, *Building IBM: Shaping an Industry and Its Technology* (MIT Press, 1995): 215.

16. Morton M. Astrahan and John F. Jacobs, "History of the Design of the SAGE Computer—The AN/FSQ-7," *Annals of the History of Computing* 5, no. 4 (1983): 340–349.

17. The $30 million for each Q-7 computer is a figure cited in Claude Baum's book *The System Builders: The Story of SDC* (System Development Corporation, 1981). It is also cited in other sources, among them Paul Edwards' *The Closed World: Computers and the Politics of Discourse in Cold War America* (MIT Press, 1996): 101–102. Emerson Pugh, head of a major technical history project at IBM, wrote that IBM received $500 million in revenue from the SAGE computer project in the 1950s. See Pugh, *Memories That Shaped an Industry: Decisions Leading to IBM System/360* (MIT Press, 1984): 126. This seeming incongruity might be the result of when delivery of many of the computers occurred (after FY 1959).

18. Kenneth Flamm, *Creating the Computer: Government, Industry, and High Technology* (Brookings Institution, 1988): 88–89; IBM annual report, 1958. Bob Evan of IBM indicated –in a private communication with Flamm—that IBM's leaders believed such a project might take IBM "too close to the operational aspects of war time."

19. Martin Campbell-Kelly, *From Airline Reservations to Sonic the Hedgehog: A History of the Software Industry* (MIT Press, 2003): 38. It is not clear whether MIT's leaders did not want the responsibility for Lincoln Laboratory or whether Lincoln Laboratory's leaders did not want to seek the primary programming contract (or both).

20. H. D. Benington, "Large Computer Programs," *Annals of the History of Computing* 5, no. 4 (1983): 350–361.

21. At its start, RAND was wholly supported by the Air Force, which was its primary funder for many years. After its first decade, its funding broadened (gradually at first).

22. J. R. Goldstein, "RAND: The History, Operations, and Goals of a Non-Profit Corporation," presentation to Harvard Business School alumni, University Club, Los Angeles, February 23, 1961 (revised April 1961): 3; System Development Corporation Records, Burroughs Corporation Records, CBI (hereafter SDC Records). At this time, Goldstein was vice president of the RAND Corporation.

23. Goldstein (1961): 5–11. Individual contracts could show a profit, but receipts had to be plowed back into other work of the RAND Corporation. There are several high-quality histories of the RAND Corporation, most notable among them Martin J. Collins' *Cold War Laboratory: RAND, the Air Force, and the American State, 1945–1950* (Smithsonian Institution Scholarly Press, 2002) and Bruce L. R. Smith's *The RAND Corporation: A Case Study of a Nonprofit Advisory Corporation* (Harvard University Press, 1969).

24. William Aspray, *John Von Neumann and the Origins of Modern Computing* (MIT Press, 1990): 92–93.

25. F. N. Marzocco, "The Story of SDD" (n.d.), SDC Records.

26. Ibid.

27. Willis H. Ware, *RAND and the Information Evolution: A History in Essays and Vignettes* (RAND Corporation, 2008): 94–98.

28. C. Robert Wieser, "The Cape Cod System," *Annals of the History of Computing* 5, no. 4 (1983): 362; Benington (1983): 353.

29. Redmond and Smith (2000).

30. "Notes of SDD Staff Meeting No. 1" (March 21, 1957); "Notes of SDD Staff Meeting No. 2" (May 20, 1957). SDC Records. System Development Division's earliest staff meetings prominently discussed the capacity problem once the major part of the SAGE programming project was complete.

31. Ware (2008): 94–98.

32. "IBM 701 Customers," IBM Corporate Archives. https://www-03.ibm.com/ibm/history/exhibits/701/701_customers.html (accessed March 13, 2015).

33. M. Mitchell Waldrop, *The Dream Machine: J. C. R. Licklider and the Revolution that Made Computing Personal* (Penguin, 2001). Waldrop provides a useful biography of Licklider and articulates his visionary role in funding research that laid the groundwork for the modern personal computer, its graphical user interface, and the Internet.

34. J. C. R. Licklider oral history interview, conducted by William Aspray and Arthur Norberg, Cambridge, Massachusetts, October 28, 1988 (CBI).

35. Arthur L. Norberg and Judy O'Neill, *Transforming Computer Technology: Information Processing for the Pentagon, 1962–1986* (Johns Hopkins University Press, 1996).

36. William Biel certainly did not match Licklider's technical achievement or fame, but he became a vice president of SDC.

37. Baum (1981): 23.

38. Ware (2008): 97–98.

39. Baum (1981): 26.

40. System Development Division annual report, 1956 (SDC Records). As each of the three buildings was completed, more SDC programmers relocated to Santa Monica, or off to SAGE sites during 1957 and 1958.

41. Baum (1981): 41.

42. System Development Corporation Articles of Incorporation (November 23, 1956), SDC Records.

43. Baum (1981): 28.

44. Bernd Ulmann, *AN/FSQ-7: The Computer That Shaped the Cold War* (Oldenbourg, 2014): 210.

45. Ibid.: 52.

46. Ibid.: 49.

47. Ibid.: 49–51.

48. Ibid.: 47.

49. Ibid.: 51.

50. John F. Jacobs, *The SAGE Air Defense System: A Personal History* (MIITRE Corporation, 1986): 105–106.

51. Baum (1981): 43–44.

52. Benington (1983): 352.

53. Ibid.

54. Ibid.: 355.

55. Ibid.: 356. Excerpts from Benington's 1956 ONR symposium paper were republished in his 1983 *Annals* article.

56. Baum (1981): 33.

57. Ibid.: 48.

58. Ibid.: 52.

59. Jacobs (1986): 108.

60. Frederick P. Brooks Jr., *The Mythical Man-Month: Essays on Software Engineering* (Addison-Wesley, 1975): 26. This details challenges with scaling up programming labor, especially late in a project. "One cannot," Brooks famously articulated, "get workable schedules using more men and fewer months. More software projects have gone awry for lack of calendar time than from all other causes combined." This piece of wisdom certainly applied to the programming of SAGE, as delivery of the master systems programs were a year off the original schedule, despite grossly exceeding initial labor force projections and continual hiring. Also, there was a long-term need for debugging and refinement of code.

61. Edward Hudson, "New Air Defense Opened in Jersey: First Step in SAGE System of Electronic Computers Dedicated at McGuire," *New York Times*, June 27, 1958.

62. Baum (1981): 65–86.

63. Jules I. Schwartz, "The Development of JOVIAL" *ACM SIGPLAN Notices* 13, no. 3 (1978); 203–214.

64. Clark Weissman, "Security Controls in the ADEPT-50 Time-Sharing System," in *Proceedings of the Fall Joint Computer Conference* (1969): 119–133.

65. Ibid.

66. Willis Ware, "Security and Privacy in Computer Systems," in *Proceedings of the Spring Joint Computer Conference* (1967): 279–282.

67. Schwartz (1978): 203–214.

68. Ibid.

69. Schwartz (1978): 204; Baum (1981): 50.

70. Baum (1981): 97–102.

71. Ibid.: 104.

72. http://www.csc.com/about_us/ds/40546/40550-five_decades_of_success (accessed 4/8/2015).

73. Baum (1981): 111.

74. Ibid.: 104.

75. SDC annual report, 1968 (SDC Records).

76. Ibid.

77. Baum (1981): 187.

78. Ibid.: 195–217.

79. Ibid.: 259.

80. Marcia Blumenthal, "Burroughs Seeks to Buy SDC for $98 Million," *Computerworld*, August 25, 1980: 61. Its brief achievement of a 50/50 mix of DoD and non-DoD revenue, achieved in 1976, was soon lost as DoD revenue once again became the overwhelming majority of its business.

81. Baum (1981): 282.

82. Though SDC was the software contractor for SAGE, IBM's maintenance of Q-7 and Q-8 systems involved significant exposure to hardware, software, and networking—systems integration.

83. Paul Ceruzzi, *Internet Alley: High Technology in Tysons Corner, 1945–2005* (MIT Press, 2008): 104–105.

84. "Computer Systems Specialists" [advertisement], *New York Times*, May 12, 1963.

85. The first anti-ballistic-missile systems, developed in the late 1950s, were incapable of effectively intercepting ICBMs. The partially effective systems of the 1960s were central to the Strategic Arms Limitation Talks (SALT I) in 1969. The resulting Anti-Ballistic Missile (ABM) Treaty of 1972 set strict limits on ABM systems.

86. Davis Dyer, *TRW: Pioneering Technology and Innovation Since 1900* (Harvard Business School Press, 1998): 167–172.

87. http://www.afspc.af.mil/shared/media/document/AFD-100405-061.pdf (accessed 4/8/2015)

88. Walter Bauer oral history interview, conducted by Arthur L. Norberg, Woodland Hills, California, May 16, 1983 (CBI); Richard L. Foreman, *Fulfilling the Computer's Promise: The History of Informatics, 1962–1982* (Informatics General Corporation, 1985): 1-4, 1-5, 1-8. R-W probably had more programmers than any other company other than SDC and IBM in the late 1950s. Bauer remembers the number as 400, but Foreman wrote 250, at the R-W center. The Foreman book is oddly organized—a compilation of data, source material, long block quotes, biographies, and narrative. Some things are clearly wrong in his book—it places the Ramo-Wooldridge and Thompson Products merger to create TRW in 1956, it was in 1958—but it is nevertheless a useful source. It, too, places a heavy emphasis on Informatics' transition to software products, though the first part of the book does have some good content on services the firm provided before it began to offer products.

89. "The New Ramo-Wooldridge Laboratories" [advertisement], *Datamation* 6, no. 1 (1960): 49.

90. Walter Bauer oral history (interview conducted by Arthur L. Norberg, Woodland Hills, California, May 16, 1983) (CBI).

91. Ibid.

92. Ibid.

93. Walter Bauer oral history interview, conducted by Luanne Johnson via telephone, June 16, 1995 (Computer History Museum).

94. Jeffrey R. Yost, "Werner Frank," in R. Daniel Wadhwani, ed., *Immigrant Entrepreneurship: German-American Business Biographies, 1720 to the Present*, volume 5 (German Historical Institute, 2013).

95. Bauer oral history (1983): 16.

96. Charles A. Phillips, "The Government: A $1.5 Billion Factor," *Datamation* 8, no. 1 (1962): 23.

97. Werner Frank oral history, conducted by Jeffrey R. Yost, Mountain View, California, February 14, 2006 (CHM).

98. Jeffrey R. Yost, "Computer Industry Pioneer: Erwin Tomash (1921–2012)," *IEEE Annals of the History of Computing* (April–June 2013): 4–7.

99. Ibid.

100. Foreman (1985): 1–13.

101. Frank oral history (2006).

102. Ibid.

103. Walter F. Bauer, "Why Multi-Computers?" *Datamation* 8, no. 9 (1962): 51–55.

104. Foreman (1985): 1–17, 19.

105. Ibid.: 1–20, 21; Frank oral history (2006).

106. Frank oral history (2006).

107. Ibid.

108. "Why Must Progress in Computer Usage Plunge Ahead at a Snail's Pace?" [advertisement by Informatics, Inc.], *Datamation* 10, no. 12 (1964): 57.

109. Martin Campbell-Kelly. *From Airline Reservations to Sonic the Hedgehog* (MIT Press, 2003).

110. Goetz oral history, conducted by Jeffrey R. Yost, May 3, 2002 (CBI).

111. Bauer, "Informatics Acquisition by Sterling Software: Unsolicited Offer, Takeover Attempt, and Merger," *IEEE Annals of the History of Computing*, July–September 2006: 32–40.

Chapter 5

1. Louis Galambos, *Competition and Cooperation: The Emergence of a National Trade Association* (Johns Hopkins University Press, 1966): viii. Galambos, who presents the case of cotton textiles as part of a larger associational movement in need of further study, relates trade associations to medieval guilds and generally presents them as stabilizing forces particularly important in the face of rising competition in twentieth-century industrial America. While I agree, I believe certain trade associations in certain industries at certain times were as much defined by facilitating or advancing battles of different firms or industry segments separated by size, activities, patents and licensing, positions within value chains, or a host of other divides. In this sense, they could be both stabilizing, and if not fully destabilizing, at least divisive within industries. This is evident in the battle between employee-based computer services firms and independent contractor computer services brokerages (set up briefly in this chapter but more fully explored in chapter 9, which addresses the industry association the NACCB), or in the early automobile industry where the Selden Patent litigation gave birth to the Manufacturers Mutual Association (MMA) and its successor the Association of Licensed Automobile Manufacturers (ALAM). The MMA was created initially to fight the Selden patent holding Electric Vehicle Company, but evolved to become the Selden patent defending ALAM (the most powerful early automobile trade association until the defeat of the patent and the organization's dissolvent in 1911) that battled non-member companies and their American Motor Car Manufacturers Association (AMCMA); James J. Flink, *The Automobile Age* (MIT Press, 1990): 54–55.

2. Unlike public corporations, trade associations are not required to file reports with the Security and Exchange Commission or distribute annual reports. Relative to companies, they are less frequently the subject of trade press literature or business journalism. Their very purpose is to facilitate and advance collaboration and sharing among like businesses, and many of them carefully skirt the boundary between lawful cooperation and unlawful collusion. Legal considerations, principals' lack of self-recognition of themselves as historical actors and of their organizations' historical relevance, the possible possession of member company confidential or sensitive data, common reorganizations as industries evolve, and frequent home office relocations (especially for smaller associations) all present pitfalls to the long-term retention of trade association records. Further, collection development authorities at libraries, museums, and other repositories have tended to focus more on records of well-known companies than trade associations.

3. Burt Grad and Luanne Johnson founded and co-ran the Software History Center (SHC) from its origin in 1997 until 2005. That year SHC merged with the Computer History Museum and became the museum's Software Industry Special Interest Group (which Grad and Johnson headed from 2005 through 2015). In addition to holding SHC meetings on ADAPSO history (which I was fortunate to participate in as a mod-

erator and oral historian), they also held important meetings on the history of the software products and (IT) professional services industries (which I also participated in). Grad and Johnson funded (through money they received from the Charles Babbage Foundation) historian Thomas Haigh's examination of ADAPSO—research that resulted in three concise, important pieces he published (in his *IEEE Annals* Biographies Department) on the trade association, offering periodization to three different eras—ADAPSO's service bureau origins, time-sharing and software products, and its increased political and lobbying role with professional services: "ADAPSO and the Service Bureau Industry, 1961–1968" [biographies], *IEEE Annals of the History of Computing* 26, no. 1 (2004): 78–85; "ADAPSO, Time-Sharing Firms and Software Companies" [Biographies], *IEEE Annals of the History of Computing* 27, no. 1 (2005): 67–73; "ADAPSO, Regulated Competition, and Professional Services: 1976–1986" [biographies], *IEEE Annals of the History of Computing* 27, no. 2 (2005): 89–92. The Charles Babbage Institute's ADAPSO Records (1960–1999) contain 17.7 cubic feet of material detailing early meetings, organizational publications, commissioned industry studies, and a host of other historically valuable information. ADAPSO changed names to the Information Technology Association of America (ITAA) in 1991—ITAA was the donor of the records.

4. Further, the leading systems integration firm, System Development Corporation, was (for over a decade) a nonprofit corporation—ADAPSO required (full) member organizations to be for-profit.

5. The earliest programming services and systems integration firms tended to concentrate in areas of greatest computer adoption—Los Angeles (the center of the aerospace industry), Washington (with the massive adoption of computers by the federal government), and New York City (the largest city in the country and its center of banking and other industries). This was true of C-E-I-R, CUC, CSC, PRC, Informatics, and others. Also consulting firms and independent consultants located most often in these three metropolitan regions—including Arthur Andersen Administrative Services, Diebold, Inc., Canning, Sisson, and Associates, Daniel McCracken, Robert Patrick, and others. While there were many service bureaus in these major cities, on a relative basis, service bureaus were far more spread out throughout the nation.

6. Thomas Haigh, "ADAPSO and the Service Bureau Industry, 1961–1968" [biographies], *IEEE Annals of the History of Computing* 26, no. 1 (2004): 79.

7. Before settling on ADAPSO, at the emerging group's earliest meetings in the 1960s the organization went by the name Data Actuating Technical Association. See "Informal Meeting of the Data Actuating Technical Association," April 27, 1960, ADAPSO Records, 1960–1999, CBI.

8. Sizable programming services or systems integrators generally had at least one facility with a computer that provided services and therefore qualified as full

Notes to Chapter 5

members—an example is C-E-I-R, which was principally a programming services firm (that did some work on client machines) with multiple bureaus.

9. Haigh (2004): 79–82.

10. ADAPSO, 1961 Directory of Data Processing Services Centers, ADAPSO Records, 1960–1999 (CBI).

11. Bernard Goldstein oral history (interview conducted by David Allison, Washington, May 3, 2002) (CBI).

12. ADAPSO, 1965–1966 Directory of Data Processing Services Centers, ADAPSO Records, 1960–1999 (CBI).

13. Jerome Dreyer oral history (interview conducted by Thomas Haigh, Pittsburgh, May 1, 2004) (Computer History Museum).

14. SHARE, Inc. meeting talks/discussion and rosters of members from the early 1960s show no programmatic content linked directly to service bureaus or the computer services business and very few participants in SHARE, Inc. in these years from computer services companies. SHARE, Inc. did attract a handful of computer services managers and executives from larger computer services enterprises such as C-E-I-R, Inc., Computer Sciences Corporation, Electronic Data Systems, etc. (particularly by the late 1960s), but there was not an organized group or recognizable content that overlapped with the type of technical and business/managerial information on services shared at ADAPSO meetings. Surveyed SHARE, Inc. Proceedings (1960–1963 and 1968–1969), Boxes 3 and 5, SHARE, Inc. Records, 1955–1994 (CBI).

15. Romuald Slimak, "Opening Address," Proceedings of ADAPSO Management Symposium, New York, January 20, 1961: 5 (ADAPSO Records, 1960–1999, CBI).

16. William W. Eaton, "The Organizational Structure of Chain Centers" [and discussion transcript], Proceedings of ADAPSO Management Symposium, New York, January 20, 1961: 33–35 (ADAPSO Records, 1960–1999, CBI).

17. "Members Enrolled January 21–June 7, 1961," Proceedings of ADAPSO Management Symposium, New York, January 20, 1961: 42 (ADAPSO Records, 1960–1999, CBI).

18. The better consultants tended to establish expertise through successful publication—this was true of Diebold, Canning and Sisson, Patrick, and McCracken for their important books and articles. Canning also launched the successful journal *EDP Analyzer*.

19. "ADAPSO Services," Proceedings of ADAPSO Management Symposium, New York, January 20, 1961: 14 (ADAPSO Records, 1960–1999, CBI).

20. A. M. Lount, "Business Ethics in Service Center Operation," Proceedings of ADAPSO Management Symposium, New York, January 20, 1961: 21–29 (ADAPSO Records, 1960–1999, CBI).

21. P. J. Cranmer, "Service Bureau Management-A Panel, Part III Service Bureau Contracts," Proceedings of ADAPSO Management Symposium, New York, January 20, 1961: 21–29 (ADAPSO Records, 1960–1999, CBI).

22. Malcolm D. Smith, "The Significance of COBOL to the Service Center," Proceedings of ADAPSO Management Symposium, St. Louis, May 5, 1961 (ADAPSO Records, 1960–1999, CBI).

23. David Green, "Uses of Operations Research Techniques in the Solution of Management Problems," Proceedings of ADAPSO Proceedings Management Symposium, Los Angeles, October 16, 1961: 4–16; Richard Hill "Summary," Proceedings of ADAPSO Management Symposium, Los Angeles, October 16, 1961: 17; James T. Pettit, "The Data Processing Hardware Selection Problem" [and discussion transcripts of all three], Proceedings of ADAPSO Management Symposium, Los Angeles, October 16, 1961: 18–47.

24. R. W. Johnson, "How to Operate a Service Bureau for Profits—Minimizing Running Costs," ADAPSO Eight Management Symposium "Schedule for Future Symposia," Toronto, May 23–24, 1964: 66 (CBI). In 1964 individual sessions were printed as separate booklets rather than an overall proceedings.

25. ADAPSO Proceedings Management Symposium (New York January 20, 1961); ADAPSO Proceedings Management Symposium (St. Louis, May 5, 1961); ADAPSO Proceedings Management Symposium (Los Angeles, October 16, 1961); ADAPSO Proceedings Management Symposium (Chicago, May 14, 1962); ADAPSO Proceedings Management Symposium (New York, February 19, 1962); ADAPSO Proceedings Management Symposium (San Francisco, January 21, 1963). ADAPSO Records, 1960–1999. CBI, UMN.

26. J. A. Cross, T. C. Scrymgeour, and G. W. Dick, "How to Operate a Service Bureau for Profits—Protecting Client Records," ADAPSO Eight Management Symposium (Toronto, May 23–24, 1964). CBI, UMN.

27. Cross, Scrymgeour, and Dick (1964); "Pentagon Fire Destroys Air Records and Computers: Area of Pentagon Is Swept by Fire," *New York Times*, July 3, 1959.

28. Extensive holdings, if not full runs, of these various publications are held in the Charles Babbage Institute's ADAPSO collection.

29. For a brief discussion of ERMA and MICR and two important and lengthy chapters on the history of computing in banking, see James W. Cortada, *The Digital Hand*, volume II: *How Computers Changed the Work of American Financial, Telecommunications, Media, and Entertainment Industries* (Oxford University Press, 2006).

30. Robert H. Long, "Data Processing Service Opportunities in Banking," *ADAPSO Management Guidon* (March 1965): 9.

31. Ibid.: 11.

Notes to Chapters 5 and 6

32. Luanne Johnson, ed., ADAPSO Reunion Transcript, May 2–4, 2002 (iBusiness Press, 2003): 62.

33. Johnson, ed. (2003): 61.

34. Haigh (2004): 82–83.

35. For a discussion on ADR and Autoflow, see Martin A. Goetz oral history (interview conducted by Jeffrey R. Yost, Washington, May 3, 2002) (CBI).

36. On the Cullinane Corporation, see John Cullinane oral history (interview conducted by Jeffrey R. Yost, Boston, July 29, 2003) (CBI).

37. Patrick McGovern oral history (interview conducted by David Morrow, Boston, August 4, 2000) (accessed May 7, 2015, from IDC successor IDG's website: https://www.idg.com).

38. DesJardins (1967): 1–12, ADAPSO Records, 1960–1999 (CBI).

39. Though early symposiums and later small side meeting of CEOs (roundtables) continued the ADAPSO tradition of sharing information, companies tended to be guarded on problems, strategies, and financials, but could benefit greatly from industry aggregate data.

40. Robert B. DesJardins, *An Economic Analysis of the Data Processing Services Industry* (ADAPSO, 1967), ADAPSO Records, 1960–1999 (CBI).

41. Robert B. DesJardins, *Second Economic Analysis of the Data Processing Services Industry* (ADAPSO, 1967); DesJardins, *Third Annual Industry Study* (ADAPSO, 1969); DesJardins. *Fourth Annual Industry Study* (ADAPSO, 1970), ADAPSO Records, 1960–1999 (CBI).

42. Johnson, ed. (2003): 181 [panel discussion].

43. Ibid.: 297.

44. Ibid.: 191.

Chapter 6

1. In its first decade (the 1960s), facilities management could also involve a facilities management provider renting computing equipment (or time on that equipment) not owned by the client (and, of course, balancing loads for computing resources between clients).

2. These long-term contracts typically had a fixed monthly charge, plus a set rate based on the variable volume (generally of transactions). As with any long-term contract, at times (with changing needs), new elements had to be negotiated or renegotiated.

3. Sometimes adding computers might significantly reduce cost if no computer time was ever wasted, but that could rarely if ever be fully achieved.

4. Thomas P. Hughes, *Networks of Power: Electrification in Western Society, 1880–1930* (Johns Hopkins University Press, 1983). With electrical networks, differential pricing has long been used as a tool to try to even out loads.

5. Charles M. Lawson, "A Survey of Computer Facility Management," *Datamation*, July 1962: 29–31.

6. Waste is based on total hours in a day and days in a week. Many respondents to SDC's survey cited round-the-clock computer room operations.

7. The term "facility management" or "facilities management" had referred to internal management previously, but as EDS grew, and spawned imitators, the usage of the term increasingly referred to outsourcing the data processing department work to a "facilities management" specialist. An example of earlier usage of the term for internal managing of computing facilities is Lawson (1962).

8. Perot and EDS did create a new field with facilities management, though there were precursors of military installations outsourcing to contractors—including facilities engaged in either data processing or calculation using data processing or computational equipment. EDS created this model for the computer industry, and for the serving of corporate, and over time, a vast range of organizational clients (in the federal government, local government, nonprofits, and so on). See Martin Campbell-Kelly. *From Airline Reservations to Sonic the Hedgehog: A History of the Software Industry* (MIT Press, 2013): 62–63.

9. Martin Campbell-Kelly, William Aspray, Nathan Ensmenger, and Jeffrey R. Yost. *Computer: A History of the Information Machine*, third edition (Westview, 2014): 138; Elmer C. Kubie, "Recollections of the First Software Company," *IEEE Annals of the History of Computing* 16, no. 2 (1994): 65–71; Gerd Meissner, *SAP: Inside the Secret Software Power* (McGraw-Hill, 2000); Catherine Fredman and Gideon Gartner, *About Gartner: The Making of a Billion-Dollar IT Advisory Firm* (Lemonade Heroes, 2014); Michael Lewis. *The New New Thing: The Silicon Valley Story* (Norton, 1999).

10. Meissner (2000); Fredman and Gartner (2014); Lewis (1999).

11. David Yoffie and Michael Cusumano, *Competing on Internet Time: Lessons from Netscape and Its Battle with Microsoft* (Free Press, 1998); Lewis (1999).

12. Doron P. Levin, *Irreconcilable Differences: Ross Perot versus General Motors* (Little, Brown and Co., 1989): 28. Noted business journalist and specialist on the automobile industry published this engaging and important work on Ross Perot and his battles with General Motors. The biographical information on Perot and discussion of the start of EDS and its early evolution is particularly useful. "Why Perot Thrived in Fertile Kansas," *New York Times*, November 6, 1992.

13. Levin (1989): 24–26.

14. Ibid.: 28.

15. Ibid.: 29.

16. Ibid.: 30.

17. "Offsite Computer Departments. Making Friends in Dallas," *Datamation* 10, no. 8 (1964): 19.

18. Ibid.

19. Levin (1989): 30.

20. "EDS Takes Over PepsiCo Data Processing," *Datamation* 12, no. 11 (1966): 86; "Pepsi-Cola Adds the Obvious: Maker of Potato Chips, Fritos," *New York Times*, February 25, 1965.

21. Levin (1989): 30.

22. "Perot's Stock Falls $445 Million in a Day," *New York Times*, April 23, 1970.

23. Arthur M. Louis, "The Fastest Richest Texan Ever," *Fortune* 78 (November 1968): 168–170.

24. John Nordheimer, "Billionaire Texan Fights Social Ills," *New York Times*, November 28, 1969.

25. Stuart Auerbach, "Perot Medicare Bonanza Revealed," *New York Times*, September 29, 1971. Levin (1989). Levin drawn from in this and the prior paragraph. For discussion of EDS hiring ex-military see Levin.

26. Wyatt Wells, "Certificates and Computers: The Remaking of Wall Street, 1967 to 1971," *Business History Review* 74 (summer 2000): 193–235.

27. R. I. Petruschell et al., *Reducing Costs of Incomplete Stock Transactions: A Study of Alternative Trade Completion Systems* (RAND Corporation, 1970).

28. Wells (2000).

29. Levin (1989): 39–47.

30. Ibid. "Perot Completes F. I. DuPont Deal," *New York Times*, May 15, 1971.

31. DuPont Walston Failure Is First Defeat for Perot," *New York Times*, January 22, 1974.

32. John Brooks, *The Go Go Years: The Drama and the Crashing Finale of Wall Street's Bullish 60s* (Wiley, 1999).

33. Levin (1989): 39–47.

34. Electronic Data Systems annual report, 1972.

35. Electronic Data Systems annual reports, 1972 and 1973.

36. Electronic Data Systems annual reports, 1975 and 1981.

37. Ronald H. Coase, "The Nature of the Firm," *Economica* 4, no. 16 (1937): 386–405; Oliver E. Williamson. *The Economic Institutions of Capitalism: Firms, Markets, and Relational Contracting* (Free Press, 1985).

38. Levin (1989): 39–47.

39. Electronic Data Systems annual report, 1977.

40. Levin (1989): 61.

41. Ken Follett. *On Wings of Eagles* (Signet, 1984).

42. Levin (1989): 63–64.

43. Electronic Data Systems annual reports, 1981 and 1982.

44. "Electronic Data Gets Army Contract," *New York Times*, April 21, 1982.

45. *Opportunities for Investment in the Computer Services Industry* (INPUT, 1977), ADAPSO Records, CBI; Electronic Data Systems annual reports, 1981 and 1982.

46. Gary Reich, "The Innovator," *New York Times*, April 25, 1985.

47. Paul Ingrassia and Joseph B. White, *Comeback: The Fall and Rise of the American Automobile Industry* (Simon & Schuster, 1995).

48. Levin (1989).

49. John Holusha, "GM Chief Returns Fire in Skirmish with Perot," *New York Times*, November 28, 1986.

50. Benjamin Klein, Robert G. Crawford, and Armen A. Alchian, "Vertical Integration, Appropriable Rents, and the Competitive Contracting Process," *Journal of Law and Economics* 21, no. 2 (1978): 297–326.

51. Ronald H. Coase, "The Acquisition of Fisher Body by General Motors," *Journal of Law and Economics* 43, no. 1 (2000): 15–32.

52. Reich (1985).

53. Holusha (1986).

54. Levin (1989).

55. William Glaberson, "Head of GM Sees End to Perot Controversy," *New York Times*, January 29, 1987.

56. "Electronic Data Official a Key to Navy Contract," *New York Times*, August 12, 1987.

57. Gary Jacobson, "EDS Sees Success After Purchase by GM—But at a Cost," *Dallas Morning News*, December 9, 2012.

58. Ibid.

Chapter 7

1. Martin Greenberger, "The Computers of Tomorrow," *Atlantic Monthly* 213, no. 5 (1964): 63–67.

2. Robert M. Fano, "The MAC System: The Computer Utility Approach," *IEEE Spectrum* 2, no. 1 (1965): 56–64; Fred Gruenberger, *Computers and Communications—Toward a Computer Utility* (Prentice-Hall, 1967).

3. Greenberger (1964).

4. Ibid.

5. Martin Greenberger, ed., *Computers and the World of the Future* (MIT Press, 1962).

6. Both acronyms were used (to show deference to both the time-sharing research group and the artificial intelligence research group), but Multiple Access Computer was most fitting in the 1960s, as it was the primary basis for receiving Project MAC funding.

7. Greenberger (1964).

8. Auerbach Corporation, *A Jointly Sponsored Study of Commercial Time-Sharing Services*, volumes I and II, 1968 (CBI).

9. Fernando J. Corbató oral history (interviews conducted by Arthur L. Norberg, Cambridge, Massachusetts, April 18, 1989 and November 14, 1990) (CBI); Emerson W. Pugh, Lyle R. Johnson, and John H. Palmer, *IBM's 360 and Early 370 Systems* (MIT Press, 1991): 362.

10. Martin Campbell-Kelly and Daniel Garcia-Swartz, "Economic Perspectives on the History of the Computer Time-Sharing Industry, 1965–1985." *IEEE Annals of the History of Computing* 30, no. 1 (2008): 16–36.

11. Bolt Beranek and Newman, Inc., preceded both GEIS and Tymshare in offering time-sharing services on a very small scale for science and engineering users, but it was a modest sized research and engineering firm focused on pioneering technology, not computer services.

12. These two pioneering articulations are discussed and analyzed in Fernando J. Corbató, "Time Sharing," in Anthony Ralston, Edwin D. Reilly, and David Hemmendinger, eds., *Encyclopedia of Computer Science*, fourth edition (Nature Publishing Group, 2000): 1779–1780.

13. Robert M. Fano, The MAC System: A Progress Report, MAC-Technical Report-12 (October 9, 1964), Fernando J. Corbató Papers, Box 18, MIT Archives.

14. Fernando J. Corbató, Marjorie Merwin-Daggett, and Robert C. Daley, "An Experimental Time-Sharing System," in *Proceedings of the 1962 Spring Joint Computer Conference* 21 (AFIPS, 1962): 335–344.

15. John McCarthy et al., A Time-Sharing System for a Small Computer, unpublished report [n.d., probably 1963], Fernando J. Corbató Papers, Box 13, MIT Archives.

16. Arthur L. Norberg and Judy O'Neill, *Transforming Computer Technology: Information Processing for the Pentagon, 1962–1986* (Johns Hopkins University Press, 1996): 102.

17. "Chronicle of the Early Years," *SPECTRUM: A Special Edition. Commemorating the Twentieth Anniversary of General Electric Information Services Company* (December 1985), Corporate History Collection, Computer History Museum.

18. Ibid.

19. Warner Sinback oral history (interview conducted by Luanne Johnson, Mountain View, California, September 24, 2004) (CHM).

20. Bernard Goldstein oral history (interview conducted by David Allison, Washington, May 3, 2002) (CBI).

21. "Chronicle ..." (1985).

22. Thomas O'Rourke oral history (interview conducted by Luanne Johnson, Saratoga, California, March 13, 2002) (CHM).

23. Ibid.

24. Ibid.

25. Ibid.; Sinback oral history.

26. LaRoy Tymes oral history (interview conducted by Luanne Johnson, Cameron Park, California, June 11, 2004).

27. O'Rourke oral history.

28. Butler Lampson oral history (interview conducted by Jeffrey R. Yost, Cambridge, Massachusetts, December 11, 2014) (CBI).

29. Rebecca M. McNown, "Tymshare, Inc.: 1965–1970." (August 17, 1970), Corporate Histories Collection, Computer History Museum. This unpublished paper was a short report by a Tymshare, Inc. employee and part-time university student.

30. O'Rourke oral history.

31. Butler Lampson, Wayne Lichtenberger, and Mel W. Pirtle, "A User Machine in a Time-Sharing System." *Proceedings of the IEEE* 54, no. 12 (1966): 1766–1774.

32. Ann Hardy oral history (interview conducted by Jeffrey R. Yost, Palo Alto, April 3, 2012) (CBI).

33. "Tymshare History: Notes from Ann Hardy" (unpublished notes, provided to author before interview with Hardy).

34. Ann Hardy oral history.

35. O'Rourke oral history.

36. Ibid.

37. Ibid.

38. McNown (1970).

39. Ann Hardy oral history.

40. O'Rourke oral history.

41. Ibid.

42. Ronald W. Braniff oral history (interview conducted by Jeffrey R. Yost, San Francisco, April 5, 2012) (CBI).

43. Auerbach Corporation, volume II (1968).

44. David Schmidt oral history (interview conducted by Jeffrey R. Yost, El Dorado Hills, California, April 5, 2012) (CBI).

45. Tymshare, Inc., annual report, 1970 (Corporate History Collection, CHM).

46. "Tymshare Offer Sold Out." *Wall Street Journal*, September 25, 1970.

47. O'Rourke oral history.

48. Ann Hardy oral history; Norman Hardy oral history (interview conducted by Jeffrey R. Yost, Portola Valley, California, April 4, 2012) (CBI).

49. Norman Hardy oral history.

50. Ibid.

51. Tymes oral history interview.

52. Ibid.

53. Ibid.

54. Ibid.

55. Ibid.

56. Tymshare, Inc. annual report, 1976, University Libraries, University of Minnesota (hereafter UL, UMN).

57. O'Rourke oral history.

58. Tymshare, Inc. annual report, 1976 (UL, UMN).

59. "TYMNET Asks to Be Value-Added Common Carrier." *Computerworld* 10, no. 21 (1976): 20; "Tymshare Says FCC Clears Unit to Operate as Carrier," *Wall Street Journal*, December 28, 1976.

60. Tymshare, Inc. annual report, 1976 (UL, UMN).

61. Auerbach Corporation, volume I (1968).

62. Campbell-Kelly and Garcia-Swartz (2008).

63. Tymshare, Inc. annual report, 1970 (CHM).

64. Goldstein oral history; Warren Prince oral history (interview conducted by Jeffrey R. Yost, Sunnyvale, California, April 4, 2012) (CBI).

65. "Tymshare Will Acquire United Data Centers for $6.1 Million in Stock," *Wall Street Journal*, August 9, 1974.

66. Goldstein oral history.

67. Tymshare, Inc. annual report, 1976 (UL, UMN).

68. Corporate Histories Database, Corporate History Collection, Computer History Museum (http://corphist.computerhistory.org/corphist/view.php?s=themes&id=5, accessed July 24, 2015). This online resource was created by Luanne Johnson.

69. Norman Hardy oral history.

70. Tymes oral history.

71. http://amturing.acm.org/award_winners/engelbart_5078811.cfm (accessed July 24, 2015).

72. Lynn Sanden oral history (interview conducted by Luanne Johnson, Cameron Park, California, June 11, 2004) (CHM).

73. Ibid.

74. Ann Hardy oral history.

75. O'Rourke oral history.

76. "McDonnell Gets Approval on Acquisition of Tymshare," *Wall Street Journal*, March 29, 1984.

77. "McDonnell Douglas Sues British Telecom on Tymnet Purchase," *Wall Street Journal*, January 30, 1990.

78. Auerbach Corporation, volume I (1968).

79. Andrew Pollack, "General Electric Buying 3 Software Companies," *New York Times*, October 2, 1981.

80. Claudia H. Deutsch, "G.E. Is Selling Off Most of Electronic Business," *New York Times*, June 25, 2002.

81. Auerbach Corporation, volume II (1968).

82. Stanley Penn, "Computer Competition: Justice Agency Mulls Move to Force IBM Out of the Time-Sharing Services Business," *Wall Street Journal*, October 1, 1968.

83. "IBM's Consent Decree and Time-Sharing," *Datamation* 12, no. 2 (1966): 101.

84. "IBM Studying Separate Prices for Servicing," *Wall Street Journal*, December 9, 1968.

85. Auerbach Corporation, volume II (1968).

86. "Control Data Settlement Seen as Setback by Others Suing IBM; Key File Destroyed," *Wall Street Journal*, January 22, 1973.

87. Campbell-Kelly and Garcia-Swartz (2008).

88. Ibid.

Chapter 8

1. On maintenance, and maintenance labor, see Kevin L. Borg, *Auto Mechanics: Technology and Expertise in Twentieth Century America* (Johns Hopkins University Press, 2007). On maintenance of industrial production machinery, see Pierre Claude Reynard, "Unreliable Mills: Maintenance Practices in Early Modern Papermaking," *Technology and Culture* 40 (April 1999): 237–262. Little has been published on the topics of technology maintenance and maintenance workers.

2. Martin Campbell-Kelly, *From Airline Reservations to Sonic the Hedgehog: A History of the Software Industry* (MIT Press, 2003): 41–45.

3. Thomas J. Misa, *Digital State: The Story of Minnesota's Computing Industry* (University of Minnesota Press, 2013). Using approximately one fifth of his book's chapter on Control Data to discuss services, Misa provides a brief and important history of this side of that firm, especially the emergence and expansion of its data centers. Nearly all other works in the quite small historical literature on CDC focuses on supercomputing, and ignore or substantially marginalize services (as well as CDC's large and important business in peripherals).

4. "Endicott School Graduates Its 10,000th EAM Customer Engineer," *Business Machines* (November 20, 1956): 10–11 (IBM Corporate Archives).

5. "CEs Turned Author Write FE Manuals," *IBM News: Field Engineering Division* (February 25, 1969): 1 (IBM Corporate Archives).

6. IBM Corporate Archives. https://www-03.ibm.com/ibm/history/witexhibit/wit_decade_1940.html (accessed December 15, 2015).

7. IBM Corporate Archives. https://www-03.ibm.com/ibm/history/witexhibit/wit_decade_1930.html (accessed December 15, 2015).

8. "IBM Graduates Class of Young Women for System Services Work," *Business Machines* (November 12, 1936): 1 (IBM Corporate Archives).

9. *Outlook: Information Systems Group National Marketing Division* [special issue 25 SE] (IBM, May/June 1985): 4–6 (IBM Corporate Archives).

10. "25 Young Women Pioneered Field," *IBM Data Processing News* (July 22, 1960): 1 (IBM Corporate Archives).

11. Glenn E. Meyers, "IBM Field Engineering Experiences: A Personal Memoir," *IEEE Annals of the History of Computing* 21, no. 4 (1999): 72–73.

12. Ibid.: 73–74.

13. Ibid.: 73–75.

14. "How the New Division Will Work," *IBM News: Field Engineering Division* (March 17, 1963): 1, 3 (IBM Corporate Archives).

15. "FE Realigns Areas to Ensure Top Customer Service," *IBM News: Field Engineering Division* (January 10, 1968): 1; "FE Data Center Comes of Age, Handles Entire IR Processing," *IBM News: Field Engineering Division*, December 16, 1968: 3.

16. "CEs Help Develop System/360 Programs," *IBM News: Field Engineering Division* (March 10, 1966): 1, 3 (IBM Corporate Archives).

17. "FE Gears Up for Expanded Role in Programming Services," *IBM News: Field Engineering Division* (November 25, 1969): 1, 3 (IBM Corporate Archives).

18. W. H. Evers and S. S. Thakur, "Programmed Automatic Customer Engineer (PACE) Dispatch," *IBM Journal of Research and Development*, July 1969: 357–365.

19. "Five 'Satellite' Centers Speed Emergency Parts in Major Cities," *IBM News: Field Engineering Division* (September 25, 1969): 1 (IBM Corporate Archives).

20. IBM Archives. http://www-03.ibm.com/ibm/history/history/year_1959.html (accessed December 15, 2015).

21. Claude Baum, *The System Builders: The Story of SDC* (System Development Corporation): 50.

22. Emerson W. Pugh, *Building IBM: Shaping an Industry and Its Technology* (MIT Press, 1995): 219.

23. David Mindell, *Between Human and Machine* (Johns Hopkins University Press, 2002): 315.

24. Martin Campbell-Kelly, William Aspray, Nathan Ensmenger, and Jeffrey R. Yost, *Computer: A History of the Information Machine*, third edition (Westview, 2014): 153–156; Campbell-Kelly (2003): 44.

25. SABRE, "Manpower Requirements, File and Program Specifications," June 30, 1960 (CBI); Frederick P. Brooks Jr., *The Mythical Man-Month: Essays in Software Engineering* (Addison-Wesley, 1975). It is unclear whether the same lesson Frederick Brooks insightfully drew from the OS/360 development project could have been learned from a reflective post-project analysis of SABRE (adding programmers, throwing additional man-months into the effort with new personnel late in the project, compounding problems and adding to the programming challenges and time). What is known is man-month estimates rose as the SABRE project progressed and lines of code far exceeded original plans—by a factor of 5.

26. Martin Rothstein, "American Airlines 'SABRE': The Analyst's Viewpoint," October 9, 1962 (CBI).

27. Robert V. Head, "Getting SABRE Off the Ground," *IEEE Annals of the History of Computing* 24, no. 4 (2002): 34.

28. Ibid.: 35.

29. Campbell-Kelly (2003): 44.

30. Memo from W. B. Emore to J. Siegfried, June 26, 1959 (CBI).

31. Campbell-Kelly (2003): 44–45.

32. *Outlook*, 3–6.

33. Ibid.

34. "25 Young Women Pioneered Field."

35. *Outlook*, 4.

36. "Loraine McLennan Recalls Birth of Systems Service," *IBM Data Processing News* (July 22, 1960) (IBM Corporate Archives).

37. *Outlook*, 4–6.

38. Ibid., 6.

39. Ibid.

40. "Announce Systems Engineering," *IBM Data Processing News* (December 5, 1960): 1 (IBM Corporate Archives).

41. Shirley Benzer, "SE Education Paves the Road for Growth," *Data Processing News*, July 7, 1961: 3.

42. *Outlook*, 6.

43. "Systems Engineers Hold Symposium," *Business Machines* (October 1961): 16 (IBM Corporate Archives).

44. *Outlook*, 6.

45. Ibid.

46. Ibid., 7.

47. Arnold H. Lubasch, "Hearing Set on IBM Case Dismissal," *New York Times*, March 20, 1982.

48. *Outlook*, 6–7.

49. Joshua Mills, "IBM to Sell Its Military Unit to Loral," *New York Times*, December 14, 1993.

50. Arthur L. Norberg, *Computers and Commerce: A Study of Technology and Management at Eckert-Mauchly Computer Company, Engineering Research Associates, and Remington Rand, 1946–1957* (MIT Press, 2005).

51. Misa (2013): 100–105.

52. Jeffrey R. Yost, *The Computer Industry* (Greenwood, 2005): 82.

53. Control Data Corporation annual reports, 1959, 1960, and 1961 (CBI).

54. Yost (2005): 86–88.

55. Control Data Corporation annual report, 1960 (CBI).

56. R. C. Gunderson to William C. Norris, "Input to the Control Data History File," July 11, 1980 (CBI)

57. Ibid.

58. Ibid.

59. Control Data Corporation annual report, 1962 (CBI).

60. Thomas J. Misa, ed., *Building the Control Data Legacy: The Career of Robert M. Price* (Charles Babbage Institute, 2012): 38–42.

61. R. C. Gunderson to William C. Norris (1980): 2.

62. Misa (2012): 1–30.

63. Ibid.

64. J. E. Davis, R. B. Smith, and D. H. Wentworth to William C. Norris, "PSD History," July 17, 1980 (CBI).

65. R. F. Buelow, J. D. Kee, and O. K. Staton to William C. Norris, "History of Engineering Services," July 3, 1980 (CBI).

66. Ibid.

67. Ibid

68. Ibid.

69. Control Data Corporation annual report, 1963 (CBI).

70. B. T. von Schmidt-Pauli to A. T. Bassett, July 7, 1980 (CBI).

71. Control Data Corporation annual report, 1964 (CBI).

72. "Control Data to Buy C-E-I-R for Less Than Announced," *Wall Street Journal*, October 2, 1967.

73. Control Data Corporation annual report, 1968 (CBI).

74. Ibid.

75. Control Data Corporation annual report, 1969 (CBI).

76. Control Data Corporation annual report, 1968.

77. Ibid.

78. Moore's Law—an observation by Gordon Moore (which later evolved to become a prediction and in part a self-fulfilling prophecy that the capacity of integrated circuits would double every 12 to 18 months)—accentuated the perception that software was not keeping up with advances in circuitry and processing power.

79. D. C. States, J. T. Lynch, and L. G. Kinney to William C. Norris, "CDI History," July 11, 1980 (CBI).

80. Ibid.

81. Ibid.

82. Ibid.

83. Ibid.

84. C. R. McEwen to G. G. Smith, "PLATO History," October 6, 1980 (CBI).

85. Ibid.

86. "Software Products Task Force Report" [draft], August 29, 1969 (CBI).

87. Ibid.

88. D. H. Wentworth to William C. Norris, "History of Unbundling," July 17, 1980 (CDC-CR, EHPR).

89. Ibid.

90. Control Data Corporation annual reports, 1970, 1971, 1972 (CDC-CR, CBI).

91. C. K. Bardos to William C. Norris, July 7, 1980. Bardos attached "Control Data Total Services" report (October 30, 1970): 1–4. Quote from 4.

92. The division had multiple name changes, but a constancy of mission in meeting the federal government's defense and Big Science needs by serving departments, agencies, and defense contractors.

93. "History of Military Products Division Control Data Corporation, 1958–1970s," CDC-CR, EHPR, box 3, folder 1. The logistics needs of the Air Force were larger than that of the largest US corporations.

94. Control Data Corporation annual reports, 1971 and 1975 (CDC-CR, CBI).

95. Control Data Corporation annual report, 1974 (CDC-CR, CBI).

96. Control Data Corporation annual report, 1973 (CDC-CR, CBI).

97. Control Data Corporation annual report, 1975 (CDC-CR, CBI).

98. Davis, Smith, and Wentworth (1980): 10.

99. Ibid.: 9.

100. Jeffrey R. Yost, "Materiel Command and the Materiality of Commands: An Historical Examination of the US Air Force, Control Data Corporation, and the Advanced Logistics System," in Arthur Tatnall, ed., *History of Computing: Learning from the Past* (Springer, 2010): 89–100; Willis H. Ware oral history (interview conducted by Jeffrey R. Yost, Santa Monica, August 11, 2003) (CBI).

101. Ware oral history.

102. Hans Neukom, "Ubisco and CDC: Analysis of a Failure," *IEEE Annals of the History of Computing* 31, no. 2 (2009): 31–43.

103. Control Data Corporation annual report, 1977 (CDC-CR, CBI).

104. Misa (2012): 306.

105. Control Data Corporation annual reports, 1984, 1986, 1987, 1988, 1989 (CDC-CR, CBI).

Chapter 9

1. Michael Brick, "Man with Grudge Against Tax System Crashes Plane into Texas IRS Office," *New York Times*, February 19, 2010.

Notes to Chapter 9

2. http://www.democraticunderground.com/discuss/duboard.php?az=view _all&address=389x7740464 (accessed January 14, 2016).

3. Harvey Shulman, "Our Low-Tech Tax Code," *New York Times*, February 21, 2010.

4. Harvey Shulman oral history (interview conducted by Jeffrey R. Yost, Mountain View, California, March 30, 2007) (Computer History Museum).

5. Shulman "Our Low Tech Tax Code."

6. Ibid.

7. Shulman oral history; Shulman, "Our Low Tech Tax Code."

8. Shulman oral history.

9. Shulman, "Our Low Tech Tax Code."

10. Shulman oral history.

11. Shulman "Our Low Tech Tax Code" (2010).

12. Business history's focus on large corporations follows the lead of the influential early business historian Alfred D. Chandler, who concentrated on large enterprises and the trend and rationale toward vertical integration—a trend that was prevalent (though far from the only model) in the first eight decades of the twentieth century. Chandler's book *The Visible Hand: The Managerial Revolution in American Business* was published by the Harvard University Press in 1977. In recent decades, networks (of suppliers) or market models for asset allocation and sourcing of inputs, which have long been present in many industries, have become increasingly common. In addition to Chandler's precedent, available archival records also makes studying large companies possible, whereas there often is a scarcity of documentary records on the work of small enterprises. Further, with greater numbers of employees, resources, and revenue they might seem more important objects of study to some scholars. For recent history, a set of oral histories can help overcome both the scarcity of source material on small enterprise and entrepreneurial women. A cluster of oral history interviews (focused on women entrepreneurs) conducted by the author are fundamental to this chapter. As are my oral history with Harvey Shulman, and Burt Grad's oral history with Fred Shulman.

13. Jeffrey R. Yost, "Programming Enterprise: Women Entrepreneurs in Software and Computer Services," in Thomas J. Misa, ed., *Gender Codes: Why Women are Leaving Computing* (Wiley and IEEE Computer Society, 2010). A major topic of my chapter in *Gender Codes* is female entrepreneurs in software—both software products and programming services. I profile the careers of the software products pioneer Luanne Johnson and programming services entrepreneurs Grace Gentry and Phyllis Murphy (overlapping in some with themes but with different text from this chapter).

14. Grace Gentry oral history (interview conducted by Jeffrey R. Yost, Berkeley, California, August 11, 2008) (CBI).

15. Richard Gentry oral history (interview conducted by Burton Grad, Mountain View, California, May 24, 2006) (CHM).

16. Grace Gentry oral history.

17. Ibid.

18. Janet Abbate, *Recoding Gender: Women's Changing Participation in Computing* (MIT Press, 2012): 119–128. In this *path-breaking* work on gender and the history of computing, Abbate details the hurdles and triumphs for women in computing within companies and academic settings. In a chapter on female entrepreneurs, she profiles Elsie Shutt and Stephanie Shirley and the programming services businesses (Computations Inc. and Freelance Programmers Ltd., respectively) they launched—Shutt in the late 1950s (in the United States) and Shirley in the early 1960s (in Great Britain). Both started as just a one-person consultant/programmer operations and then, realizing an opportunity, they grew the consulting enterprise. Abbate refers to both women hiring "employees," so these might have been solely employee-based rather than the independent contractor brokerage model pioneered by Grace Gentry. Regardless, these early companies were not known to Gentry and were not models for her business. Shutt kept things small for many years, with fewer than 15 employees, whereas Shirley continued to grow her business to include a roster of 750 by the mid 1980s. Both Shutt and Shirley apparently continued to focus on hiring women over the long term, whereas Gentry, Inc. hired predominantly women only at the start.

19. Grace Gentry oral history.

20. Ibid.

21. Ibid.

22. Ibid.

23. Harvey Shulman oral history.

24. Grace Gentry oral history.

25. Phyllis Murphy oral history (interview conducted by Jeffrey R. Yost, Burbank, California, December 18, 2009) (CBI).

26. Ibid.

27. Ibid.

28. Ibid.

29. Ibid.

30. http://www.pmurphy.com/about-us (accessed January 14, 2016).

31. Fred Shulman oral history (interview conducted by Burton Grad, Mountain View, California, March 28, 2007) (CHM).

32. Ibid.

33. Ibid.

34. Ibid.

35. Ibid.

36. Ibid.

37. Ibid.

38. Ibid.

39. Ibid.

40. Ibid.

41. Ibid. After Shulman stated a range from $15 million to $25 million as the price paid for COMSYS, he responded "Yes" to being paid roughly 40 cents on the dollar for annual revenue, placing the price at around $22 million.

42. "NACCB: A Historical Perspective" (unpublished and undated report provided to author by past NACCB president Grace Gentry, circa late 1990s).

43. Ibid.

44. Ibid.

45. Peggy Noell Smith oral history (interview conducted by Jeffrey R. Yost, Greensboro, North Carolina, January 9, 2009) (CBI).

46. "NACCB: A Historical Perspective."

47. Ibid.

48. NACCB, "1995 NACCB Operating and Compensation Survey," 1995 (survey provided to author by past NACCB president Grace Gentry).

49. "NACCB: A Historical Perspective"; "1995 NACCB Operating and Compensation Survey"; Phyllis Murphy oral history; Grace Gentry oral history; Fred Shulman oral history.

Chapter 10

1. INPUT, "US Information Services Markets, 1983–1988: Industry Specific Markets, Volume 1" (1983): 1. INPUT and other early IT market research firms concentrated

on "spending" by end users (organizations) to accurately identify the size of the industry—as opposed to sum of all revenue of participants as subcontracting could lead to some double (or even triple) counting. This set a precedent for latter research firms in identifying the size or revenue of the industry (to accurately identify, and not inflate, the overall industry revenue). In the text I use the term "revenue" throughout rather than spending, but all industry figures given are estimates that seek to avoid double counting and artificially inflating revenue.

2. INPUT (1983): 48. A 2014 global total of $954.8 billion is reported in press release by Gartner, Inc. (https://www.gartner.com/doc/3032029/market-share-analysis-it-services, accessed March 2, 2016). Different market research firms have different methods for their estimates on industry size, and these methods (along with industries and their definitional boundaries) change. Thus, the 30.8-fold growth from 1982 to 2014 is an approximation (and is based on estimates of two different firms decades apart). INPUT (which operated until 2000) and Gartner, Inc. (still active) are highly trusted names at the top of the IT market research field and thus the general trend (growth by a factor of about 30) is a best possible estimate for growth that in actuality might be slightly more or slightly less.

3. https://www.adp.com/solutions/multinational-business/services/payroll-services.aspx (accessed March 2, 2016).

4. INPUT (1983): 1. This figure of 6,000 undoubtedly only includes multi-person firms—as the number of self-employed, independent contractors in IT has long been in at least the tens of thousands.

5. http://www.gartner.com/newroom/id/1666524 (accessed February 2, 2016).

6. Ernest Holsendolph, "US Settles Phone Suit, Drops I.B.M. Case; A.T.&T. to Split Up, Transforming Industry," *New York Times*, January 9, 1982.

7. Ibid.

8. http://www.fujitsu.com/global/about/corporate/history/1992-today (accessed April 26, 2016).

9. Cognizant is based in New Jersey, but was founded by an Indian entrepreneur, grew through a partnership with India-based Satyam, and has about two thirds of its workforce in India. For that reason, it is often considered (by market research firms and journalists) to be an Indian firm and to be among the Big Five in India in IT Services.

10. Gartner Newsroom press releases. http://www.gartner.com/newsroom (accessed March 5, 2016).

11. Steve Lohr, "IBM Reports Declines in Fourth-Quarter Profit and Revenue Despite Gains in New Fields," *New York Times*, January 20, 2016.

12. http://www.cnbc.com/2015/02/26/ibms-rometty-focusing-on-high-growth-wont-forget-core.html (accessed March 2, 2016).

13. IBM annual reports, 1991–1993.

14. Jeffrey R. Yost. *The Computer Industry* (Greenwood, 2005).

15. https://www-03.ibm.com/press/us/en/pressrelease/7641.wss (accessed March 3, 2016)

16. IBM annual reports, 1984–1986, IBM Corporate Archives.

17. Louis V. Gerstner Jr., *Who Says Elephants Can't Dance: Leading A Great Enterprise Through Dramatic Change* (HarperBusiness, 2002); Lee Iacocca, *Iacocca: An Autobiography* (Bantam, 1986).

18. David Kirkpatrick and Jennifer Reese, "Breaking Up IBM: Facing Horrendous Challenges, John Akers Is Taking Blue Apart and May Even Sell Stock in the Pieces," *Fortune* (July 27, 1992). http://archive.fortune.com/magazines/fortune/fortune_archive/1992/07/27/76699/index.htm (accessed March 30, 2017).

19. "IBM Global Services: A Brief History" (May 2002), IBM Corporate Archives. https://www-03.ibm.com/ibm/history/documents/pdf/gservices.pdf (accessed March 4, 2016).

20. Ibid.

21. Ibid.

22. Ibid.

23. "IBM to Acquire PwC Consulting," IBM press release, July 30, 2002. https://www-03.ibm.com/press/us/en/pressrelease/585.wss (accessed March 6, 2016).

24. IBM annual reports, 2000 and 2011, IBM Corporate Archives.

25. Charles House and Raymond Price, *The HP Phenomenon: Innovation and Business Transformation* (Stanford Business Books, 2009). This is a useful company history that provides a strong history of HP and many of its major technologies and businesses. It, however, is deficient in ignoring HP's services business, which was large and growing quickly nearly a decade before the book was published—truly one of HP's greatest business transformations.

26. Hewlett-Packard annual reports, 2000 and 2002.

27. Hewlett-Packard annual reports, 2000, 2003, and 2005. HP's rapid ramp up in services staff and services revenue without question was remarkable, but may have been aided by how services employees and services revenue were defined by the company. Much as there is a push to classify income today as cloud (by IBM, Microsoft, HP, and others), the push to classify income as services in the 2000s likely was a goal for the company.

28. Pete Carey, "HPs Acquisition Misstep Far from the First," *San Jose Mercury News*, November 20, 2012. http://www.mercurynews.com/2012/11/20/hps-acquisition-misstep-far-from-the-first (accessed April 6, 2016).

29. Gary Jacobson, "EDS Sees Success After GM Purchase—But at a Cost," *Dallas Morning News*, December 9, 2012.

30. Ibid.

31. Matt Richtel and Andrew Ross Sorkin, "HP Acquires EDS for $13.9 Billion," *International Herald Tribune*, May 13, 2008.

32. IBM annual report, 2007; Richtel and Sorkin (2008).

33. Richtel and Sorkin (2008).

34. Damon Darlin, "Patricia C. Dunn Dies at 58; Led H.P. During Spying Case," *New York Times*, December 6, 2011.

35. Ibid.

36. http://www8.hp.com/us/en/hp-news/press-release.html?id=588867#.VrT0CF_naM8 (accessed March 3, 2016).

37. "HP to Separate into Two New Industry-Leading Public Companies." http://www8.hp.com/us/en/hp-news/press-release.html?id=1809455#.VrUZQF_naM8 (accessed March 5, 2016).

38. Michael J. de la Merced, "Hewlett-Packard Spin-Off Is Preparing to Spit Again," *New York Times*, May 25, 2016; "CSC to Combine Government Services Unit with SRA Upon Separation from CSC; Combination will Create Leading Pure-Play Government I.T. Business in the US," CSC press release, August 31, 2015 (http://www.csc.com, accessed July 1, 2016).

39. *Values, Driven, Leadership: The History of Accenture* (History Factory/Accenture, 2005): 59.

40. Ibid.

41. https://www.accenture.com/us-en/company-women.aspx (accessed March 6, 2016). IBM, also committed to gender diversity, recently reported 29 percent of their work force were women, and 25 percent of managers and executives—and IBM CEO Virginia Rometty previously led IBM's largest division, Global Services. At HP, in addition to Fiorina's CEO appointment, and later Whitman's, the company has had a strong commitment to hiring and promoting women with 30 percent of its global employees being women and about 22 percent of managers and executives by 2008. The workforce participation of women at these companies appears to be slightly above national averages. The National Council of Women in Information Technology reported that women made up 26 percent of the US professional computing occupations work force in 2014. With major advances toward balanced

gender participation in other professions (biological sciences, medicine, and law), and the substantial decline in the number of female computer science majors since its late 1980s peak, more can and should be done in K–12 schools, universities, government, small companies, and major corporations to encourage women's participation. Beyond just numbers, future studies need to record the experiences of women within these organizations. Recently an oral history project has been conducted along these lines with female programmers and managers at corporations by Thomas Misa, funded by the Alfred P. Sloan Foundation.

42. *Values, Driven, Leadership* (2005).

43. Ibid.

44. Accenture annual reports 2011–2014. Some small firms undoubtedly grew faster.

45. The French computer historian Pierre Mounier-Kuhn has published extensively and insightfully on the history of informatics/computer science in France and the French computing industries. His book *L'Informatique en France de la Seconde Guerre Mondiale au Plan Calcul: L'Emergence d'un Science* (Paris: Presses de l'Université Paris-Sorbonne, 2010) focuses on French Informatique, on the emergence and diffusion of computer science and computing technology in France. Among his important articles on the industry (and specifically on service bureaus) is "Le Traitement à Façon: Un Survol Historique," *Entreprises et Histoire* (Novembre 2005): 52–86. In this article he discusses Capgemini as well as IBM (both pre-computer and computer-based service bureaus), EDS, and others.

46. Tristan Baston-Breton, *La Saga Capgemini: L'Incroyable Histoire de L'Une des Plus Belles Francaises de l'Informatique* (Point de Mire, 1999): 22, 30–32, 40–47.

47. Ibid.: 72–73.

48. Ibid.: 140–149; "Capgemini Ernst & Young History" (n.d., probably early 2000s), http://www.fundinguniverse.com/company-histories/cap-gemini-ernst-young-history (accessed October 19, 2016).

49. Capgemini annual report, 2008 (https://www.capgemini.com). (The Capgemini annual reports cited in this chapter were accessed on the company's website in October 2016.)

50. "Capgemini Ernst & Young History."

51. Capgemini annual report, 2008.

52. Capgemini annual report, 2015.

53. http://museum.ipsj.or.jp/en/computer/dawn/0012.html (accessed March 30, 2017).

54. Source: email Correspondence with Dr. Koh Hotta (October 4, 2016). Dr. Hotta was a 35-year veteran and senior manager at Fujitsu (former Director of Software

Assurance and Software Development for Supercomputing at Fujitsu and the current Director of Fujitsu's Heritage Hall (Fujitsu history museum).

55. Source: email correspondence with Dr. Koh Hotta, October 4, 2016.

56. Fujitsu annual report, 1997. http://www.fujitsu.com (accessed March 30, 2017).

57. Ibid.; Lawrence M. Fischer, "Fujitsu to Pay $850 Million to Acquire Rest of Amdahl," *New York Times*, July 31, 1997. http://www.nytimes.com/1997/07/31/business/fujitsu-to-pay-850-million-to-acquire-rest-of-amdahl.html?_r=0 (accessed March 30, 2017).

58. Dan Sabbagh, "The End of an Era as ICL Becomes Fujitsu," *Telegraph*, June 21, 2001.

59. Naoyuki Akikusa and Yuji Hirose, "Fujitsu's Internet Business Strategy," in *Proceedings of the Ninth International Symposium on Semiconductor Manufacturing* (2000): 3.

60. "Fujitsu to Construct New Facilities at Its Tatebayashi and Akashi System Centers," press release, January 27, 2015. http://www.fujitsu.com/global/about/resources/news/press-releases/2015/0127-01.html (accessed March 30, 2017).

61. Ibid.

62. Anirban Sen, "21st Century Will Be India's: Virginia Rometty, IBM," *Economic Times* (India), February 3, 2016. http://articles.economictimes.indiatimes.com/2016-02-03/news/70314193_1_virginia-rometty-watson-ibm-cloud (accessed March 9, 2016).

63. IBM annual reports, 2012 and 2014. https://www.bloomberg.com/news/articles/2015-02-24/ibm-employee-count-falls-for-second-year-in-transition-to-cloud (accessed March 3, 2016).

64. Patrick Thibodeau, "In a Symbolic Shift, IBM India Workforce Likely Exceeds US," *Computerworld*, November 29, 2012. http://www.computerworld.com/article/2493565/it-careers/in-a-symbolic-shift--ibm-s-india-workforce-likely-exceeds-u-s-.html (accessed March 3, 2016).

65. Ibid.

66. Ross Bassett, *The Technological Indian* (Harvard University Press, 2016).

67. Ibid.: 245–268.

68. Ibid.

69. Dinesh C. Sharma, *The Outsourcer: The Story of India's IT Revolution* (MIT Press, 2015): 53–69; Bassett (2016). Sharma's important study insightfully analyzes the long history of India building research and educational infrastructure in information technology. He emphasizes the significance of the development of institutions

and know-how in IT that long predated economic liberalization. Both Sharma's book and Bassett's book stand out for their complementary elucidation of the long and rich historical context to India's rise in the global IT industry.

70. Sharma (2015): 69–75; James W. Cortada, *The Digital Flood: Diffusion of Information Technology Across the US, Europe, and Asia* (Oxford University Press, 2012). Cortada's chapter on India provides a rich understanding of the scale and scope of adoption of computing in this country over the past 50 years.

71. Bassett (2016): 266–267.

72. Sharma (2015): 124–127.

73. AnnaLee Saxenian, *The New Argonauts in the Global Economy: Regional Advantage in a Global Economy* (Harvard University Press, 2007).

74. http://www.fundinguniverse.com/company-histories/cognizant-technology-solutions-corporation-history (accessed March 30, 2017).

75. "HCL Technologies Ltd.," *Economic Times*, date unknown. http://economictimes.indiatimes.com/hcl-technologies-ltd/infocompanyhistory/companyid-4291.cms (accessed March 9, 2016).

76. Sharma (2015): 164–168.

77. Ibid.: 175.

78. Jeffrey R. Yost, "Materiel Command and the Materiality of Commands: An Historical Examination of the US Air Force, Control Data Corporation, and the Advanced Logistics System," in Arthur Tatnall, ed., *History of Computing: Learning from the Past* (Springer, 2010).

79. Watts Humphrey oral history (interview conducted by Grady Booch via Skype, July 17–22, 2009) (CHM).

80. Mark C. Paulk et al., "Capability Maturity Model, Version 1.1," *IEEE Software* 10, no. 4 (1993): 18–27.

81. Sharma (2015): 177–178.

82. "Some Key Facts and Events in Y2K History," *Computerworld*, January 3, 2000. http://www.computerworld.com/article/2597231/some-key-facts-and-events-in-y2k-history.html; http://www.cisco.com/c/en/us/about/government-affairs/archives/y2k.html (accessed March 11, 2016).

83. Sharma (2015): 180; http://www.cisco.com/c/en/us/about/government-affairs/archives/y2k.html (accessed March 23, 2016).

84. Thomas L. Friedman. *The World Is Flat: A Brief History of the 21st Century* (Farrar, Straus, and Giroux, 2005).

85. For a meaningful general counter to Friedman's bold assertion, see Harm J. de Blij, *The Power of Place: Geography, Destiny, and Globalization's Rough Landscape* (Oxford University Press, 2008).

86. Lauren Csorny, "Careers in the Growing Field of Information Technology Services," *Beyond the Numbers* 2, no. 9 (2013): 1, 3.

87. https://www.gartner.com/doc/3032029/market-share-analysis-it-services and http://www.gartner.com/newsroom/id/2496815 (accessed March 18, 2016).

88. http://www.gartner.com/newsroom/id/2875020 (accessed March 18, 2016).

89. http://www.tcs.com/investors/Documents/Annual%20Reports/TCS_Annual_Report_2014-2015.pdf (accessed March 18, 2016); http://finance.yahoo.com/q/pr?s=HPQ+Profile; http://finance.yahoo.com/q/pr?s=ACN+Profile (accessed March 18, 2016).

90. http://www.computerweekly.com/feature/Top-five-outsourcing-destinations-to-watch (accessed March 18, 2016).

91. http://www.zdnet.com/article/india-it-salary-benchmark-2012 (accessed March 15, 2016). This 2012 survey usefully provided IT salary by occupation at less than 5 years, 5 to 10 years, and 10 or more years of experience. I used an inflation calculator (about 28 percent) to convert 2012 dollars to 2016 dollars.

92. By *The Economist*'s "Big Mac" index, the cost of goods is more than 60 percent less on average in India than the US average. http://www.economist.com/content/big-mac-index (accessed March 18, 2016).

93. http://profit.ndtv.com/news/corporates/article-why-more-employees-are-quitting-tcs-now-than-ever-before-780080 (accessed March 18, 2016).

94. "On the Turn: India Is no Longer the Automatic Choice for IT Services and Back-Office Work," *The Economist*, January 19, 2013. http://www.economist.com/news/special-report/21569571-india-no-longer-automatic-choice-it-services-and-back-office-work-turn (accessed March 19, 2016).

95. https://www.quora.com/What-is-the-average-salary-hike-in-IT-companies-in-India (accessed March 20, 2016). Most top of range (30 percent) increases are off a low base, higher-salaried individuals often get average increases of upper single digits, somewhat in line with inflation.

96. "Get Off of My Cloud" appears on the Rolling Stones' LP *December's Children (and Everybody's)* (London Records, December 1965). It was released as a single before the release of the LP.

97. Joni Mitchell, "Both Sides Now" (Reprise Records, 1969).

98. A few examples from literature and the visual arts: Johann Wolfgang von Goethe, "Poem on the Clouds" (1865); Percy Bysshe Shelly "The Clouds" (1820);

Rainer Maria Rilke, "Clouds" (n.d.); Titian, "Bacchus and Ariadne" (1520–1523); Ansel Adams, "Clouds, White Pass" (1941).

99. Eric Masanet, Arman Shehabi, and Jonathan Koomey, "Characteristics of Low-Carbon Data Centres," *Nature Climate Change* 3 (July 2013): 627–630.

100. https://www.carbontrust.com/about-us/press/2013/08/carbon-trust-unlaces-carbon-bootprint-of-watching-football (accessed March 18, 2016).

101. "The Rise of the Sharing Economy," *The Economist*, March 9, 2013.

102. Nir Kshetri, "Privacy and Security Issues in Cloud Computing: The Role of Institutions and Institutional Evolution," *Telecommunications Policy* 37 (2013): 373.

103. The "wild, wild West" metaphor for the Internet has persisted, especially in the face of increasing criminal and state-sponsored cyber-attacks. President Obama used the metaphor in a talk on cyber-attack/cyber-defense at Stanford University in 2015. See Nicole Perlroth and David E. Sanger, "Obama Calls for New Cooperation to Wrangle the New 'Wild West' Internet," *New York Times*, February 14, 2015.

104. Patrick Thibodeau, "One in Four Cloud Providers Will Be Gone by 2015," *Computerworld*, December 11, 2013. http://www.computerworld.com/article/2486691/cloud-computing/one-in-four-cloud-providers-will-be-gone-by-2015.html(accessed March 30, 2017).

105. "The Cheap, Convenient Cloud," *The Economist*, April 18, 1015.

106. Without question B2C providers of platforms for social or business networking (Facebook and LinkedIn) are providing cloud computing. Facebook is the one firm whose revenue would place it above all of the enterprise cloud services leaders. It, however, is absent from discussion in market research cloud industry articles, which often draw from proprietary market research reports, or the press releases on industry growth/projections market research firms put out. Clearly, social networking and B2C appears outside of the numbers presented for the cloud computing industry by leading market research firms—it is its own space, where advertising provides the vast majority of revenue. Google, of course, is a big player in B2C cloud with Google+, Gmail, cloud storage, and other software products and platforms. Other firms lead the enterprise cloud space, but Google is expanding rapidly in this area.

107. Kenji E. Kushida, Jonathan Murray, and John Zysman, "Cloud Computing: From Scarcity to Abundance," *Journal of Competition and Trade* 15 (2015): 5–19.

108. Salesforce.com has signed on many large enterprises in recent years.

109. http://www.gartner.com/newsroom/id/920712 (accessed March 18, 2016).

110. https://www.technologyreview.com/s/406593/servers-for-hire (accessed March 18, 2016).

111. David Streitfeld, "Amazon Delivers Some Pie in the Sky," *New York Times*, December 3, 2013.

112. http://fortune.com/2015/04/23/amazon-web-services-5-billion and http://www.cio.com/article/2921180/cloud-computing/amazon-opens-up-about-aws-revenues.html (accessed February 10, 2017).

113. http://fortune.com/2016/01/25/vendor-cloud-numbers-a-mystery and https://qz.com/821060/amazon-web-services-amzn-is-now-a-11-billion-a-year-cloud-computing-business (accessed February 10, 2017). (As one financial journalist points out there may be an incentive for companies with major software and services businesses to recognize revenue as "cloud" to show momentum in this fast-growing competitive field. This is less likely with Amazon owing to the distinctions between its main business, e-commerce, and cloud services.)

114. "Cloud Computing: The Sky Is the Limit," *The Economist*, October 17, 2015). The cloud segment is growing fast, and revenue numbers become dated quite quickly. Companies also have an incentive to recognize revenue as in or tied to the cloud, as the major firms in this arena are all vying for both the perception and reality of being the number-one provider of cloud services.

115. "Partly Cloudy," *The Economist*, October 17, 2015.

116. Barb Darrow, "Shocker! Amazon Remains the Top Dog in Cloud by Far, but Microsoft, Google, Makes Strides," *Fortune*, May 19, 2015.

117. Louis Columbus, "Roundup of Cloud Forecasts and Market Estimates Q3 Update 2015," *Forbes*, September 27, 2015.

118. "Gartner Says Worldwide Cloud Infrastructure-as-a-Service Spending to Grow 32.8 Percent in 2015," Garner Newsroom [press release].

119. Christina Rexrode, "Amazon Cloud Services Is Aimed at Big Banks," *Wall Street Journal*, February 24, 2016.

120. Jay Greene, "Google Chips Away at Amazon's Cloud Business," *Wall Street Journal*, February 24, 2016.

121. Ibid.

122. http://www.nytimes.com/2016/01/29/technology/microsoft-earnings.html (accessed March 18, 2016)

123. Jay Greene, "Microsoft Marshals Cyberattack Defenses," *Wall Street Journal*, February 9, 2016.

124. http://fortune.com/2014/01/23/lessons-from-the-death-of-a-tech-goliath (accessed March 30, 2017). Siebel Systems was the fastest-growing firm in US history, with a five-year growth rate exceeding 780,000 percent in the second half of the 1990s, according to Deloitte.

125. http://fortune.com/2015/08/21/oracle-salesforce-hookup and http://www.informationweek.com/software/enterprise-applications/oracle-adds-cloud-adapter-for-salesforcecom/d/d-id/1113461 (accessed March 18, 2016).

126. http://fortune.com/2014/01/23/lessons-from-the-death-of-a-tech-goliath (accessed March 30, 2017).

127. http://investor.salesforce.com/about-us/investor/investor-news/investor-news-details/2015/Salesforce-Announces-Fiscal-2015-Fourth-Quarter-and-Full-Year-Results/default.aspx (accessed March 30, 2017).

128. Louis Columbus, "By 2018, 62% of CRM Will Be Cloud Based," *Forbes*, June 20, 2015.

129. http://www-03.ibm.com/press/us/en/pressrelease/47838.wss (accessed March 30, 2017).

130. https://www-03.ibm.com/press/us/en/pressrelease/45022.wss (accessed March 30, 2017).

131. http://www.dell.com/learn/us/en/vn/secure/2015-10-12-dell-emc-transaction (accessed March 18, 2016).

132. http://www.dell.com/learn/us/en/vn/secure/2015-10-12-dell-emc-transaction and http://www.wsj.com/articles/dell-to-buy-emc-for-67-billion-1444649012 (accessed April 21, 2016).

133. Claire Zillman, "Accenture's Millennial Hiring Spree," *Fortune*, March 15, 2016.

Conclusion

1. Among the global industries that are significantly larger are agriculture, health care, and automobiles.

2. The important work of Martin Campbell-Kelly (both his single-authored scholarship and his co-authored scholarship with Daniel Garcia-Swartz) stands as an exception—two chapters in *From Airline Reservations to Sonic the Hedgehog*, Campbell-Kelly's book on the software products industry, focus entirely or largely on services, and Campbell-Kelly and Garcia-Swartz's excellent, concise, and engagingly written history of the entire IT industry, *From Mainframes to Smartphones*, is insightful. Because the survey by Campbell-Kelly and Garcia-Swartz covers the six-decade history of the entire global computer industry—computers/hardware, software, and services—it gives more attention to the industry's hardware side. It is also more an economic history than business history. In contrast, this book is more focused on business history, and offers case studies of numerous IT firms and their evolving strategies—including some in industry sectors largely ignored by other scholars (such as consulting, facilities management, data processing services, and independent contractor brokerages).

3. To a certain degree, as a turnaround artist, former CEO Lou Gerstner became a major public figure as IBM was building services into being a major profit center in the early to mid-1990s. Over the past half-decade, Virginia Rometty also has had a high profile. She has taken undue criticism for financial results when, to her credit, she has focused more on longer-term strategic reorientation of the firm toward more profitable areas. Another exception is a recent entrant, Amazon's AWS cloud division and the firm's founder and CEO—Jeff Bezos certainly has become well known as the leader of this e-commerce and now cloud services giant (a company that most people still think of solely as an e-commerce retailer).

4. Ellen Ullman, *Close to the Machine: Technophilia and Its Discontents* (City Lights, 1997).

5. http://www.nytimes.com/2012/12/09/technology/air-force-stumbles-over-software-modernization-project.html?_r=0 (accessed March 28, 2016).

6. JoAnne Yates, *Structuring the Information Age: Life Insurance and Technology in the 20th Century* (Johns Hopkins University Press, 2008); James W. Cortada, *The Digital Hand*. volumes 1–3 (Oxford University Press, 2003, 2005, 2007); Thomas J. Misa and Jeffrey R. Yost, *FastLane: Managing Science in the Internet World* (Johns Hopkins University Press, 2016).

7. Jeffrey R. Yost, "Manufacturing Mainframes: Component Fabrication and Component Procurement at IBM and Sperry-Univac, 1960–1975," *History and Technology* 25, no. 3 (2009): 219–235.

8. Alfred D. Chandler, *The Visible Hand: The Managerial Revolution in American Business* (Harvard University Press, 1977). Chandler himself cites IBM as an example. Many other historians have identified IBM as "Chandlerian"; see, e.g., Naomi R. Lamoreaux, Daniel M. G. Raff, and Peter Temin, "Beyond Markets and Hierarchies: Toward a New Synthesis of American Business History," *American Historical Review* (April 2003): 404–433.

9. Yost (2009).

10. Lamoreaux, Raff, and Temin (2003).

11. Jeffrey R. Yost, *The Computer Industry* (Greenwood , 2005).

12. James P. Womack, Daniel T. Jones, and Daniel Roos, *The Machine That Changed the World: The Story of Lean Production—Toyota's Secret Weapon in the Global Car Wars That Is Now Revolutionizing World Industry* (Free Press, 1990).

13. Don Clark, Dana Cimilluca, and Robert McMillan, "EMC Takeover Marks the Return of Michael Dell," *Wall Street Journal*, October 13, 2015. http://www.wsj.com/articles/dell-to-buy-emc-for-67-billion-1444649012 (accessed January 18, 2016).

14. Richard Barras, "Towards a Theory of Innovation in Services," *Research Policy* 15, no. 4 (1986): 161–173.

Notes to Conclusion

15. Nicholas G. Carr, "IT Doesn't Matter," *Harvard Business Review* (May 2003): 41–49.

16. Nicholas G. Carr, *Does IT Matter?: Information Technology and the Corrosion of Competitive Advantage* (Harvard Business School Press, 2004).

17. Nicholas Carr, *The Big Switch: Rewiring the World, From Edison to Google* (Norton, 2008).

18. Carr (2004): 5. He cites the 1998–2000 Compass World IT Strategy Census—a survey looking to the future that was conducted in 1997 and published in 1998.

19. Thomas Landauer, *The Trouble with Computers: Usefulness, Usability, and Productivity* (Bradford, 1996); Erik Brynjolfsson and Adam Saunders, *Wired for Innovation: How Information Technology Is Reshaping the Economy* (MIT Press, 2013).

20. Carr (2004, 2008). In the former, little mention is made of security. In the latter, security makes several bit appearances in discussions of malicious bots and hacks, but it is not well integrated into Carr's argument as to what this might mean for the commodity cloud or for the IT as a utility era he foresees. In general, a degree of customization and diversity makes things harder for malicious attackers. And corporate and government spending on technology, but also on education, behavior, and processes, will have to increase dramatically (as it has already begun to do). I co-led a major history of computer security project sponsored by the National Science Foundation (2012–2015), and am currently writing a book on the history of computer security under contract for the MIT Press.

21. Yost (2009).

Bibliographic Notes

This book analyzes the complex, rapidly evolving, and highly influential computer/IT services industry through a series of company case studies. The cases were selected to best understand key companies, as well as the origin and evolution of the half dozen segments of the IT services industry. Given the book's heavy concentration on the strategies of and developments at computer services firms, it is a work squarely at the intersection of the history of technology and business history, and contributes to historiography in both of these fields. Drawing extensively on archival materials and sixteen carefully selected oral history interviews by the author, the book also speaks to issues in social and cultural history by exploring professionalization, gender, labor, and users. It is the first scholarly book on the history of the computer services trade—a remarkable industry that was and is critical to the use of computers by organizations and the shaping of information technology worldwide.

Primary Sources

Most of the cases studies in this book were made possible by extensive analysis of archival materials. This includes the cases of Diebold and Associates/Diebold Group, Inc., International Business Machines (IBM), C-E-I-R, Inc., System Development Corporation, the Association of Data Processing Service Organizations (ADAPSO), Tymshare, Inc., the National Association of Computer Consultant Businesses (NACCB), and Control Data Corporation. These and several of the other cases are enhanced by the oral histories I conducted that are now publicly available at the Charles Babbage Institute. Oral histories (at CBI and CHM) conducted by other historians and researchers also proved highly useful. Published primary sources, especially articles contemporaneous to developments in trade publications, company publications, and national newspapers also were quite useful—particularly *Datamation, Computerworld, EDP Analyzer, Business Machines, IBM News, Think,* the *New York Times,* and the *Wall Street Journal.* Likewise, company annual reports were a significant source of financial and other information.

Archival Collections

Association of Data Processing Service Organization Records (Charles Babbage Institute—CBI)
Burroughs Corporation Records (CBI)
C-E-I-R, Inc. Records (CBI)
Control Data Corporation Records (CBI)
Corporate Histories Project Collection (Computer History Museum—CHM)
Diebold Group, Inc., Client Reports (CBI)
Fernando Corbató Papers (MIT Archives)
International Business Machines Corporate Archives (IBM)
SHARE, Inc. Records (CBI)
Sperry-Univac Records (Hagley Museum and Library)
System Development Corporation Records (CBI)

Oral Histories (Interviewer in parentheses)

CBI Oral Histories

Walter F. Bauer (A. L. Norberg)
Ronald W. Braniff (J. R. Yost)
Richard G. Canning (J. R. Yost)
Fernando J. Corbató (A.L. Norberg)
John J. Cullinane (J. R. Yost)
George W. Dick (W. Aspray)
Grace Gentry (J. R. Yost)
Martin A. Goetz (J. R. Yost)
Bernard Goldstein (D. K. Allison)
Ann Hardy (J. R. Yost)
Norman Hardy (J. R. Yost)
Cuthbert C. Hurd (C. R. Fillerup)
Butler Lampson (J. R. Yost)
Frank R. Lautenberg (P. Ceruzzi)
J. C. R. Licklider (W. Aspray and A. L. Norberg)
Harry M. Markowitz (J. R. Yost)

Bibliographic Notes

Phyllis Murphy (J. R. Yost)
Warren Prince (J. R. Yost)
Herbert W. Robinson (B. Bruemmer)
David Schmidt (J. R. Yost)
Peggy Noell Smith (J. R. Yost)
Willis H. Ware (J. R. Yost)
Sam Wyly (D. K. Allison)

CHM Oral Histories

Walter F. Bauer (L. Johnson)
Jerome L. Dreyer (T. Haigh)
Werner Frank (J. R. Yost)
Richard Gentry (B. Grad)
Peter Harris (T. Haigh)
Watts S. Humphrey (G. Booch)
Thomas J. O'Rourke (L. Johnson)
Lynn Sanden (L. Johnson and A. Hardy)
Fred Shulman (B. Grad)
Harvey Shulman (J. R. Yost)
Warner Sinback (L. Johnson)
LaRoy Tymes (A. Hardy and L. Johnson)

Secondary Sources

Few scholars have studied the history of the computer services industry. Martin Campbell-Kelly's excellent book, *From Airline Reservations to Sonic the Hedgehog: A History of the Software Industry* (MIT Press, 2003), has chapters on "The Origins of the Software Contractor, the 1950s," and "Programming Services, the 1960s." In these two chapters, systems integration and programming services are more the focus than consulting, data processing/service bureaus, or facilities management. The book does not examine the services industry's later evolution (1970s and beyond), and thus, the two chapters principally serve as a prehistory to the book's main topic, the software products trade. Campbell-Kelly and Daniel D. Garcia-Swartz's *From Mainframes to Smartphones: A History of the International Computer Industry* (Harvard University Press, 2015) is an insightful and concise book on all three of the major IT global industries—the computer, software products, and services trades—and is structured in a way that the relative attention to the three industries is in that order. These two scholars also published an important article ("Economic Perspectives on the History of the Computer

Time-Sharing Industry, 1965–1985" *IEEE Annals of the History of Computing* 30, no. 1 [2008]: 16–36) on the economic history of time-sharing, indicating how the time-sharing industry persisted into the early personal computer era before its decline. On ADAPSO, Thomas Haigh wrote three "Biographies Department" contributions (cited in chapter 5) in *IEEE Annals of the History of Computing* that usefully explore the evolution of this significant trade association—especially enlightening is Haigh's analysis of the trade group's battle with the banking industry over what many ADAPSO members believed were anti-competitive practices by large banks.

The book takes a fundamentally different approach to examining IBM's history than all of the many existing books and articles on the company. Prior historical scholarship on IBM has focused on IBM hardware, where the company recorded the vast share of its revenue through much of its more than hundred year history. The iconic mainframes—the IBM 701, IBM 650, IBM 1401, and IBM System/360—are center stage in this literature. My book reinterprets IBM as fundamentally involved in services—for punch card tabulation machines and then computers—from its origin in 1911 to the present. In substantial part, IBM has always been a services company. Prior to 1970 much of this services work (maintenance, programming, systems integration, and so on), which was performed by IBM customer engineers, systems engineers, and other technical staff, was not priced or charged, it was "bundled" to support the company's hardware business. Between 1970 and 1988, in implementing "unbundling," a meaningful portion of services was charged to customers. In 1989 IBM began to expand its data center operations and soon launched its Global Services Division—charging for all services. With this, services became IBM's largest business and the company became the global leader in IT services in revenue—distinctions it holds to this day. Regardless of whether IBM charged for services, this work was always critical to IBM's competitive advantage. IBM's hardware, of course, was important and there has been some high quality scholarship on the company; most notably, Steve Usselman has published a number of outstanding articles and book chapters on IBM's history. His scholarship on IBM has concentrated on the firm's strategies and its position with respect to its global competitors, as well as on political economy. With regard to the latter, he provides compelling analysis of antitrust and IBM leaders' long-term dialogue with the US Department of Justice (DOJ), including the firm's defense against lawsuits brought by the DOJ. While Usselman's focus has not been services, his important scholarship stands out in being very much informed by both software and services in IBM's different eras. A very useful survey—which is focused primarily on IBM hardware—is Emerson Pugh's *Building IBM: Shaping an Industry and Its Technology* (MIT Press, 1995). Robert Sobel's *IBM: Colossus in Transition* (Times Books, 1981) also concentrates heavily on hardware but is less comprehensive than Pugh's study (and while it has a short bibliography, it does not have notes/citations).

Paul Ceruzzi published an engaging book on an IT region's development over six decades, *Internet Alley: High Technology in Tysons Corner, 1945–2005* (MIT Press, 2008). Not only is this Washington, DC, metro region a central point for routing Internet traffic, it also is an area peppered with major facilities of computer services firms focused on defense contracting, including Computer Sciences Corporation, Planning Research Corporation, and CACI. While Ceruzzi briefly discusses these companies, he offers minimal contextualization of their place in the broader IT services industry. Ceruzzi has a different goal, which he successfully accomplishes, characterizing technology and a changing regional landscape and culture. On India's services industry two recent works make substantial contributions, Ross Bassett's *The Technological Indian* (Harvard University Press, 2016) and Dinesh C. Sharma's *The Outsourcer: The Story of India's IT Revolution* (MIT Press, 2015). Bassett concentrates on analysis of the long-term influence of MIT and its Indian graduates on education, science and technology research, and industry in India. Sharma also examines the deep roots of India's technology, showing how recent developments are the product of many decades of the nation's embrace of technology and its development of underlying institutional infrastructure. On the history of gender and computing there are three standout works: Janet Abbate's *Recoding Gender: Women's Changing Participation in Computing* (MIT Press, 2012), Thomas J. Misa's edited volume *Gender Codes: Why Women Are Leaving Computing* (Wiley 2010), and Marie Hicks' *Programmed Inequality: How Britain Discarded Women Technologists and Lost Its Edge in Computing* (MIT Press, 2017). All three provide useful discussion, analysis, and framings. Nathan Ensmenger's *The Computer Boys Take Over: Computers, Programmers, and the Politics of Technological Expertise* (MIT Press, 2010), is the leading study of the history of programming and the programming profession. More broadly, there are many strong scholarly books and articles on computer history (with minimal or no direct content on IT services) that provide important context, as well as some works on other technologies or industries that connect thematically, hence the numerous citations to secondary works throughout the book.

Index

Accenture, 6, 9, 15, 23, 29, 113, 231, 234–235, 242–244, 247, 250–251, 259–260, 263, 268, 270, 274, 281–282
Akers, John, 237
Amazon Web Services (AWS), 266–267
Amdahl, Gene, 153, 248–249
Amdahl Corporation, 248–249
Andersen Consulting, 234, 241, 243, 244, 246
Army-Navy/Fixed Special eQuipment-7 computers (AN/FSQ-7), 91–92, 96–97, 181
Arthur Andersen and Company, 6–7, 9, 23, 24–29, 31, 41–42, 232
Association of Data Processing Service Organizations (ADAPSO), 117–134
  becoming multidivisional, 127–128
  charter members, 123
  competition from banks, 125–126
  founding, 117–122
  industry identity and statistics, 129–133
Automatic Data Processing, 6, 8, 10, 45, 55–62, 113, 126, 131, 137, 144, 223, 231–232
Automatic Payrolls, Inc., 8, 46, 55–57, 277, 283
Automation: Advent of the Automatic Factory, 30
Automobile industry, 271

Babbage, Charles, 47
Bauer, Walter, 108–114
Benton, Charles, 71
Bezos, Jeff, 265–266
Biel, William, 93–96
Birkenstock, James, 53
Braniff, Ronald, 162
British Petroleum Corporation, 83
Burroughs Corporation, 2, 21, 36, 51, 104–106, 109–110, 130, 145, 151, 231–232, 234, 242, 252
Business Process Outsourcing (BPO), 257

Campbell-Kelly, Martin, 4, 7, 85, 113, 155
Canning, Richard, 23, 37–43, 277
Canning, Sisson, and Associates, 23, 37–42
Cape Cod System, 93
Capgemini, 6, 15, 235, 244–251, 263, 269, 271, 274
  Daimler-Benz, 245
  expansion 244–251
  founding, 244–245
Carbon and climate change, 262–263
Carr, Nicholas, 283–285
Charles Babbage Institute, 6, 10, 12, 13, 31, 117, 247–248

Cloud computing, 1, 3, 5, 15–16, 154, 175, 231, 234–235, 242, 249–251, 258–272
Coase, Ronald, 145
Cognizant, 7, 235, 254, 255, 270
Compatible Time Sharing System. *See* Massachusetts Institute of Technology, Compatible Time Sharing System
Computer Consulting, 1, 6–10, 19–43, 64, 109, 132, 186–187, 198, 206, 211–228, 237–246, 276, 281
Computer Sciences Corporation (CSC), 6, 9, 10, 63, 69, 72, 84–88, 104, 106, 109, 112–113, 223, 232, 242, 272, 282
Computer Usage Corporation (CUC), 6–7, 10, 63, 68–72, 86–87, 148, 213, 272, 282
 Computer Usage Facilities Management Division, 68–69
 early clients, 65–68
 initial public offering, 69
 origins, 64–65
 relations with IBM, 68–70
Computer utility, 13, 102, 285
Comshare, 160–161
COMSYS, 14, 218, 221–225, 228
Control Data Corporation (CDC), 9, 13, 33, 46, 55, 61, 83, 120, 132, 137, 148, 155, 168, 175, 189–209
 bundling and unbundling, 195, 202–204
 CDC 1604, 191, 198
 CDC 6600, 191
 C-E-I-R, Inc., acquisition of, 196, 203
 CO-OP, 198
 CYBERNET, 196–197
 data centers, 191–198
 educational services, 198–202
 federal systems, 205
 financials, 190, 205
 founding, 189–191
 PLATO, 201–202, 207
 restructuring and renaming, 197–208
 "Total Services" marketing directive, 204
 unbundling (*see* Control Data Corporation, bundling and unbundling)
Corbató, Fernando, 68, 101, 155–156
Council for Economic and Industrial Research (C-E-I-R, Inc.), 6–7, 10, 13, 63, 72–84, 85–87, 111, 113, 122–123, 125, 155, 196, 203, 213, 277
 financials, 80
 founding, 72–73
 mergers and acquisitions, 79–82
 sale of, 84
Crandall, Richard, 133, 161
Crawford, Perry, 182
Cray, Seymour, 189–191
Cunningham, Peter, 129, 131, 231
Customer Relationship Management (CRM) Software. *See* Salesforce.com

Dartmouth Time Sharing System, 157–159
Data Processing Digest, 40–41
Data processing services, origins of, 45–62
DATA-TECH, 79
Dell, Inc., 269, 278
Dick, George, 82–83
Diebold, John, 7, 9, 23, 29–31, 37, 43, 213
Diebold and Associates/Diebold Group
 early clients, 32–37
 founding, 30–31
 sale of, 37
DuPont Glore Forgan, Inc., 143–146

Eckert, J. Presper, 19–21, 26
Eckert-Mauchly Computer Corporation, 20–21, 53, 189

Index

EDP Analyzer, 42
Educational services, 37–40, 69, 83, 198–202
Electronic Data Systems, 135–151
  becoming GM division, 146–151
  brokerage processing, 141–146
  financials, 142
  founding, 139
  initial public offering, 140
  Medicaid processing, 147
  Medicare processing, 140
  overseas expansion, 147–148
Electronic Numerical Integrator and Computer (ENIAC), 19, 65, 276
EMC, 269
Evans, William H., 118–120, 131

Facilities management, 8, 12–13, 70–72, 119, 127–128, 131, 133, 135–151, 158, 245–246, 248, 275, 279
FACT (compiler), 87
Fano, Robert, 153, 156, 175
F. I. DuPont Glore Forgan and Company, 143–146
Fiorina, Carly, 240
Forrester, Jay, 21, 90–91, 253
Frank, Werner, 109–110
Fujitsu, 6, 9, 15, 231, 234–235, 244, 247–250, 263, 269, 271, 284
  datacenters, 249–250
  "everything on the internet" campaign, 249
  FACOM 128A Scientific Computer, 247
  founding, 247
  international partnerships, 248–249
  PROPOSE, 247–248

Gates, William, 5, 274
Gender, 59, 65–66, 70, 161–163, 170, 179, 185–186, 213–221, 226–228, 240–242, 282–283

General Electric, 9, 12, 26–27, 32, 64–66, 68–69, 77, 97, 103, 108, 127, 154, 156, 158–159, 161, 178–179, 185–189, 211, 213–214, 226, 232, 243, 274, 282–283
General Electric Information Systems (GEIS), 154, 156, 158, 160, 173, 175, 197, 226, 232, 234, 279
General Motors, 5, 138, 148–150
Gentry, Grace, 213–219, 221, 224, 226, 278
Gentry, Inc., 214–218
Gentry, Richard, 214–218
Gerstner, Louis, 209, 236–237
Glickauf, Joseph, 25–28
Goldstein, Bernard, 120, 132, 158, 168–169
Google, 267
Greenberger, Martin, 153–154, 175
Griffin, Katherine, 222–223, 225
Grosch, Herbert, 76–77, 154, 156
Grosch's Law, 154

Hardy, Ann, 161–163, 170
Hardy, Norman, 163–164, 166, 169
Hart, Milledge, 139, 141–144
HCL Technologies, 9, 234, 255, 259
Head, Robert V., 183
Healthcare.gov, 3
Hewlett-Packard, 6, 151, 231, 233–235, 239–243
  acquisitions, 240–243
  expansion to services business, 240–242
  founding, 239
  restructuring, 242–243
Higgins, Joseph, 25–28
Hill, Richard, 109
Honeywell, 69, 84–87, 130, 145, 154, 158, 191, 201, 224, 233, 278, 280
Hopper, Grace, 153
Humphrey, Watts, 253

Hurd, Cuthbert, 53, 64, 68, 70–71, 111, 137
Ikeda, Toshio, 247
Independent contractor brokerages, 211–228
India, computer services in, 6–7, 9, 15, 62, 151, 234–235, 241, 244, 247, 254–259, 270, 274, 284
Informatics, Inc., 106–114
Infosys, 6, 247, 251, 254–255, 257, 259, 270
Infrastructure as a service (IaaS), 264–269
INPUT, 129, 231
International Business Machines (IBM), 6–11, 18, 21–22, 24, 25, 28, 34, 35, 36, 45–61, 64–67, 76–79, 86–89, 91–94, 96–97, 99, 101–105, 109–110, 118, 120–121, 125–126, 128, 130–132, 135, 137–141, 145, 151, 155–158, 160–164, 169, 174–175, 177–189, 196, 198, 202–203, 206, 209, 213–215, 223, 231, 233–239, 241–244, 247, 250, 253, 255, 256, 263, 267–269, 270, 276, 278, 280–281, 284
 Advanced System Development Division, 183
 bundling and unbundling, 173, 180, 186–188
 "Call 360" (time-sharing system), 173
 Customer Engineers (CEs), 13, 178–181
 Federal Systems Division, 182, 185, 188–189, 278
 Field Services Division, 178–181
 founding of (as C-T-R), 46–48
 Global Services Division, 238–239, 250, 259
 IBM 650, 54
 IBM 701, 53
 IBM 704, 75
 IBM 1401, 55, 253
 IBM System/360, 55, 60, 68, 105, 145, 155, 173, 177, 180, 189, 243, 248, 350
 IBM Systems Journal, 187
 India, operations in, 235, 250–251, 253
 restructuring (in 1980s and early 1990s), 237–238
 SABRE, 13, 33, 106, 114, 173, 181–185, 206, 278
 Service Bureau Corporation, 45, 50–55, 57, 62, 64, 67, 118–120, 124, 126, 131–132, 137, 141, 155, 160, 173, 175, 188, 206
 STRETCH, 68, 81–82, 161–163
 Systems Engineers (SEs), 185–189, 193, 238, 280
 System Services Women Corps, 179, 185–186
 tabulation field services, 48–53
 Technical Computing Bureau (TCB), 64, 67
 unbundling (*see* International Business Machines, bundling and unbundling)
 Watson (analytics division), 269

John Diebold and Associates. *See* Diebold and Associates/Diebold Group
Jones, Fletcher, 84
Jules Own Version of the International Algorithmic Language (JOVIAL), 102–103

Kampf, Serge, 244
Kanodia, Lalit, 252
Kappler, Melvin, 93–95, 98, 103
Kemeny, John, 153
Kohli, F. C., 252
Kubie, Elmer, 64–67, 71–72, 137
Kurtz, Thomas, 157–158

Index

Lampson, Butler, 161
Lautenberg, Frank, 58–61
Lawrence Radiation Laboratory, 161, 163–164, 262
Lehman, John, 82–83
Leontief, Wassily, 72–74
Lesser, Barbara, 65–66
Licklider, J. C. R., 94–95, 101, 157
Lincoln Laboratory, 91–96, 98

Management consulting, 6, 9, 24–26, 29, 39, 206, 238, 246
Markets versus hierarchies, 146–151, 278–280
MARK IV, 113
Markowitz, Harry, 76–77
Marquez, Tom, 139
Massachusetts Institute of Technology (MIT), 77, 90–91, 93–94, 101, 103, 107–109, 113, 122, 127, 129, 138, 150, 155–156, 163, 182, 251–253, 265
  Compatible Time Sharing System, 153–154
  Multics, 68–69, 101, 154–156, 252
  Project MAC, 68–69, 154–157, 163, 252
Mauchly, John, 19–21, 26
McCarthy, John, 156
Melahn, Wesley, 103
Meyerson, Mort, 141, 143–144, 147, 149, 151
Microsoft, 5, 15, 234, 236, 260, 264, 267, 274, 279, 282
Milligan, Margaret, 40–41
MITRE Corporation, 81, 94, 109, 255
Moore, Frederick, 72
Moore's Law, 15
Motorola, 256
Multics. *See* Massachusetts Institute of Technology, Multics

Murphy, Phyllis. *See* Phyllis Murphy and Associates
Myers, Glenn, 179–180

Nadella, Satya Narayana, 5
National Association of Computer Consultant Businesses (NACCB), 14–15, 211, 218, 220–221, 224–228
National Cash Register Corporation, 21, 38, 40, 122–123
Norris, William, 189–191, 194, 198, 200–203, 207, 274
Nutt, Roy, 84

Orchard-Hays, William, 76–77, 79
Osborn, Roddy, 27–28

Patrick, Robert, 84
Perot, H. Ross, 5, 12, 71, 72, 135–151, 158, 237, 240, 255, 279
Phyllis Murphy and Associates, 14, 213, 218–221, 224, 227–228
Planning Research Corporation, 103–104, 106–107, 223, 351
Platform as a service (PaaS), 264–267
Postley, John, 112
Price, Robert, 193–194, 200, 203, 207
Programming services
  origins and early history, 63–88
Project Genie, 157
Project MAC. *See* Massachusetts Institute of Technology, Project MAC

Ramo, Simon, 107–108
Ramo-Wooldridge, 107–109
RAND Corporation, 89, 92–94, 97, 109, 142, 207
  Systems Research Laboratory, 93–94
Remington Rand Corporation, 21–22, 32–33, 51, 53, 91, 110, 189, 245
Universal Automatic Computer (UNIVAC), 22, 26–27, 38, 53

Robinson, Herbert W., 73–77, 79–83, 89, 111
Rometty, Virginia, 235, 250–251, 270

Salesforce.com, 259, 263, 265, 267–269, 271, 281
Satyam, 254
Schaefer, Ellen Kerksieck, 185–186
Scientific Data Systems (SDS), 157–158, 160–163, 167
Section 1706, 212–213, 218, 220, 223–225
Semi-Automatic Business Research Environment (SABRE). See International Business Machines, SABRE
Semi-Automatic Ground Environment (SAGE), 10–11, 77, 82, 90–102, 109, 112, 114, 154, 156–158, 167, 181–182, 185, 188, 209, 213, 217, 261, 267, 275, 277, 281–283
Service Bureau Divison/Corporation. See International Business Machines, Service Bureau Corporation
SHARE, Inc., 11, 84, 109–110, 121
Sheldon, John, 64–65, 72, 137
Shulman, Fred, 221–228
Shulman, Harvey, 211, 218, 221–224, 226–228
Sinback, Warner, 158–159
Sisson, Roger, 37–38, 40–41, 277
Smith, C. R., 182
Smith, R. Blair, 182
Smith, Roger, 5, 124, 138, 149–151
Software as a service (SaaS), 264–268
Software Engineering Institute, 255–256
  Capability Maturity Model (CMM), 255–256
Southern Pacific Railroad Internal Network (SPRINT), 261
Spacek, Leonard, 25–26
Sperry Rand Corporation, UNIVAC Services Centers (USC), 118, 123–124

Stack, Andrew Joseph, 211
Stein, Howard, 222–224, 225, 228
Strategic Air Command Control System (Air Force SACCS), 101–102
System Development Corporation, 8, 11, 65, 88–89, 93–106, 114, 124, 136–137, 156, 181–182
  financials, 104
  origins, 89–96
  SAGE programming, 94–100
  System Development Division, 95–96
  transition to for-profit corporation, 105
  "university of programmers," 98–100
Systems integration, 1–4, 8–11, 13, 63, 88–114, 118, 124, 128, 137, 151, 182, 184, 189, 206, 234, 237–238, 240, 245–246, 258–260, 274–283

Tata Consultancy, 6–7, 234, 247, 255, 257, 259, 270, 274, 284
Taub, Henry, 57–61
Thompson Ramo-Wooldridge (TRW), 108
Time sharing, 1, 9, 12, 13, 15, 55–56, 68–69, 83–84, 87, 95, 101, 104–105, 118, 127–128, 130–133, 153–175, 178, 197, 203, 206, 226, 231–234, 260–265, 279, 285, 288, 350
Trade associations, 117–134
Trunk, Rheena, 65–66
Tymes, LaRoy, 163–164, 166
TYMNET, 158, 163–164, 166–172
Tymshare, Inc., 9, 12, 113, 132, 133, 153, 156, 158–173, 175, 197, 213, 234, 260
  acquired by McDonnell Douglas, 172
  acquisitions, 169–172
  financials, 167
  international, 169
  network (see TYMNET)
  origins, 159–160, 170

Index

Ullman, Ellen, 275
United Data Centers, 158
Universal Automatic Computer. *See* Remington Rand, UNIVAC
University Computing Corporation, 86–87

Ware, Willis, 91, 96, 98
Watson (system/division), 250–251, 268
Watson, Thomas J., 46, 48–51, 54, 64, 70
Watson, Thomas J., Jr., 53–54, 64, 70, 141, 185–187, 237, 274, 278
Whitman, Meg, 242
Williamson, Oliver, 145
Wipro, 6, 9, 234, 247, 251, 254–257, 259, 270, 274
Women, in computer services. *See* Gender
Wooldridge, Dean, 108
Wyly, Sam, 86